T0320429

Process Integration for Resource Conservation

To achieve environmental sustainability in industrial plants, resource conservation activities such as material recovery have begun incorporating process integration techniques for reusing and recycling water, utility gases, solvents, and solid waste. *Process Integration for Resource Conservation* presents state-of-the-art, cost-effective techniques, including pinch analysis and mathematical optimization, for numerous conservation problems. The second edition of this best-seller adds new chapters on heat integration and retrofitting of resource conservation networks and features multiple optimization examples via downloadable MS Excel spreadsheets.

- Emphasizes the goal of setting performance targets ahead of detailed design following the holistic philosophy of process integration
- Explains various industrial examples step by step and offers demo software and other materials online
- Features a wealth of industrial case studies
- Adds chapters on heat integration, combined heat and power, heat-integrated water network, and retrofit of resource conservation network
- Adds new optimization examples and downloadable MS Excel files on superstructural approaches and automated targeting models for direct reuse, recycle, and regeneration

Ideal for students preparing for real-world work as well as industrial practitioners in chemical engineering, the text provides a systematic guide to the latest process integration techniques for performing material recovery in process plants. The book features a solutions manual, lecture slides, and figure slides for adopting professors to use in their courses.

Green Chemistry and Chemical Engineering

Series Editor: Dominic C.Y. Foo
University of Nottingham, Malaysia

Magneto Luminous Chemical Vapor Deposition
Hirotsugu Yasuda

Particle Technology and Applications
Sunggyu Lee and Kimberly H. Henthorn

Carbon-Neutral Fuels and Energy Carriers
Nazim Z. Muradov and T. Nejat Veziroğlu

Oxide Semiconductors for Solar Energy Conversion: Titanium Dioxide
Janusz Nowotny

Water for Energy and Fuel Production
Yatish T. Shah

Managing Biogas Plants: A Practical Guide
Mario Alejandro Rosato

The Water-Food-Energy Nexus: Processes, Technologies, and Challenges
I. M. Mujtaba, R. Srinivasan, and N. O. Elbashir

Hemicelluloses and Lignin in Biorefineries
Jean-Luc Wertz, Magali Deleu, Séverine Coppée, and Aurore Richel

Materials in Biology and Medicine
Sunggyu Lee and David Henthorn

Resource Recovery to Approach Zero Municipal Waste
Mohammad J. Taherzadeh and Tobias Richards

Hydrogen Safety
Fotis Rigas and Paul Amyotte

Nuclear Hydrogen Production Handbook
Xing L. Yan and Ryutaro Hino

Efficiency and Sustainability in the Energy and Chemical Industries: Scientific Principles and Case Studies, Third Edition
Krishnan Sankaranarayanan

Process Integration for Resource Conservation, Second Edition
Dominic C.Y. Foo

For more information about this series, please visit: https://www.routledge .com/Green-Chemistry-and-Chemical-Engineering/book-series/ CRCGRECHECHE

Process Integration for Resource Conservation

Second Edition

Dominic C.Y. Foo

CRC Press
Taylor & Francis Group
Boca Raton   London   New York

CRC Press is an imprint of the
Taylor & Francis Group, an **informa** business

Designed cover image: shutterstock

Second edition published 2025
by CRC Press
2385 NW Executive Center Drive, Suite 320, Boca Raton FL 33431

and by CRC Press
4 Park Square, Milton Park, Abingdon, Oxon, OX14 4RN

CRC Press is an imprint of Taylor & Francis Group, LLC

© 2025 Dominic C.Y. Foo

First edition published by CRC Press 2012

Library of Congress Cataloging-in-Publication Data
Names: Foo, Dominic C. Y., author.
Title: Process integration for resource conservation / by Dominic Foo.
Description: Second edition. | Boca Raton : CRC Press, 2025. | Series: Green chemistry and chemical engineering | Includes bibliographical references and index. | Summary: "To achieve environmental sustainability in industrial plants, resource conservation activities such as material recovery have begun incorporating process integration techniques for reusing and recycling water, utility gases, solvents, and solid waste. Process Integration for Resource Conservation presents state-of-the-art, cost-effective techniques, including pinch analysis and mathematical optimization, for numerous conservation problems. The Second Edition of this best seller adds new chapters on heat integration and retrofitting of water networks and features multiple optimisation examples via downloadable MS Excel spreadsheets. Emphasizes the goal of setting performance targets ahead of detailed design following the holistic philosophy of process integration Explains various industrial examples step by step and offers demo software and other materials online Features a wealth of industrial case studies Adds chapters on heat integration, combined heat and power, heat-integrated water network, and design of retrofit water network Adds new optimisation examples and downloadable MS Excel files on superstructural approaches and automated targeting models for direct reuse, recycle, and regeneration Ideal for students preparing for real-world work as well as industrial practitioners in chemical processing, the text provides a systematic guide to the latest process integration techniques for performing material recovery in process plants. The book features a solutions manual, lecture slides, and figure slides for adopting professors to use in their courses"-- Provided by publisher.
Identifiers: LCCN 2024019469 (print) | LCCN 2024019470 (ebook) | ISBN 9781032003962 (hbk) | ISBN 9781032003931 (pbk) | ISBN 9781003173977 (ebk)
Subjects: LCSH: Chemical processes. | Mathematical optimization.
Classification: LCC TP155.7 .F63 2025 (print) | LCC TP155.7 (ebook) | DDC 660/.28--dc23/eng/20240515
LC record available at https://lccn.loc.gov/2024019469
LC ebook record available at https://lccn.loc.gov/2024019470

ISBN: 978-1-032-00396-2 (hbk)
ISBN: 978-1-032-00393-1 (pbk)
ISBN: 978-1-003-17397-7 (ebk)

DOI: 10.1201/9781003173977

Typeset in Palatino
by KnowledgeWorks Global Ltd.

Access the Support Material: Routledge.com/9781032003931

Contents

Part I Material Recovery

Part II Heat Integration

Series Preface

Toward the late 20th century, the development of various environmentally friendly processes, techniques, and methodologies saw significant growth in the scientific community. The main driving force for such growth was due to the rising awareness of sustainable development, more stringent environmental regulation, and increasing costs of raw materials and waste treatment. After several decades of development, we now broadly term these environmentally friendly processes, techniques, and methodologies as *green/clean technologies*.

In the 21st century, the global community has experienced many extreme weather events such as prolonged draught, extreme heat, tornadoes, and wildfires. The scientific community believes that these extreme weather events are closely linked to climate change, and they are expected to increase in frequency and intensity in the future. Following the Paris Agreement and Glasgow Climate Pact, there is now an international commitment to limit the rise of global average temperature to well below 2°C, and pursuing efforts to limit the temperature increase to 1.5°C above pre-industrial levels. Hence, it is believed that green/clean technologies have a much bolder role to play in combating the global climate change in the coming years.

It is also worth mentioning that the United Nation Sustainable Development Goals (UNSDG) that were launched in 2015 define 17 important goals to transform the world by 2030. It is believed that some of these goals may be addressed with the development of green/clean technologies. These include:

- Goal 6—Clean Water and Sanitation: Ensure access to water and sanitation for all
- Goal 7—Affordable and Clean Energy: Ensure access to affordable, reliable, sustainable, and modern energy
- Goal 12—Responsible Consumption and Production: Ensure sustainable consumption and production patterns
- Goal 13—Climate Action: Take urgent action to combat climate change and its impacts

The *Green Chemistry and Chemical Engineering* book series by CRC Press/ Taylor & Francis focuses on the subset of green technologies dedicated to address the "2E" agenda, i.e., *environment* and *energy*. It involves the development of materials (e.g., catalysts, nanomaterials), methodologies (e.g., process optimization, footprint reduction, artificial intelligent) and processes (e.g., waste treatment) that will bring forth solutions to address pressing problems such as:

- Greenhouse gas management and reduction
- Sustainable water production
- Wastewater treatment and recycling
- Circular economy and waste reduction
- Renewable energy
- Sustainable use of energy resources

I am hopeful that this *Green Chemistry and Chemical Engineering* series will serve as a de facto source of reference materials and practical guides for academics and industrial practitioners looking to advance the discipline and aims of green chemistry and chemical engineering.

Dominic C. Y. Foo
Series editor, Green Chemistry and Chemical Engineering

Foreword

The ubiquitous use of natural resources throughout the process industries is one of the major hurdles limiting the extent of sustainability worldwide. Tremendous amounts of raw materials and energy are used in manufacturing chemical products. As industry endeavors to conserve natural resources, two critical questions need to be answered:

1. What are the targets for minimum consumption of mass and energy?
2. How to methodically achieve these targets in a cost-effective manner?

This book provides an integrated set of frameworks, methodologies, and tools to answer these two questions.

The author is a renowned leader in the area of resource conservation through process integration, and I had the pleasure of closely observing his professional growth, global leadership, and remarkable contributions that have been documented through excellent publications. In this book, Dr. Foo manages to elegantly transform the theories and concepts into effective educational tools, exciting reading materials, and very useful applications. Various techniques are used, including graphical, algebraic, and mathematical tools. The addressed problems span a wide range, including direct recycle, inplant modifications, and waste treatment for continuous and batch systems. Numerical examples are used to further explain the tools and to demonstrate their effectiveness. The second edition of the book maintains the key strengths of the first edition and adds new and emerging tools and case studies. The inclusion of spreadsheet-based examples is an additional advantage of the second edition that will help coupling the integration methods with standard process engineering tools.

Overall, this is an excellent contribution that will benefit numerous researchers, students, and process engineers and will serve the cause of sustainability worldwide.

Mahmoud El-Halwagi, PhD
Professor and Holder of Bryan Research
and Engineering Chair
Texas A&M University
College Station, Texas

It is no exaggeration to say that the integration of environmental considerations is one of the most significant changes to occur in the basic chemical
engineering curriculum in recent decades. In fact, the same can probably
be said for other engineering disciplines as well. The major environmental
issues of the 20th century are now routinely discussed by lecturers in university classrooms worldwide. Although climate change looms over the world
as the major sustainability crisis of this era, we still face diverse environmental challenges, ranging from the well-established ones such as biodiversity
loss, to the emergent challenges such as plastic pollution.

The state-of-the-art approach combines so-called end-of-pipe treatment
technologies with pollution prevention strategies to ensure that engineering
systems are designed to be sustainable, legally compliant, and economically
viable. While end-of-pipe technologies such as gas scrubbers and wastewater treatment plants make use of physical and chemical processes to reduce
the environmental hazards of existing waste streams, pollution prevention
approaches make use of a broader and less well-defined set of technologies—
both "hard" and "soft"—to achieve reductions in environmental impacts.

This book deals with the efficient use of utilities in industrial systems
through systematic recycle and reuse of process streams. Many applications
deal with efficient water utilization, which is fast becoming a critical aspect of
process plant operations in many parts of the world, and is likely to become
even more so as the effects of global climate change make themselves felt at
the local level. Economizing on water resources will surely be an important
climate change adaptation strategy in many countries in the coming decades.
However, many of the principles described here are readily extended toward
problems involving efficient use of other resources, including utility gases
and industrial solvents. It is quite instructive how analogous structures can
be solved by common techniques even though the physical nature of the
underlying industrial processes is fundamentally different. Furthermore,
efficient use of purchased utilities yields economic benefits that may offset
some, or all, of the costs incurred in implementing these pollution prevention strategies. Finally, an added benefit is that enhancing the efficiency of an
industrial system generally reduces the quantity of waste streams that need
to be treated after the fact.

The main contribution of this textbook is that it brings together a family
of systematic design tools that can be used to determine the most cost-effective measures to implement recycle and reuse of process streams in industrial plants. These techniques are drawn from the state-of-the-art in process
integration literature, mostly from the author's own research over the past
decade, and presents it in a form suitable for a wide range of audiences, from
advanced undergraduate students to practicing engineers from the process
industries.

It has been nearly two decades since Prof. Dominic C. Y. Foo, then an ambi-
tious, freshly minted PhD at the beginning of his academic career, described
to me his idea for this book at a conference banquet dinner, and more than

a decade since the publication of the original edition. Today, with Prof. Foo having established himself as a leading researcher in the field of process integration, and, as the ever-growing public concern about sustainability, in general, and the consequences of climate change, in particular, continues to induce changes in the chemical engineering curriculum, the second edition of this book comes as a timely update to teach the next generation of chemical engineers how to "save the planet" more systematically and intelligently.

Raymond R. Tan, PhD
Full Professor and University Fellow
Department of Chemical Engineering
De La Salle University, Manila, Philippines

Preface

Resource conservation and *waste minimization* have been major concerns in the process industry in the past three decades. This could be due to several reasons. Among these, the increase of public awareness toward environmental sustainability is probably the most influential factor. Hence, ever-stringent emission regulations are observed in many developed and developing countries. To maintain business sustainability and to fulfill social responsibilities, many industrial sectors have taken initiatives to improve emission quality by reducing their emission load. Besides, industrial sectors also enjoy economic benefits from this initiative, since reduced waste generation means more efficient use of resources, which in turn helps reduce overall production costs.

Concurrently, the development of various systematic design tools within the *process systems engineering* (PSE) community has seen significant achievements in the past decades. It should also be noted that most recent scholarly activities in the PSE community have been treating chemical processes as an integrated system rather than focusing on the individual units as in the past. This is indeed a good move, as it is now generally recognized that an optimum unit operation designed independently does not necessarily lead to an optimum overall process. Hence, taking a holistic approach toward the synthesis of an optimum process flowsheet is indeed important. In this aspect, the specialized PSE techniques of *process integration* have much to offer. Process integration may be defined as *a holistic approach to process design, retrofitting, and operation which emphasizes the unity of the process* (El-Halwagi (1997; El-Halwagi and Foo, 2014). The technique has its roots in energy recovery system design due to the first world oil crisis back in the 1970s. It was then extended to various waste recovery systems in the late 1980s. Between the mid-1990s to early 2010s, various process integration tools were developed, focusing on the recovery of material resources, such as water, utility gas, solvent, and solid wastes. However, most of these techniques are published in technical papers, targeted for academic researchers in the PSE community. Hence, this book is meant to respond to the need to disseminate the state-of-the-art techniques developed in the past three decades to a wider audience, especially those looking for cost-effective techniques for various resource conservation problems. Both *pinch analysis* and *mathematical optimization* techniques are covered. A significant change of this second edition of the book is the extensive use of spreadsheet software, i.e., Microsoft Excel. Almost all engineers have this spreadsheet software on their PC/laptop nowadays, which makes it easier for them to carry out the analysis on resource conservation using process integration methods described in this book.

As the book aims to bridge the gap between academic and industrial practitioners, various industrial examples and case studies collected from public literatures were included (see the list in the following page). Hence, the book is useful for upper-level undergraduate or postgraduate students in chemical, process, or environmental engineering, who will soon face the "real world" upon graduation. On the other hand, industrial practitioners will also find the various industrial examples as good reference points in their efforts toward implementing resource conservation initiatives in their plants. The main emphasis of the book is *to set performance targets ahead of detailed design*, following the philosophy of process integration. With the performance targets identified, one will get away with the question of "Is there a better design?" that is always raised in any technical project.

One of the main aims of this book is to enable readers self-learning. Hence, most of the examples are explained in a methodical manner to facilitate independent reading. Besides, electronic calculation files for examples and problems are also made available on publisher's website (www.routledge.com/9781032003931) for the ease of use for readers. Also made available on the book website is the revised version of the prototype software for material conservation, i.e., RCNet (first reported by Ng et al. (2014)], along with its user guide.

I hope you enjoy reading the book and have fun in setting targets!

References

El-Halwagi, M. M. 1997. *Pollution Prevention Through Process Integration: Systematic Design Tools*. San Diego: Academic Press..

El-Halwagi, M. M. and Foo, D. C. Y. 2014. Process synthesis and integration. In: Seidel, A. and Bickford, M. eds., *Kirk-Othmer Encyclopedia of Chemical Technology*. John Wiley & Sons.

Ng, D. K. S., Chew, I. M. L., Tan, R. R., Foo, D. C. Y., Ooi, M. B. L., and El-Halwagi, M. M. 2014. RCNet: An optimisation software for the synthesis of resource conservation networks, *Process Safety & Environmental Protection*, 92(6), 917–928.

Example and Case Studies

Throughout this book, various hypothetical and industrial examples have been utilized for conceptual illustration, as well as problem for exercises. A complete list of these examples (labeled as "E") and problems (labeled as "P") are given here for better reference of readers of specific interest.

Water Minimization

Literature Examples

Wang and Smith (1994)—P3.3, P5.9, E6.1, E6.2, E6.4, P6.2, E8.3, P8.16, E9.7, E9.8, P9.1,

Polley and Polley (2000)—P3.4, P4.12, P5.6, P6.1, E8.1, E8.4, P8.1, P8.14, E9.6, P9.2, E12.3, P12.1

Savelski and Bagajewicz (2001)—P4.4, P8.17

Wang and Smith (1995)—P4.12

Jacob *et al.* (2002)—P4.3, P8.10

Sorin and Bédard (1999)—P5.3, P6.6, P8.12, P9.8

Gabriel and El-Halwagi (2005)—P9.12

Zero fresh resource network—E3.5, P3.7, E4.3, P4.7, P8.9

Zero waste discharge network (Foo, 2008)—E3.6, P3.7, E4.4, P4.7, P5.2, P8.8

Industrial Case Studies

Acrylonitrile (AN) production case study—E2.1, E2.2, E3.1, E3.2, E3.4, E4.1, E5.1, E5.2, P5.5, E6.3, Chapter 7 (Case study), P8.2, E9.1, E9.3, E9.4, E9.5

Alkene production from shale gas (Foo, 2019)—P2.5, P3.8, P4.14, Chapter 5 (Industrial Case Study), P6.8, P8.21, P9.12

Alumina plant (Deng et al., 2009)—P6.4, P9.10, P9.11

Bioethanol production—P4.10, P9.10

Brick manufacturing—P3.9, P6.12, P8.19

Bulk chemical production (Foo, 2010)—P3.5, P5.4, P8.6, P9.4

Chlor-alkali plant—P4.15, P9.7

Coal mine water system—P5.7, P6.3, P8.13

Kraft pulping process 1 (El-Halwagi, 1997)—P2.2, P4.1, P5.8, P6.13, P8.15

Kraft pulping process 2 (Parthasarathy and Krishnagopalan, 2001)—
P2.8, P4.11, P5.10, P6.10, P8.20

Organic chemical production (Hall, 1997; Foo, 2008)—P3.6, P8.11

Palm oil mills (Chungsiriporn *et al.*, 2006)—P2.7, P4.8, P8.4, P9.6

Paper milling process (Foo *et al.*, 2006a)—P4.5, P6.7, P9.9, P12.2

Specialty chemical production process (Wang and Smith, 1995)—P2.6,
P4.6, P6.5, P6.9, P8.7, P9.5

Steel plant (Tian et al., 2008)—P4.9, P8.5

Textile plant (Ujang *et al.*, 2002)—P2.4, P3.2, P9.3

Tire-to-fuel process (El-Halwagi, 1997; Noureldin and El-Halwagi,
1999)—P2.1, P3.1, P.5.1, P7.3, P8.3

Tricresyl phosphate process (El-Halwagi, 1997)—P2.3, P4.2, P6.11, P8.18

Multiple Contaminant Cases

Wang and Smith (1994)—E9.2

Corn refinery (Bavar et al., 2018)—P9.15

Oil refinery 1 (Mughees & Al-Hamad, 2015)—P9.14

Oil Refinery 2 (Sujo-Nova et al., 2009)—P9.17

Sugar refinery (Balla et al., 2017)—P9.16

Utility Gas Recovery Problems

Literature Examples

Refinery hydrogen network 1 (Hallale and Liu, 2001)—E2.4, E3.3, P4.16,
P5.11, E6.6, P8.24, P9.19, E14.4

Refinery hydrogen network 2 (Foo and Manan, 2006)—P4.20, P8.26, P9.16

Oxygen-consuming process (Foo and Manan, 2006)—P3.11, P8.23, P9.18

Industrial Case Studies

Refinery hydrogen network 3 (Alves and Towler, 2002)—P2.11, P4.19,
P5.13, P6.14, P8.25, P9.20, E12.4

Refinery hydrogen network 4 (Umana et al., 2014)—P2.10, P4.18, P6.16,
P8.27, P12.3

Refinery hydrogen network 5 (Deng et al., 2020)—P5.12, P12.5

Aromatic plant (Shariati et al., 2013)—P2.12, P6.17, P9.21

Magnetic tape manufacturing process (El-Halwagi, 1997)—P2.9, P3.10, P4.17, P6.15, P8.22

Property Integration Problems

Metal degreasing process (Kazantzi and El-Halwagi, 2005)—E2.5, E3.7, P4.21, P5.14, E7.1, E8.2, P8.28

Kraft papermaking process (Kazantzi and El-Halwagi, 2005)—E2.6, E3.8, E4.5, E5.3, P8.29, P9.22

Microelectonics manufacturing (Gabriel *et al.*, 2003; El-Halwagi, 2006)—P3.13, P8.30, P9.23

Wafer fabrication process (Ng *et al.*, 2009)—P2.13, P4.23, P5.15, P6.19, P7.2, P8.31, P9.25

Palm oil mill (Ng *et al.*, 2009)—E5.4, P6.18

Vinyl acetate manufacturing (El-Halwagi, 2006)—P2.14, P3.12, P4.22, P9.24

Paper recovery study (Kit et al., 2011)—P3.14

Paper recovery study (Kit et al., 2011)—P4.24

Biorefinery (Martinez-Hernandez, 2018)—P5.16, P9.26

Pre-treatment Networks

Tan et al. (2010)—Case 1 and 2a–E6.7

Tan et al. (2010)—Case 2b–P6.20

Tan et al. (2010)—-Case 3–P6.21

Inter-plant Resource Conservation Networks
Literature Examples

Inter-plant water networks (Bandyopadhyay et al., 2010)—E10.1, E10.2, E10.3, P10.1, P10.2

Inter-plant water networks (Olesen and Polley, 1996)—P10.4

Inter-plant hydrogen network 1—P10.5

Inter-plant hydrogen network 2 (Deng et al., 2017)—P10.6

Industrial Case Studies

Integrated pulp mill and bleached paper plant (Lovelady *et al.*, 2007; Chew *et al.*, 2008) Section 10.4

Inter-plant water networks in catalyst plant (Feng et al., 2006)—P10.3

Inter-plant water networks in eco-industrial park (Lovelady and El-Halwagi, 2009)—P10.7

Batch Material Networks

Literature Examples

Wang and Smith (2005b)—E11.1–11.3

Kim and Smith, 2004—P11.1

Chen and Lee, 2008—P11.2

Property network (Ng *et al.* 2008; Foo, 2010)—11.6

Industrial Case Studies

Agrochemical production (Majozi *et al.*, 2006)—P11.3

Polyvinyl chloride (PVC) resin manufacturing (Chan *et al.*, 2006)—P11.5

Fruit juice production (Almato et al., 1999)—P11.4

Combined Heat and Power (CHP)

Sulfuric acid recovery plant—Section 14.5

Acknowledgments

I am indebted to many organizations and institutions for the completion of this book. I would like to express my gratitude to the various funding agencies that supported my scholarly activities in the past decade. These include the Ministry of Science, Technology and Innovation (MOSTI) of Malaysia, the World Federation of Scientists, as well as the University of Nottingham Research Committee.

Special thanks are due to Allison Shatkin and Sonia Tam, editors at CRC Press/Taylor & Francis Group, who have been very helpful in supporting the making of this book. I am grateful to LINDO Systems Inc., who have provided the demo versions of LINGO software that accompanies this book.

I am also indebted to several people who have played important roles in my life. Without them, I would not have been able to achieve success in my professional career. I am especially grateful to my high school teacher, Boon Nam Khoo, who taught me O-level physics and inspired me to pursue my professional career as a chemical engineer. I should also be grateful to all my formal teachers in Chi Wen Primary and Secondary Schools (where I had the elementary and middle-school education); as well as Datuk Mansor Secondary School (where I completed high school education).

I am also grateful to all my lecturers at Universiti Teknologi Malaysia, where I completed the undergraduate and graduate studies in Chemical Engineering. Special thanks are due to Professor Ramlan Abdul Aziz, the former and founding Director at the Institute of Bioproduct Development, who inspired me tremendously during my early days as a postgraduate student. His dedication to work and the motivation to help others are the core values that I have tried to emulate. He has also given me a lot of career and self-development opportunities since the initial stages of my career. I should also acknowledge Professor Zainuddin Abdul Manan and Dr Kamarul Ariffin who laid the foundation of my process integration knowledge.

I am indebted to my supportive colleagues at the Department of Chemical and Environmental Engineering, University of Nottingham Malaysia (UNM). During the academic year 2010–2011, when I was working on the first edition of this book, many colleagues helped me by temporarily relieving my academic load in teaching and project supervision. Without their help, this book would not have been a reality. Special thanks are due to Professors Denny K.S. Ng (now with Sunway University Malaysia), Hon Loong Lam, and Nishanth who formed the Sustainable Process Integration Group where excellent research ideas were laid, apart from sharing teaching load for various undergraduate and postgraduate subjects in the department. It was through undergraduate teaching where the various assignments for students were prepared, which later became examples and problems in this book.

I am honored to be able to work with a group of collaborators who became very close friends over the years. Professor Raymond Tan of De La Salle University, Philippines, is one those in the top of the list. We started our collaboration as early as when I was still a PhD student back in 2004. The idea of writing a book was also nurtured over a conference dinner during our early days of friendship (see the foreword by Professor Tan). Besides, I have been very fortunate to work with Professor Mahmoud El-Halwagi of Texas A&M University since 2003. He is an inspiring mentor who once told me "contribution will always prevail" during the early, difficult years of my research career. To this day, I am still using his quote to tell the novices in the academic and research community that "your detractors cannot stop you; they can only slow you down." I should also mention that his first two process integration books have been a great source of inspiration in the writing of this book.

I would also like to acknowledge my other close collaborators for their continuous support. These include Professor Cheng-Liang Chen (Taiwan), Professor Santanu Bandyopadhyay (India), Professor Thokozani Majozi and Adeniyi Jide Isafiade (both from South Africa), the late Professor Jiří Klemeš (Czech Republic), Professors Feng Xiao (China), Professor Robin Smith and Dr. Nick Hallale and Dr. Michael Short (United Kingdom), the late Professor Jacek M. Jeżowski (Poland), Professor Paul Stuart and Dr. Alberto Alva-Argaez (Canada), Professor Denny Ng, Professor Rosli Yunus, Dr. Sivakumar Kumaresan, Dr. Yin Ling Tan, Dr. Cheng Seong Khor, Dr. Mimi Haryani Hashim, Dr. Jully Tan, Dr. Lee Pheng Chee, Dr. Zulfan Adi Putra, Mike B. L. Ooi, Mustafa Kamal, and Choon Lai Chiang (all from Malaysia), Professors Valentin Pleşu and Vasile Lavric (Romania), Professor Vasiliki Kazantzi (Greece), Professor Petar Varbanov (Hungary), Professor Abeer Shoaib (Egypt), Professors ChangKyoo Yoo and Jin-Kuk Kim (Korea).

I am also very fortunate to have been working with a group of excellent young researchers throughout my academic career. Some of them came to my research group as short-term visiting scholars, and now have established themselves as excellent scholars. They include Dr. Seingheng Hul (Cambodia), Dr. Jui-Yuan Lee (Taiwan), Professor Chun Deng, Tianhua Wang, Yuhang Zhou, Wei Li, Zengkun Wen, Dr. Jin Sun (all from China), Sophanna Nun (Cambodia), Dr. Gopal Chandra Sahu (India), Choon Hock Pau (Malaysia), Edward L. Z. Yong (Brunei), and Dr. Grzegorz Poplewski (Poland). Besides, the undergraduate and graduate students whom I have guided their projects on resource conservation are worth mentioning. They include Mahendran Subramaniam, Jun Hoa Chan, Shin Yin Saw, Liangming Lee, Melwynn K. Y. Leong, Ming Hann Lim, Sinthi Laya Thillaivarrna, Victor Francis Obialor Onyedim, Dr. Diban Pitchaimuthu, Dr. Rex T. L. Ng, Hoong Teng Loh, Chee Yan Wong, Yan Beng Ong, Zi Ming Yeo, Eric Chee, Wei Li Soh, Xin Yi Chua, Wan You Kho, Tamelasan Ramanath, and Jia Chun Ang. My academic career would not have been as exciting without the contributions of these young people!

Lastly, I am very grateful to my wonderful family. I am indebted to my wife, Cecilia C. S. Cheah, for her constant support. Her dedication and love toward our family and kids, Irene X. H. Foo, Jessica X. H. Foo, and Helena M. H. Foo, have been crucial in supporting me to continuously excel in my professional career. I would also like to mention the invaluable support I have received from my parents, Tian Juan Foo and Mathilda A. N. Siaw, and from my siblings, Agnes H. H. Foo and Jude C. W. Foo.

Author

Dominic C.Y. Foo is a Professor of Process Design and Integration at the University of Nottingham, Malaysia, and is the Founding Director of the Centre of Excellence for Green Technologies. He is a Fellow of the Academy of Sciences Malaysia (ASM), a Fellow of the Institution of Chemical Engineers (IChemE), a Fellow of the Institution of Engineers Malaysia (IEM), Chartered Engineer (CEng) with the Engineering Council UK, Professional Engineer (PEng) with the Board of Engineer Malaysia (BEM), ASEAN Chartered Professional Engineers (ACPE), as well as the Past President for the Asia Pacific Confederation of Chemical Engineering (APCChE). Professor Foo served on International Scientific Committees for many important international conferences (e.g., PRES, FOCAPD, ESCAPE, PSE, SDEWES, etc.). Professor Foo is the Editor-in-Chief for *Process Integration and Optimization for Sustainability*, Subject Editor for *Process Safety & Environmental Protection*, and editorial board members for many other renowned journals. He is the winners of the Innovator of the Year Award 2009 of IChemE, Young Engineer Award 2010 of IEM, Outstanding Young Malaysian Award 2012 of Junior Chamber International, Outstanding Asian Researcher and Engineer 2013 (Society of Chemical Engineers, Japan), Vice-Chancellor's Achievement Award 2014 (University of Nottingham), Top Research Scientist Malaysia 2016 (ASM), and top 2% world-renowned scientist according to Stanford List since 2020. He examines the thesis and conducts professional workshops for academics and industrial practitioners worldwide.

How to Make Use of This Book

Instructors for Process Design Courses

Incorporating sustainability concepts into the undergraduate syllabus has been a common practice in chemical engineering in the past decade. Hence, topics such as environmentally benign design, heat recovery system, and resource conservation are among those that have attracted the attention of academics who conduct process design courses at undergraduate and graduate levels. This may also be partially due to the influence of industrial practitioners.

Nevertheless, balancing the topics to be covered in an undergraduate process design course is indeed a challenge. In most universities, process design is taught at the final year (senior—year 4 level) of an undergraduate chemical engineering program along with the capstone design project. In some cases, only limited lecture series are carried out throughout the design project (i.e., no formal taught course elements). Hence, the course instructors will have to balance between the necessary materials needed in a typical design project (e.g., process synthesis, equipment design, process simulation, etc.) while adding some extra elements where they feel relevant.

Tables 0.1 and 0.2 show a list of suggested topics to be included in a typical undergraduate process design course, along with sections that may be excluded. For an introductory level, it is suggested to cover the basic concept of resource conservation network (RCN) with direct water reuse/recycle alone (water recovery is easiest for understanding and also for implementation), with suggested chapters in Table 0.1. This typically requires a duration of one to two weeks (assuming three contact hours per week). For an intermediate-level course that runs between three to five weeks, or

TABLE 0.1

Suggested Topics for Undergraduate Chemical Engineering Process Design Course (Introductory Level)

Chapters (Ch)	Sections That May Be Excluded
Ch 2: Data extraction	Sections 2.4 (refinery hydrogen network) and 2.5 (property integration), if focus is on water recovery
Ch 3: Graphical targeting for direct reuse/recycle	Sections 3.3 (multiple resources), 3.4 (threshold problems), and 3.5 (property integration)
Ch 8: Network design and evolution	Sections 8.3 (design of material regeneration network) and 8.4 (network evolution techniques)

TABLE 0.2

Suggested Topics for Undergraduate Chemical Engineering Process Design Course (Intermediate Level)

Chapters (Ch)	Sections That May Be Excluded
Ch 2: Data extraction	—
Ch 3: Graphical targeting for direct reuse/recycle	Section 3.4 (threshold problems)
Ch 4: Algebraic targeting for direct reuse/recycle	Section 4.3 (threshold problems)
Ch 5: ATM for direct reuse/recycle	Section 5.4 (ATM for bilateral problems)
Ch 6: ATM for material regeneration network	Section 6.3 (mass exchanger as regeneration unit)
Ch 8: Network design and evolution	Section 8.4 (network evolution techniques)
Ch 9: Superstructural approach[a]	Section 9.5 (multi-objective optimization)
Ch 13: Heat exchanger network[a]	Section 13.6 (utility selection)

[a] These few chapters are a matter of choice, based on the preference of instructors.

specialized/optional modules that run throughout the entire semester (e.g., for graduate level), various chapters of the book may be covered following the suggested topics in Table 0.2. Students will be trained to synthesize an RCN with reuse/recycle and regeneration schemes in this intermediate-level course. If the instructor feels that the students should learn some basic optimization techniques, Chapter 9 may be included. Note again that there are sections that may be excluded due to time constraints (See Tables 0.1 and 0.2).

Based on my own experience, the most effective way of teaching resource conservation topics in this book is through the "hands-on" approach with practical/workshop sessions. These are needed in order for the students to get to experience the "know-how" in solving an RCN problem. Hence, it is best to allocate at least 50% of the lecture time in allowing students to practice through the examples in the book. Instructors are encouraged to make use of the various teaching materials (PowerPoint slides, worksheets, calculation files) that accompany the book in conducting the course/workshop.

Instructors for Professional Courses

As resource conservation is a good means to achieve sustainability for the process industry, it has attracted lots of attention among industrial practitioners

TABLE 0.3

Suggested Topics for Optional Courses

Half-day Workshop	One-day Workshop	Two-day Workshop
Ch 2; Data extraction	Ch 2: Data extraction	Ch 2: Data extraction
Ch 3; Graphical targeting for direct reuse/recycle	Ch 3; Graphical targeting for direct reuse/recycle	Ch 3: Graphical targeting for direct reuse/recycle
Ch 8: Network design and evolution	Ch 4: Algebraic targeting for direct reuse/recycle[a]	Ch 4: Algebraic targeting for direct reuse/recycle
Ch 9: Superstructural approach[a]	Ch 12: Superstructural approach[a]	Ch 5: ATM for direct reuse/recycle
		Ch 8: Network design and evolution
		Ch 6: ATM for material regeneration
		Ch 7: Process changes[a]
		Ch 9: Superstructural approach

[a] These few chapters are a matter of choice, based on the preference of instructors. Refer to Tables 0.1 and 0.2 for sections that may be excluded for each chapter.

over the last decades. I have myself conducted many successful workshops in various countries on resource conservation. The typical suggested topics are listed in Table 0.3 for professional courses of different durations.

1

Introduction

Resource conservation is one of the most important elements for establishing sustainability in modern society. The growth of the global population, overextraction, mismanagement of natural resources, climate change, and global water availability are some of the factors that need serious consideration to achieve the goal of sustainability. By practicing resource conservation, we address various Sustainable Development Goals (SDGs) of the United Nations. In particular, material recovery such as water reuse/recycle are well suited to support SDG6 (clean water and sanitation) and SDG12 (responsible consumption and production). On the other hand, energy recovery techniques address SDG7 (clean and affordable energy) and SDG13 (climate action).

To this end, we are seeking a generic family of techniques that address the various resource conservation problems of a similar nature.

1.1 Motivating Examples

Many regard resource conservation as a simplistic problem, with a straightforward solution. Some plant operation personnel tend to leave this problem to the consultants in the belief that they will provide them with quick and sensible solutions. However, it is always good to have a benchmark in order to compare with the solution proposed by an external consultant.

Let us begin this discussion with several motivating examples. Figure 1.1 shows an acrylonitrile (AN) plant where the wastewater treatment facility is operated at full hydraulic capacity (El-Halwagi, 1997, 2006). Since water is a by-product of the reactor, any increase in AN production will lead to a larger wastewater flowrate and hence overload the existing wastewater treatment facility (see detailed description in Example 2.1).

To debottleneck the wastewater treatment facility of the AN plant, different parties may adopt different solution strategies. If the problem is approached by a wastewater treatment consultant, a typical solution would be to construct a larger wastewater treatment plant to handle the desired wastewater throughput. However, this involves substantial capital investment and operating cost, which may not even justify the desired production expansion.

DOI: 10.1201/9781003173977-1

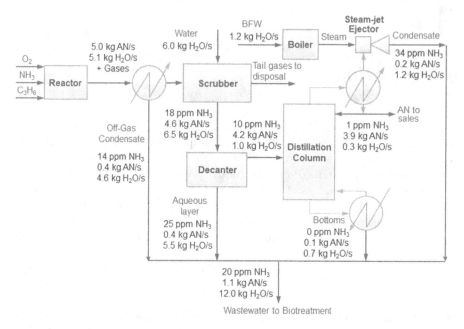

FIGURE 1.1
AN production plant. (From El-Halwagi, 1997. With permission.)

On the other hand, a process or utility engineer of the AN plant who has been instructed by the plant manager to look into the debottlenecking problem may opt for different options. He or she may want to carry out wastewater recycling in order to reduce the wastewater flowrate to the treatment facility. He or she may then call up various suppliers to look for a good filtration unit (or something similar) to purify the terminal wastewater stream for recycling purposes. This seems to be the most common industrial practice, especially for those who claim it to be a straightforward solution. However, this may not be as straightforward as it seems to be. One would certainly need to worry about the various contaminants present in the terminal wastewater stream, since it is contributed by various wastewater streams from different sections of the plant. Next, when the engineer presents the proposal to the plant manager, the plant manager may in turn question the engineer on the choice of the filtration unit, as there could be other types of purification units that might be cheaper and worth exploring. He or she may also question if any of the individual wastewater streams could be directly recovered before any treatment; and if yes, what would be the maximum recovery for the plant? The engineer has more questions in his or her mind. He or she will have to sort out the various suggestions to figure out an economically viable solution. In fact, if one were to explore the various options possible for water recycling in this problem, there are easily a dozen of them, ranging from various direct water recovery options as well as those associated with the use of purification units (see El-Halwagi, 2006 for details). Note that each of these

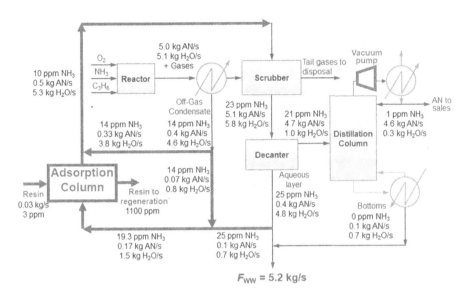

FIGURE 1.2
Optimum water recovery option for AN case.

solutions is unique and needs detailed investigation. Obviously, industrial practitioners have no such luxury of time to inspect each solution in detail.

The optimum solution for this case is given in Figure 1.2. The solution includes an installation of a resin adsorption unit that purifies the effluent from the decanter and off-gas condensate, before it is recovered to the scrubber. Note also that a big portion of the off-gas condensate is also recovered to the scrubber directly. Besides, the boiler that runs the steam-jet ejector of the distillation column is replaced by a vacuum pump, which eliminates the condensate from the steam-jet ejector. This solution is generated by a systematic approach[1] and is not immediately obvious to the engineer through an ad hoc approach.

Let us now look at a second case study, where the solvent is utilized in a metal degreasing process (Kazantzi and El-Halwagi, 2005). In order to reduce the solvent waste of the process, suggestions are made to recover the solvent from the condensate streams (Figure 1.3; see detailed description in Example 2.5).

The optimum solution for this case is given in Figure 1.4. For this case, the operating condition of the thermal processing unit is modified to allow direct recovery of the condensate streams (without any purification unit). Similar to the AN case, the solution for this case is generated via a systematic design tool[2] and is not obvious to even an experienced engineer.

For a third example, let us look at Figure 1.5, which shows a refinery hydrogen network. It is desirable to reduce both the fresh hydrogen feed and the waste gas; the latter is currently purged and is used as fuel.

---

[1] See the generation of solution for this example in Examples 3.1, 6.3 and Section 7.2.
[2] See discussion of this case study in Examples 3.1, 6.3 and Example 7.1

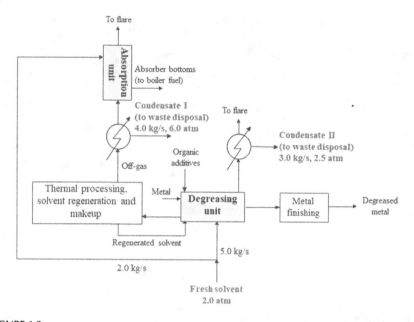

FIGURE 1.3
A metal degreasing process. (From Kazantzi and El-Halwagi, 2005. Reproduced with permission. Copyright © 2005. American Institute of Chemical Engineers [AIChE].)

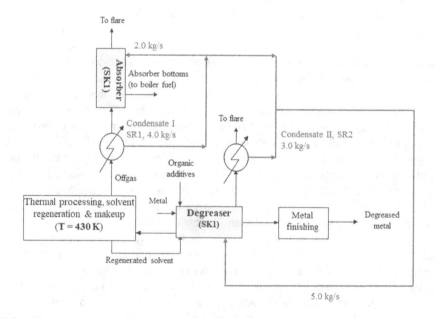

FIGURE 1.4
Solvent recovery scheme for a metal degreasing process. (From Kazantzi and El-Halwagi, 2005. Reproduced with permission. Copyright © 2005. American Institute of Chemical Engineers [AIChE].)

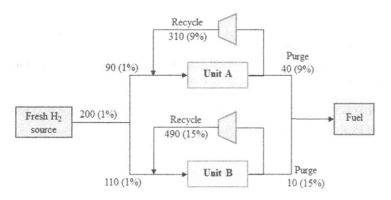

FIGURE 1.5
A refinery hydrogen network (stream flowrate given in million standard cubic feet/day-MMscfd; its purity is given in parenthesis). (From Hallale and Liu, 2001. Reproduced with permission. Copyright © 2001. Elsevier.)

The minimum cost solution for the hydrogen recovery network is given in Figure 1.6. As shown, apart from direct recovery of hydrogen sources, a pressure swing adsorption (PSA) unit is used to purify some hydrogen sources for recovery. Again, this does not seem to be an obvious solution, even for an experienced process engineer. However, the recovery scheme in Figure 1.6 may be generated by a series of systematic design steps.[3]

From the earlier examples, it is obvious that developing a material recovery system may not be as straightforward as commonly thought, even for relatively simple processes. The question "is there a better design?" often arises among process engineers. Furthermore, we realize that material recovery may involve various types of resources, e.g., water, utility gases, and solvents.

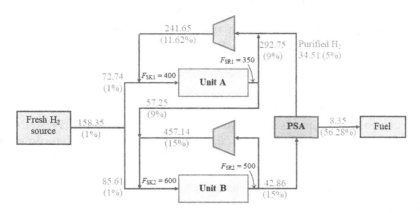

FIGURE 1.6
Hydrogen recovery for a refinery hydrogen network (stream flowrate given in million standard cubic feet/day-MMscfd; its purity is given in parenthesis).

---

[3] See the generation of solution for this example in Examples 6.6 and Section 7.2.

Hence, a systematic approach is needed to develop an optimum flowsheet for such a system. We are in need of a "quick kill" solution that addresses this problem based on some generic principles common to all material recovery systems. In a nutshell, the approach should answer the following questions that often arise in a material recovery problem:

- What is the maximum recovery potential for a given material recovery problem?
- What is the minimum waste generation for a process?
- If a purification unit is to be used, what is the minimum capacity for the unit?
- What is the minimum cost solution for a given material recovery problem?

Is there such a technique that can answer the questions given earlier? Fortunately, there is. With the development of various systematic process design techniques in the past decades, it is now possible to set rigorous performance targets for a material recovery problem. One of such promising techniques is known as *process integration*, which consists of two important elements, i.e., *process synthesis* and *process analysis* (El-Halwagi, 1997).

1.2 Process Synthesis and Analysis

Until recently, many regarded process design exercise as performing a process simulation study for a given flowsheet. This is indeed not true. A typical process design exercise should consist of both elements of *process synthesis* and *process analysis*. The term "process synthesis" was first proposed in the late 1960s by Rudd (1968). The definition for the term has evolved over the years, and one commonly acceptable definition takes the form of "the discrete decision-making activities of conjecturing (1) which of the many available component parts one should use, and (2) how they should be interconnected to structure the optimal solution to a given design problem" (Westerberg, 1987, p. 128). In a nutshell, we are aiming to produce an optimum (unknown) process flowsheet, when the process input and output streams are provided (Figure 1.7).

On the other hand, in process analysis, we aim to predict how a defined process would actually behave under a given set of operating conditions. This involves the decomposition of the process into its constituent elements (e.g., units) for performance analysis. In most cases, once a process flowsheet is synthesized, its detailed characteristics (e.g., temperature, pressure, and flowrates) are predicted using various analytical techniques. In other

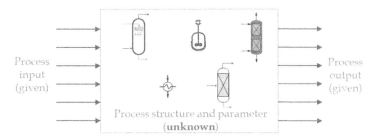

FIGURE 1.7
A process synthesis problem. (From El-Halwagi, 2006. Reproduced with permission. Copyright © 2006. Elsevier Inc.)

words, we aim to predict the process outputs for given process inputs and their flowsheet (Figure 1.8). Most often, various commercial process simulation software may be used, for both continuous (e.g., Aspen HYSYS, Aspen Plus, UniSim Design, and DESIGN II for Windows) and batch processes (e.g., SuperPro Designer).

Process synthesis and process analysis supplement each other in a process design exercise. Through process analysis, the basic conditions of a process flowsheet (e.g., mass and energy balances) may be obtained. We can then make use of the process synthesis technique to develop an optimum flowsheet. Note that in some instances, we may also reverse the task sequence of synthesis and analysis. For such a case, a process flowsheet is first constructed with the process synthesis technique; it is then analyzed to obtain its process conditions. The interactions of process synthesis and analysis in a design exercise are depicted in Figure 1.9.

Figure 1.10 shows the various mass and energy input and output streams for a processing unit. With process synthesis and analysis, the mass-energy dimensions for a chemical process are better understood. This is essential in setting performance targets for a material recovery problem.

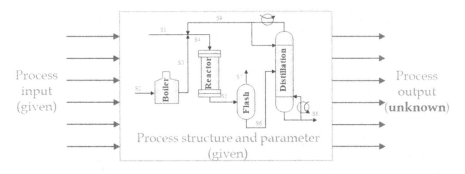

FIGURE 1.8
A process analysis problem. (From El-Halwagi, 2006. Reproduced with permission. Copyright © 2006. Elsevier Inc.)

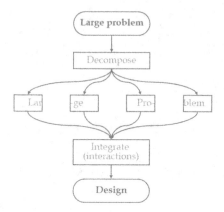

FIGURE 1.9
Interactions of process synthesis and analysis. (From Westerberg, 1987. Reproduced with permission. Copyright © 1987. Wiley-VCH Verlag GmbH & Co. KGaA.)

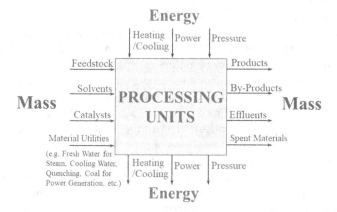

FIGURE 1.10
Mass-energy dimensions for a processing unit. (From El-Halwagi, 1997. With permission.)

We next discuss the main technique that can be used to set rigorous performance targets, i.e., *process integration*.

1.3 Process Integration: A Brief Overview

El-Halwagi (1997, 2006) defined process integration as *a holistic approach to process design, retrofitting, and operation, which emphasizes the unity of the process*. To date, three main branches of process integration problems have been established, i.e., *heat, mass*, and *property integration*. They are briefly described as follows.

The first trend in process integration was dedicated to *heat integration* problems. It was first introduced in the early 1970s and gained great attention during the first oil crisis a few years back then. The technique was first established for the optimal synthesis of *heat exchanger network* (HEN; Hohmann, 1971; Linnhoff and Flower, 1978; Umeda et al., 1978), followed by the efficient design of various energy-intensive processes, e.g., distillation and utility system. With the extensive development of heat integration techniques in the 1980s and 1990s, this trend has reached a mature stage in recent years. Many reviews (e.g., Gundersen and Naess, 1988; Linnhoff, 1993; Furman and Sahinidis, 2002) and textbook materials (Linnhoff et al., 1982; Douglas, 1988; Shenoy, 1995; Smith, 1995, 2016; Biegler et al., 1997; Turton et al., 1998; Kemp and Lim, 2020; Klemeš et al., 2010; El-Halwagi, 2017; Seider et al., 2019) are readily available.

Following the analogy between heat and mass transfer, *mass integration* was then established in the late 1980s for the synthesis of a mass exchange network (MEN, El-Halwagi and Manousiouthakis, 1989). The technique was further extended into other subcases of MEN synthesis problems, e.g., reactive MEN, combined heat and reactive MEN, and waste-interception network. References for this trend are available in review articles (El-Halwagi, 1998; El-Halwagi and Spriggs, 1998; Dunn and El-Halwagi, 2003) and textbooks (El-Halwagi, 1997, 2006, 2017; Seider et al., 2016). It is worth mentioning that the main driving force for the development of these techniques was the waste minimization initiative.

Apart from mass integration problems that are concentration-based, *property integration* has been established as a new trend in the last decade. The technique is based on property and functionality tracking of stream quality, since many chemical processes are characterized by chemical or physical properties rather than composition (Shelley and El-Halwagi, 2000; El-Halwagi, 2006, 2017).

Within the domain of mass and property integration, there exist some special cases of subareas that focus on material recovery, which has been initiated since the mid-1990s. These include the introduction of *water minimization* (Wang and Smith, 1994) and refinery hydrogen network problems (Towler et al., 1996). At a later stage, the techniques were extended to cover the recovery of various other materials, such as solvents, solids, etc. (see Kazantzi and El-Halwagi, 2005; Foo et al., 2006). In recent years, process integration researchers have collectively termed this special area of interest as *resource conservation network* (RCN) synthesis, which addresses the optimum use of various material resources, such as water, utility gases (e.g., hydrogen and nitrogen), solvent, paper, plastic waste, etc. Some recent archives have been made available in review articles (Bagajewicz, 2000; Foo, 2009; Gouws et al., 2010; Jeżowski, 2010; Kamat et al., 2022), handbook (Klemeš, 2022) and textbook materials too (Mann and Liu, 1999; Smith, 2016; El-Halwagi, 2006, 2017; Klemeš et al., 2010; Majozi, 2010). This is also the main subject of this book, which aims to compile the most state-of-the-art techniques developed in the past decade in RCN synthesis.

Besides, there are also other emerging trends of process integration that are worth mentioning, e.g., carbon management network (Tan and Foo, 2007; Foo and Tan, 2016, 2020), supply chain production planning (Singhvi and Shenoy, 2002; Foo et al., 2008b), process scheduling (Foo et al., 2007), human resource planning (Foo et al., 2010), and financial analysis (Zhelev, 2005; Bandyopadhyay et al., 2016).

We might now wonder how the aforementioned process integration problems are interconnected. In fact, there exist some similarities in all these problems. In summary, all these problems may be described in terms of stream *quantity* and *quality* aspects. For instance, in HEN and other heat integration problems, we rely on the minimum approach temperature (quality) for feasible heat transfer (quantity). On the other hand, for MEN and concentration-based RCN problems, the impurity concentration (quality) will ensure that feasible mass transfer (quantity) takes place. The same principle applies to other problems, as summarized in Table 1.1 (Sahu and Bandyopadhyay, 2011).

It is also worth noting that within the process integration community, there exist two distinct approaches to address the various RCN problems, i.e., *insight-based pinch analysis* and *mathematical optimization* techniques. Both of these techniques have their strengths and limitations. The insight-based techniques are well accepted in providing good insights for the designers. They can be used to set performance targets for a given problem before the detailed design of an RCN. However, they can only handle a single quality constraint at one time. In contrast, mathematical optimization techniques do

TABLE 1.1

Quantity and Quality Aspects of Various Process Integration Problems

Quantity	Quality	Examples/Problems	References
Heat	Temperature	Heat exchange network	Linnhoff et al. (1982)
		Heat integration	Smith (1995, 2016)
Mass	Concentration	Mass exchange network	El-Halwagi and Manousiouthakis (1989)
		Water minimization	Wang and Smith (1994)
		Refinery hydrogen network	Towler et al. (1996)
Mass	Properties	Property-based RCNs	Kazantzi and El-Halwagi (2005)
Steam	Pressure	Cogeneration	Dhole and Linnhoff (1993)
Energy	CO_2	Carbon-constrained energy planning	Tan and Foo (2007)
Mass	Time	Supply chain management	Singhvi and Shenoy (2002)
Time	Time	Process scheduling	Foo et al. (2007)
		Human resource planning	Foo et al. (2010)
Energy	Time	Isolated power system	Arun et al. (2007)

Source: From Sahu and Bandyopadhyay, 2011.

TABLE 1.2

Comparison of Insight-Based and Mathematical Optimization Techniques

Approaches	Strengths	Limitations
Insight-based pinch analysis techniques	• Provide good insights for designers • Ability to set performance targets ahead of design	• Handle single quality constraints at one time
Mathematical optimization techniques	• Handle multiple quality constraints simultaneously • Handle complex cases (e.g., process constraints and detailed stream matching) relatively well	• Poor insights for designers • Global optimality may not be guaranteed for certain types of problems

not provide such good insights for designers, but they are powerful in solving complex cases (e.g., process constraints, detailed stream matching, etc.). Comparisons of these approaches are summarized in Table 1.2.

Besides, for the insight-based techniques, one may also categorize the nature of the techniques as graphical or algebraic methods. While both methods essentially obtain the same results, graphical techniques are more powerful in providing good insights for designers as compared with the algebraic techniques. However, the latter are normally more computationally effective, which facilitates the calculation steps, especially for large and complex problems. Besides, the numerical nature of the algebraic techniques also makes them a good platform for software interactions (e.g., process simulation). A summary of the comparison of both methods is given in Table 1.3.

Note that the insight-based technique is a two-step approach. In Step 1, performance targets for a given problem are first identified based on the quality and quantity aspects as discussed earlier. In the case of RCNs, the performance targets correspond to their minimum flowrates of fresh resources and waste discharge. In Step 2, the network structure that achieves the performance targets in Step 1 is then identified; this is commonly termed as the *network design* stage.

It is also worth noting that in recent years, the trend to combine both insight-based and mathematical optimization approaches is becoming more

TABLE 1.3

Comparison of Insight-Based and Mathematical Optimization Techniques

Techniques	Strengths	Limitations
Graphical	• Better insights for designers	• Tedious solution for complex problems • Inaccuracy problems
Algebraic	• Computational effectiveness • Ease for large and complex problems • Interaction with other software, e.g., process simulators, spreadsheets	• Less insights for designers

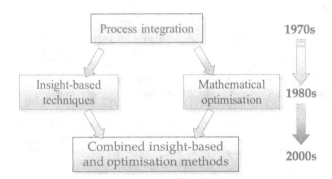

FIGURE 1.11
Development in process integration research in recent years. (Courtesy of Professor Robin Smith, University of Manchester.)

apparent (Smith, 2000). This is different from the early 1980s where developments were mainly focused on the individual approaches (Figure 1.11). As a result, some hybrid approaches (e.g., Alva-Argáez et al., 1998, 1999; Ng et al., 2009a,b,c) have been developed in recent years.

All insight-based, optimization, and hybrid techniques for RCN synthesis are found in this book.

1.4 Strategies for Material Recovery and Types of RCNs

Several terminologies are now widely accepted for RCN synthesis in the process integration community. Following the definition of Wang and Smith (1994, 1995), when an effluent stream of a process is sent to other operations and does not reenter its original process, it is termed as *direct reuse* scheme (Figure 1.12a). On the other hand, if the effluent is permitted to reenter the operation where it was generated, the scheme is termed as *direct recycle* scheme (Figure 1.12b). When the maximum potential for direct reuse/recycle is exhausted, the effluent may be sent for *regeneration* in a *purification unit* where the effluent quality is improved for further reuse (Figure 1.12c) or recycle (Figure 1.12d). When the strategies of reuse, recycle, or regeneration are implemented within the same process/plant, we term them as *in-plant RCN*.

Upon the exhaustion of in-plant material recovery potential, we may then explore the potential of material recovery across different RCNs (Figure 1.13). In practice, these RCNs may be part of a large RCNs that are geographically segregated into different sections, or different process plants that operate with each other through *industrial symbiosis*. We shall collectively term this category of RCN problems as *inter-plant resource conservation networks*

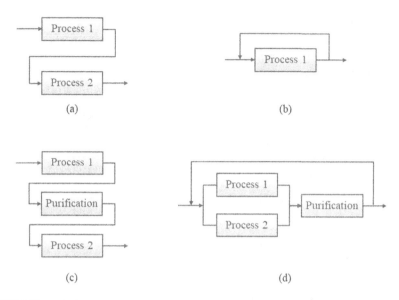

FIGURE 1.12
Strategies for in-plant resource conservation: (a) direct reuse, (b) direct recycle, (c) regeneration reuse, and (d) regeneration recycling. (From Wang and Smith, 1994. With permission.)

(IPRCNs). For an IPRCN, various reuse/recycle and regeneration strategies may be implemented for each individual RCNs, before material recovery is carried out among them.

Apart from the various RCN problems that are operated in continuous mode, there are also RCNs that are operated in batch process mode. Hence, the time dimension of this *batch material network* is of important consideration

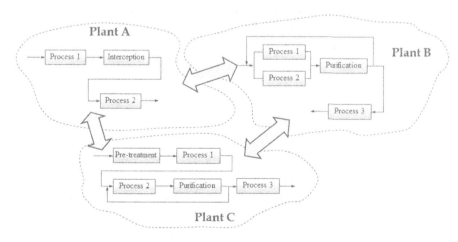

FIGURE 1.13
IPRCN where material recovery is carried out among different RCNs.

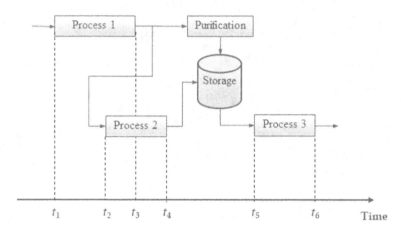

FIGURE 1.14
A batch material network.

in the analysis. Similar to the in-plant RCN problem, reuse/recycle and regeneration strategies may be implemented for a batch material network, as shown in Figure 1.14.

1.5 Problem Statements

To better understand an RCN problem, we should first understand the concept of *process sink* and *source*. For a given process where a resource is found in its outlet streams and needs to be recovered, the outlet stream is termed as a process source. On the other hand, a process sink refers to a unit where a resource is consumed (Figure 1.15).

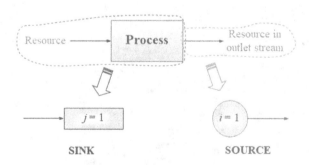

FIGURE 1.15
A material sink and source. (From Foo, 2009. With permission.)

Hence, the formal problem statement for a reuse/recycle network can be stated as follows:

- Given a number of resource-consuming units in the process, designated as process sink, or $SK = \{j = 1, 2, 3, ..., N_{SK}\}$, each sink requires a fixed flow/flowrate requirement of material (F_{SKj}) with a targeted quality index (q_{SKj}). The quality index should satisfy the following constraint:

$$q_{SKj}^{min} \leq q_{SKj} \leq q_{SKj}^{max}, \tag{1.1}$$

where q_{SKj}^{min} and q_{SKj}^{max} are the lowest and highest limits of the targeted quality. Note that the quality index of an RCN varies from one case to another, which may take the form of impurity concentration of a concentration-based RCN (e.g., ppm, mass ratio, and percentage) or property operator values for a property-based RCN (see detailed discussion in Chapter 2).

- Given a number of resource-generating units/streams, designated as process source, or $SR = \{i = 1, 2, 3, ..., N_{SR}\}$, each source has a given flow/flowrate (F_{SRi}) and a quality index (q_{SRi}). Each source can be sent for reuse/recycle to the sinks. The unutilized source(s) are sent for environmental discharge (treated as a sink).

- When the process source is insufficient for use by the sinks, external fresh resources (treated as source) may be purchased to satisfy the requirement of sinks.

  The superstructural representation for this problem is given in Figure 1.16.

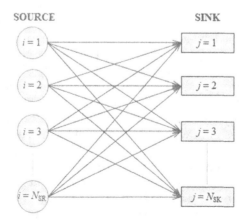

FIGURE 1.16
Superstructural representation for reuse/recycle scheme.

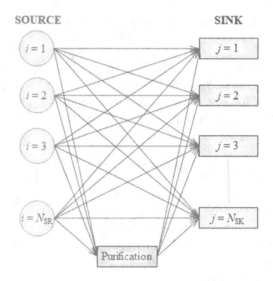

FIGURE 1.17
Superstructural representation for RCN with purification unit.

On the other hand, for material regeneration and pretreatment networks, purification units are used for purifying process sources, before they are recovered to the sinks and/or sent for final waste discharge. In this case, the superstructural representation is given in Figure 1.17, and the following is added to the problem statement:

- Given purification units of known performance, the process source is to be purified for recovery in the sinks and for meeting the environmental discharge limit (Figure 1.17).

For an IPRCN, apart from in-plant reuse/recycle/regeneration, material recovery is also carried out across different RCNs. Hence, the superstructural representations take the form presented in Figure 1.18. Note that we may differentiate them as *direct* and *indirect* integration schemes. In the latter schemes, material recovery is carried out through a *centralized utility hub,* which acts as a hub through which streams are collected before reuse/recycle. In some cases, the centralized utility hub supplies the individual RCNs with fresh resources and provides various purification units for regeneration and waste treatment facilities.

For a batch material network, where process sinks and sources are available in different *time intervals,* direct reuse/recycle and regeneration have to be carried out within the same time interval. Besides, we may also make use of storage tanks to recover process sources (or purified sources) across different time intervals. The superstructural representation for the batch material network is shown in Figure 1.19.

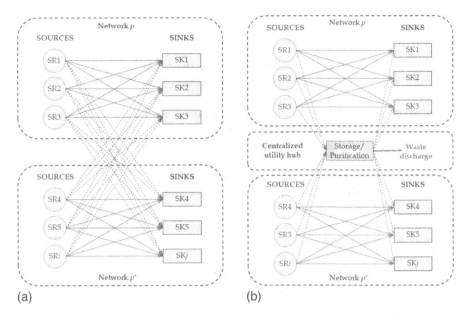

FIGURE 1.18
Superstructural model for IPRCN: (a) direct integration and (b) indirect integration schemes.
(From Chew et al., 2008.)

For all RCN problems described earlier, the objective of the RCN problem may be set to minimize the requirement of the external fresh resources and minimum cost. Note that minimizing the fresh resources also leads to minimum waste generation for an RCN. This achieves the waste minimization objective of an RCN.

This book covers all types of RCN problems described earlier.

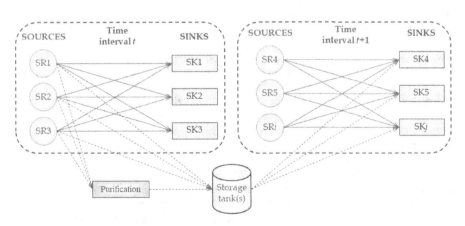

FIGURE 1.19
Superstructural model for a batch material network.

1.6 Structure of the Book

This book has 15 chapters. In Chapter 1, the overview of process synthesis and process integration is presented. The remaining chapters of the book are divided into two parts. Part I of the book focuses on material recovery. We first discuss the various data extraction principles in performing resource conservation activities in Chapter 2, as correct data extraction is the most critical step in ensuring that an optimum material RCN is synthesized. Chapters 3 through 8 in Part I present various insight-based pinch analysis techniques for material RCN synthesis, covering both stages of targeting and network design. In Chapter 9, the basic mathematical programming techniques for RCN synthesis are presented. Next, Chapters 10 through 12 present extended applications where pinch analysis and mathematical programming techniques in Chapters 3–9 are used for the synthesis of IPRCN, batch material networks, and retrofit of RCN.

Part II of the book is dedicated to energy recovery topics. Chapters 13 and 14 present the basics of heat exchange network synthesis and combined heat and power scheme, which is then followed by Chapter 15 where extended applications on heat-integrated RCN are presented.

Readers who are interested in industrial applications should read the Appendix, where the various hypothetical and industrial problems are tracked throughout the book.

References

Alva-Argáez, A., Kokossis, A. C., and Smith, R. 1998. Wastewater minimisation of industrial systems using an integrated approach, *Computers and Chemical Engineering*, 22, S741–744.

Alva-Argáez, A., Vallianatos, A., and Kokossis, A. 1999. A multi-contaminant transshipment model for mass exchange network and wastewater minimisation problems, *Computers and Chemical Engineering*, 23, 1439–1453.

Arun, P., Banerjee, R., and Bandyopadhyay, S. 2007. Sizing curve for design of isolated power systems, *Energy for Sustainable Development*, 11(4), 21–28.

Bagajewicz, M. 2000. A review of recent design procedures for water networks in refineries and process plants, *Computers and Chemical Engineering*, 24, 2093–2113.

Bandyopadhyay, S., Foo, D. C. Y., and Tan, R. R. 2016. Feeling the pinch? *Chemical Engineering Progress*, 112(11), 46–49 (November 2016).

Biegler, L. T., Grossman, E. I., and Westerberg, A. W.1997. *Systematic Methods of Chemical Engineering and Process Design*. Upper Saddle River, NJ: Prentice Hall.

Chew, I. M. L., Ng, D. K. S., Foo, D. C. Y., Tan, R. R., Majozi, T., and Gouws, J. 2008. Synthesis of direct and indirect inter-plant water network, *Industrial & Engineering Chemistry Research*, 47, 9485–9496.

Dhole, V. R. and Linnhoff, B. 1993. Total site targets for fuel, co-generation, emissions, and cooling, *Computers and Chemical Engineering*, 17(suppl), S101–S109.

Douglas, J. M.1988. *Conceptual Design of Chemical Processes*. New York: McGraw Hill.

Dunn, R. F. and El-Halwagi, M. M. 2003. Process integration technology review: Background and applications in the chemical process industry, *Journal of Chemical Technology and Biotechnology*, 78, 1011–1021.

El-Halwagi, M. M. 1997. *Pollution Prevention Through Process Integration: Systematic Design Tools*. San Diego, CA: Academic Press.

El-Halwagi, M. M. 1998. Pollution prevention through process integration, *Clean Product and Processes*, 1, 5–19.

El-Halwagi, M. M. 2006. *Process Integration*. San Diego, CA: Elsevier Inc.

El-Halwagi, M. M. 2017. *Sustainable Design Through Process Integration*, 2nd edition, Butterworth-Heinemann.

El-Halwagi, M. M. and Manousiouthakis, V. 1989. Synthesis of mass exchange networks, *AIChE Journal*, 35(8), 1233–1244.

El-Halwagi, M. M. and Spriggs, H. D. 1998. Solve design puzzle with mass integration, *Chemical Engineering Progress*, 94(8), 25–44.

Foo, D. C. Y. 2009. A state-of-the-art review of pinch analysis techniques for water network synthesis, *Industrial and Engineering Chemistry Research*, 48(11), 5125–5159.

Foo, D. C. Y. and Tan, R. R. 2016. A review on process integration techniques for carbon emissions and environmental footprint problems, *Process Safety and Environmental Protection*, 103, 291–307.

Foo, D. C. Y. and Tan, R. R. 2020. *Process Integration Approaches to Planning Carbon Management Networks*. Boca Raton, Florida, US: CRC Press.

Foo, D. C. Y., Hallale, N., and Tan, R. R. 2007. Pinch analysis approach to short-term scheduling of batch reactors in multi-purpose plants, *International Journal of Chemical Reactor Engineering*, 5, A94.

Foo, D. C. Y., Hallale, N., and Tan, R. R. 2010. Optimize shift scheduling using pinch analysis, *Chemical Engineering*, 2010, 48–52.

Foo, D. C. Y., Kazantzi, V., El-Halwagi, M. M., and Manan, Z. A. 2006. Cascade Analysis for targeting property-based material reuse networks, *Chemical Engineering Science*, 61(8), 2626–2642.

Foo, D. C. Y., Ooi, M. B. L., Tan, R. R., and Tan, J. S. 2008b. A heuristic-based algebraic targeting technique for aggregate planning in supply chains, *Computers and Chemical Engineering*, 32(10), 2217–2232.

Furman, K. C. and Sahinidis, N. V. 2002. A critical review and annotated bibliography for heat exchanger network synthesis in the 20th century, *Industrial and Engineering Chemistry Research*, 41(10), 2335–2370.

Gouws, J., Majozi, T., Foo, D. C. Y., Chen, C. L., and Lee, J.-Y. 2010. Water minimisation techniques for batch processes, *Industrial and Engineering Chemistry Research*, 49(19), 8877–8893.

Gundersen, T. and Naess, L. 1988. The synthesis of cost optimal heat exchange networks—An industrial review of the state of the art, *Computers and Chemical Engineering*, 6, 503–530.

Hallale, N. and Liu, F. 2001. Refinery hydrogen management for clean fuels production, *Advances in Environmental Research*, 6, 81–98.

Hohmann, E. C. 1971. *Optimum Networks for Heat Exchange*, PhD thesis. University of Southern California, Los Angeles, CA.

Jeżowski, J. 2010. Review of water network design methods with literature annotations, *Industrial and Engineering Chemistry Research*, 49, 4475–4516.

Kamat, S., Bandyopadhyay, S., Foo, D. C. Y., and Liao, Z. 2022. State-of-the-art review of heat integrated water allocation network synthesis, *Computers and Chemical Engineering*, 167, 108003.

Kazantzi, V. and El-Halwagi, M. M. 2005. Targeting material reuse via property integration, *Chemical Engineering Progress*, 101(8), 28–37.

Kemp, I. C. and Lim, J. S. 2020. Pinch Analysis for Energy and Carbon Footprint Reduction, 3rd edition. Amsterdam, the Netherlands: Elsevier, Butterworth-Heinemann.

Klemeš, J. J. 2022. *Handbook of Process Integration (PI) 2ⁿᵈ ed.* Cambridge: Woodhead Publishing.

Klemeš, J., Friedler, F., Bulatov, I., and Varbanov, P. 2010. *Sustainability in the Process Industry: Integration and Optimization.* New York: McGraw Hill.

Linnhoff, B. 1993. Pinch analysis: A state-of-art overview, *Chemical Engineering Research and Design*, 71, 503–522.

Linnhoff, B. and Flower, J. R. 1978. Synthesis of heat exchanger networks, *AIChE Journal*, 24, 633–642.

Linnhoff, B., Townsend, D. W., Boland, D., Hewitt, G. F., Thomas, B. E. A., Guy, A. R., and Marshall, R. H. 1982. *A User Guide on Process Integration for the Efficient Use of Energy.* Rugby, U.K: IChemE.

Mann, J. G. and Liu, Y. A.1999. *Industrial Water Reuse and Wastewater Minimization.* New York: McGraw Hill.

Majozi, T. 2010. *Batch Chemical Process Integration: Analysis, Synthesis and Optimization.* London: Springer.

Ng, D. K. S., Foo, D. C. Y., and Tan, R. R. 2009a. Automated targeting technique for single-component resource conservation networks—Part 1: Direct reuse/recycle, *Industrial and Engineering Chemistry Research*, 48(16), 7637–7646.

Ng, D. K. S., Foo, D. C. Y., and Tan, R. R. 2009b. Automated targeting technique for single-component resource conservation networks—Part 2: Single pass and partitioning waste interception systems, *Industrial and Engineering Chemistry Research*, 48(16), 7647–7661.

Ng, D. K. S., Foo, D. C. Y., Tan, R. R., Pau, C. H., and Tan, Y. L. 2009c. Automated targeting for conventional and bilateral property-based resource conservation network, *Chemical Engineering Journal*, 149, 87–101.

Rudd, D. F. 1968. The synthesis of system designs, I. Elementary decomposition theory, *AIChE Journal*, 14(2), 343–349.

Seider, W. D., Lewin, D. R., Seader, J. D., Widagdo, S., Gani, R., and Ng, K. M. (2016). *Product and Process Design Principles: Synthesis, Analysis and Evaluation*, 4th edition. New York: John Wiley and Sons.

Sahu, G. C. and Bandyopadhyay, S. 2011. Holistic approach for resource conservation, Chemical Engineering World, 2011, 104–108.

Shelley, M. D. and El-Halwagi, M. M. 2000. Componentless design of recovery and allocation systems: A functionality-based clustering approach, *Computers and Chemical Engineering*, 24, 2081–2091.

Shenoy, U. V.1995. *Heat Exchanger Network Synthesis: Process Optimization by Energy and Resource Analysis.* Houston, TX: Gulf Publishing Company.

Singhvi, A. and Shenoy, U. V. 2002. Aggregate planning in supply chains by pinch analysis, *Transactions of the Institute of Chemical Engineers, Part A*, 80, 597–605.

Smith, R.1995. *Chemical Process Design*. New York: McGraw Hill.

Smith, R. 2000. State of the art in process integration, *Applied Thermal Engineering*, 20, 1337–1345.

Smith, R. 2016. *Chemical Process: Design and Integration*, 2nd edition. West Sussex, England: John Wiley & Sons.

Tan, R. R. and Foo, D. C. Y. 2007. Pinch analysis approach to carbon-constrained energy sector planning, *Energy*, 32(8), 1422–1429.

Towler, G. P., Mann, R., Serriere, A. J.-L., and Gabaude, C. M. D. 1996. Refinery hydrogen management: Cost analysis of chemically integrated facilities, *Industrial and Engineering Chemistry Research*, 35(7), 2378–2388.

Turton, R., Bailie, R. C., Whiting, W. B., Shaeiwitz, J. A., and Bhattacharyya, D. 2014. *Analysis, Synthesis and Design of Chemical Processes*, 4th edition. Upper Saddle River, NJ: Prentice Hall.

Umeda, T., Itoh, J. and Shiroko, K. 1978. Heat exchange system synthesis, *Chemical Engineering Progress*, July 1978, 70–76.

Wang, Y. P. and Smith, R. 1994. Wastewater minimisation, *Chemical Engineering Science*, 49, 981–1006.

Wang, Y. P. and Smith, R. 1995. Wastewater minimization with flowrate constraints, *Chemical Engineering Research and Design*, 73, 889–904.

Westerberg, A. W. 1987. Process synthesis: A morphology review. In: Liu, Y. A., McGee, H. A. and Epperly, W. R. eds., *Recent Developments in Chemical Process and Plant Design*. New York: John Wiley & Sons.

Zhelev, T. K. 2005. On the integrated management of industrial resources incorporating finances, *Journal of Cleaner Production*, 13, 469–474.

Part I

Material Recovery

In Part I of this book, the readers will learn several well-established insight-based *pinch analysis* techniques for the synthesis of a material *resource conservation network* (RCN). A typical pinch analysis technique consists of two stages. In stage 1, the performance targets for an RCN are first identified based on first principles (i.e., mass balance and thermodynamic constraints)—this is called the *targeting* step. In most cases, the minimum fresh resource(s) needed for an RCN should be determined, since it translates into the minimum operating cost (and minimum fixed cost indirectly) of an RCN. In Chapter 2, various data extraction principles in performing resource conservation activities are first presented. This will guide readers to perform correct data extraction in order to ensure that an optimum material RCN is synthesized. Chapters 3 and 4 present the "traditional" graphical and algebraic pinch analysis techniques that can be used to determine the RCN performance targets. In Chapters 5 and 6, the targeting procedure is assisted with mathematical models where spreadsheet and optimization software are used to determine rigorous RCN targets. Note that all targeting exercises are performed prior to the determination of the actual structure (interconnection) of an RCN. The latter is regarded as stage 2, which is commonly termed as *network design*, where the systematic design procedure is covered in Chapter 8. In Chapter 9, the basics of the mathematical-based superstructural approach for RCN synthesis are introduced. Chapters 3 through 9 should be read sequentially as they cover basic techniques for direct material reuse/recycle, as well as regeneration.

Upon learning the basic RCN techniques in Chapters 2 through 9, readers may move to the extended applications in Chapters 10 through 12, where techniques in those earlier chapters are extended to topics such as inter-plant

RCN (Chapter 10), batch material network (Chapter 11), and retrofit of existing RCN (Chapter 12). Chapters 10 through 12 stand alone, with no interconnection among them.

To facilitate self-learning, all calculation files (spreadsheet, LINGO, etc.) of examples in Chapters 3–15 are made available to the readers on the publisher website (www.routledge.com/9781032003931). However, it is advisable that readers should redo the calculation in order to have a better understanding of the techniques.

2

Data Extraction for Resource Conservation

Data extraction is the most crucial step in any resource conservation activity. If the data for carrying out an analysis are wrongly extracted, we might end up synthesizing a suboptimum resource conservation network (RCN). Hence, the data extraction strategies in this chapter supersede various insight-based and optimization approaches for the rest of the chapters in this book.

As outlined in the problem statements in Chapter 1, there exist several material *sinks* in a process where the material *sources* may be recovered to (i.e., via direct reuse/recycle and/or regeneration). In this chapter, we focus on some basic principles on how to identify these process sinks and sources correctly, along with their limiting data for synthesizing an optimum RCN. Note that for an RCN that is based on mass integration, e.g., water and gas networks, the quality index for maximum recovery corresponds to the impurity concentration in the process sinks and sources. On the other hand, the corresponding quality index for property integration is the linearized operator for the property mixing rule.

2.1 Segregation for Material Sources

To maximize the potential for material recovery, it is of utmost importance to keep the individual source candidates segregated. However, this is not a common practice for a manufacturing process in general. In most cases, many material sources in the plant are normally mixed before being sent to the next processing unit. For instance, effluent streams from units that contain contaminated water are normally mixed in the equalizing tank before being fed to downstream wastewater treatment units. Note that some of these effluent streams may only contain a small amount of impurities (while some streams are more contaminated). Having all effluent streams mixed in the equalizing tank thus degrades the potential for water recovery. Hence, we have the following heuristic:

Heuristic 1—Keep material sources segregated to maximize their recovery potential.

DOI: 10.1201/9781003173977-3

Example 2.1 Data extraction for water minimization in acrylonitrile (AN) production

Figure 2.1 shows a process flow diagram (PFD) for AN production (El-Halwagi, 1997), one of the motivating examples in Chapter 1. AN is produced in a fluidized-bed reactor via vapor-phase ammoxidation of propylene. The reaction is single pass, with almost complete conversion of propylene. The reactor is operated at 2 atm and 450°C, with the stoichiometry shown in the following equation:

$$C_3H_6 + NH_3 + 1.5O_2 \left(\xrightarrow{\text{catalyst}} C_3H_3N + 3H_2O \right) \qquad (2.1)$$

Effluent from the reactor is cooled and partially condensed. The off-gas from the condenser is sent to a scrubber for purification, while its condensate is sent to the biotreatment facility. Freshwater is used as the scrubbing agent in the scrubber, before the tail gas may be sent for disposal. The bottom product from the scrubber is sent to a decanter, where it is separated into aqueous and organic layers. The aqueous phase is sent to the biotreatment facility, while the organic phase (which contains most AN products) is fractionated in a mild vacuumed distillation column. The latter is induced by a steam-jet ejector, where steam condensate is produced (this is sent for biotreatment too). Besides, the

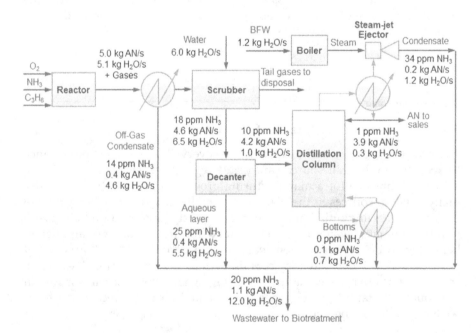

FIGURE 2.1
PFD for AN production. (From El-Halwagi, 1997. With permission.)

distillation column bottom also produces wastewater, which is sent to the biotreatment facility. Detailed material balances for the process are given in Figure 2.1.

Due to increased customer demand, the plant authority is exploring opportunities to increase the plant's overall AN throughput. However, Equation 2.1 reveals that water is a by-product of AN production. Hence, an increased AN production also leads to increased wastewater flowrate to the biotreatment facility. Furthermore, the latter is currently operated at full hydraulic capacity. Hence, any increase in production capacity is only possible upon the debottlenecking of the biotreatment facility, either by reducing the total wastewater flowrate or by installing another biotreatment unit. The latter strategy is reported to be an expensive one (El-Halwagi, 1997). Thus, it has been decided to approach the problem by reducing its total wastewater flowrate, by carrying out water recovery between its process sinks and sources.

Identify the water sinks and sources to carry out a water recovery scheme for the process. Note that for this process, ammonia (NH_3) concentration is the main concern for water recovery (i.e., the quality index in this case).

SOLUTION

From the process description, it is estimated that the two units that require water intake are boiler feed water (BFW) and scrubber. So they are designated as the *water sinks* if water recovery is to be carried out.

On the other hand, it is a common practice to take the terminal wastewater stream that is sent to the wastewater facility as a candidate for water reuse/recycle, i.e., *water source*. Note, however, that doing this does not result in the maximum water recovery potential for the process. As shown in Figure 2.1, the terminal wastewater stream has an NH_3 concentration of 20 ppm, which is actually an average concentration after the mixing of four individual wastewater streams, i.e., off-gas condensate (from condenser), aqueous layer from the decanter, distillation bottom, as well as the steam condensate from the steam-jet ejector. Among the four wastewater streams, distillation bottom is the cleanest source, as it is free from ammonia content. Hence, this stream can be fully recovered. However, when the stream is mixed with other wastewater streams (and reaches a concentration of 20 ppm), its potential for recovery is degraded.

Following Heuristic 1, the individual streams that contribute to the terminal wastewater stream are segregated as water sources for recovery:

1. Distillation bottom (0.8 kg/s; 0 ppm)
2. Off-gas condensate (5.0 kg/s; 14 ppm)
3. Aqueous layer from the decanter (5.9 kg/s; 25 ppm)
4. Steam condensate from steam-jet ejector (1.4 kg/s; 34 ppm)

2.2 Extraction of Limiting Data for Material Sink for Concentration-Based RCN

To maximize the material recovery potential for a concentration-based RCN problem (e.g., water minimization), the following heuristics are useful for material sinks:

Heuristic 2—Minimize the flowrate of a sink to reduce the overall fresh resource intake.

Heuristic 3—Maximize the inlet concentration of a sink to maximize material recovery.

It is a common fact that during plant operation, processing units are normally operated within a given range for their operating parameters. For an RCN, flowrate and inlet concentration are two important parameters for any resource-consuming processes. Similar to other operating parameters, operating flowrate and concentration are also maintained within the range that is bound by their upper and lower limits, such as that shown in Figure 2.2. For the operating flowrate, the lower limit is important to ensure that the unit (e.g., scrubber, which utilizes water as scrubbing agent) is performing according to its design specification (e.g., to purify the sour gas content to a given quality). In contrast, the upper limit is always decided based on the designed capacity of the unit (e.g., to avoid flooding in the column, etc.). A similar situation also applies for inlet concentration for the processes. In this case, the lower and upper limits correspond to the desired operating concentration range that ensures normal operating conditions for the resource-consuming units.

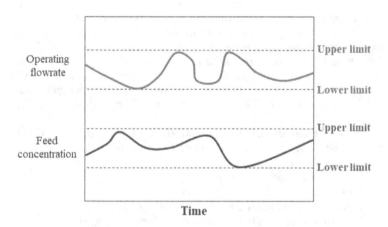

FIGURE 2.2
Historical data for plant operation. (From El-Halwagi, 2006. With permission.)

Note, however, that experience shows that the establishment of the operating range for flowrate and concentration are often not directly measurable. One will have to rely on experience or some empirical models to predict the lower and upper limits of these variables. It is always useful to work with experienced plant personnel to identify these limits. Besides, the safety factor (10%–20% of the most severe cases) should also be incorporated in choosing the operating limits. For instance, a vessel that can tolerate 100 kg/h of water throughput is always operated at its maximum limit of 90 kg/h, i.e., 90% of its design capacity. Hence, even if the upper limit has been taken for a material recovery scheme, it is still within the operating range that has incorporated the safety factor.

Following Heuristics 2 and 3 means that the lower flowrate and upper concentration limits of the process sinks are taken for material recovery. Since the flowrate and concentration values are always taken based on their operating limits, they are often known as the *limiting data*, which will enable the maximum potential to be explored for a material recovery scheme.

Example 2.2 Water minimization for AN production (revisited)

The AN process in Example 2.1 is revisited. To ensure a feasible water recovery scheme, the following process constraints on flowrate and concentration are to be complied with:

i. Scrubber

$$5.8 \leq \text{Flowrate of wash feed } \left(\text{kg/s}\right) \leq 6.2 \tag{2.2}$$

$$0.0 \leq NH_3 \text{ content of wash feed } \left(\text{ppm}\right) \leq 10.0 \tag{2.3}$$

ii. BFW

$$NH_3 \text{ content} = 0.0 \text{ ppm} \tag{2.4}$$

$$AN \text{ content} = 0.0 \text{ ppm} \tag{2.5}$$

iii. Decanter

$$10.6 \leq \text{Feed flowrate } \left(\text{kg/s}\right) \leq 11.1 \tag{2.6}$$

iv. Distillation column

$$5.2 \leq \text{Feed flowrate } \left(\text{kg/s}\right) \leq 5.7 \tag{2.7}$$

$$0.0 \leq NH_3 \text{ content of feed } \left(\text{ppm}\right) \leq 30.0 \tag{2.8}$$

$$80.0 \leq AN \text{ content of feed } \left(\text{wt\%}\right) \leq 100.0 \tag{2.9}$$

TABLE 2.1

Limiting Water Data for AN Production

Water Sinks, SK$_j$		
j Stream	Flowrate F_{SKj} (kg/s)	Concentration C_{SKj} (ppm)
1 BFW	1.2	0
2 Scrubber inlet	5.8	10

Water Sources, SR$_i$		
i Stream	Flowrate F_{SRi} (kg/s)	Concentration C_{SRi} (ppm)
1 Distillation bottoms	0.8	0
2 Off-gas condensate	5	14
3 Aqueous layer	5.9	25
4 Ejector condensate	1.4	34

Identify the limiting data for the process sinks in the AN process to carry out a water minimization study.

SOLUTION

Example 2.1 identifies that the two process sinks for a water minimization study are BFW and the scrubber inlet. Hence, we focus only on the constraints that are related to these sinks (i.e., decanter and distillation column will not be considered). For the BFW, Equation 2.4 indicates that no NH_3 content is tolerated in this sink; hence, an impurity of 0 ppm is taken as the limiting concentration of the latter. Note that its flowrate may be extracted from the PFD in Figure 2.1. On the other hand, Equations 2.2 and 2.3 indicate the operating range for the flowrate and concentration in this sink. Applying Heuristics 2 and 3 leads to the lower flowrate (5.8 kg/s) and upper concentration (10 ppm) limits being identified as the limiting data for this process sink. The limiting data for both sinks are summarized in Table 2.1.

Besides, Example 2.1 also identifies four water sources for the water minimization study, i.e., off-gas condensate, aqueous layer from the decanter, distillation bottom, and steam condensate from the steam-jet ejector. The limiting flowrate and concentration for these sources are also summarized in Table 2.1.

2.3 Data Extraction for Mass-Exchange Processes

Many unit operations in the process industry may be categorized as mass-exchange processes, e.g., washing, absorption, extraction, etc. The typical characteristic of these operations is that mass-separating agents (MSAs,

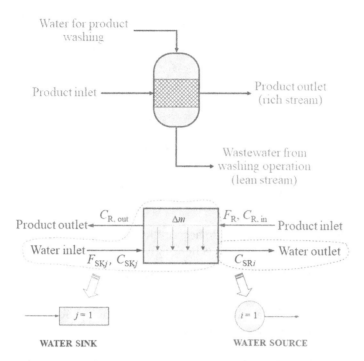

FIGURE 2.3
A product washing operation (a) for the above and (b) for the below.

e.g., water, solvent, etc.) are used to selectively remove certain components (e.g., impurities, pollutants, by-products, products) from a component-laden stream (El-Halwagi, 1997, 2006). The latter is normally termed as the *rich stream*, while the MSA is termed as the *lean stream*. Figure 2.3 shows a typical product washing operation, where water is used as an MSA (lean stream) to partially remove the impurity content from the product (rich) stream.

We may represent the product washing in Figure 2.3a as a countercurrent operation in Figure 2.3b. For most mass-exchange processes, the inlet and outlet flowrates of the operation are normally nearly the same (i.e., with negligible flowrate losses). Hence, for a product stream with flowrate F_R, to be washed in order to reduce its impurity content from inlet ($C_{R,in}$) to outlet concentrations ($C_{R,out}$), the following mass balance equation is used to determine the impurity load (Δm) to be removed from the rich stream:

$$\Delta m = F_R \left(C_{R,in} - C_{R,out} \right). \tag{2.10}$$

Since water is used as an MSA for this operation, the inlet of the operation is hence designated as the water sink, while its outlet as the water source (see Figure 2.3b). Similarly, due to the negligible flowrate losses for the water sink and source, their flowrates are normally kept uniform (i.e., $F_{SKj} = F_{SRi}$).

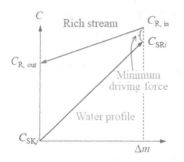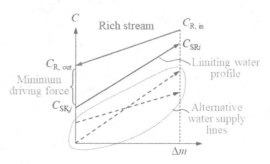

FIGURE 2.4
A product washing operation represented in concentration versus load diagram. (From Wang and Smith, 1994. Reproduced with permission. Copyright ©1994. Elsevier.)

A material balance for the MSA to account for the impurity load *(Am)* that is transferred from the rich stream is hence given as

$$\Delta m = F_{SKj}\left(C_{SRi} - C_{SKj}\right). \tag{2.11}$$

A convenient way of analyzing the mass-exchange processes is the concentration versus load diagram (Wang and Smith, 1994). Such a diagram for a single washing operation in Figure 2.3 is shown in Figure 2.4. As shown, inlet and outlet concentrations of the rich stream as well as those of the water stream (known as the *water profile*) are plotted versus the impurity load transferred between them. If pure freshwater (i.e., zero impurity) is used in this operation, the water profile is drawn from the origin, such as that in Figure 2.4a. To minimize the freshwater flowrate, the slope of the water profile is kept to the maximum, so that the water stream will exit at the highest possible outlet concentration. Note that the latter value is normally separated from the inlet concentration of the rich stream by the minimum driving force, to ensure feasible mass transfer between the two streams. This is shown in Figure 2.4a.

However, doing this only ensures the use of the minimum freshwater flowrate in the washing operation but does not include the consideration of water reuse/recycle, since the inlet concentration is set at zero. Reduction of freshwater is only possible when the washing operation allows the recycling of effluent from the same unit or the reuse of effluent from other units (when there is more than one water-using process). For such a case, the inlet concentration of the washing process should be raised to a higher value, such as that shown in Figure 2.4b. Similar to the outlet concentration, the inlet concentration should take the maximum possible value that is bounded by the minimum mass transfer driving force at the lean end of the concentration versus load diagram to ensure feasible mass transfer, as shown in Figure 2.4b. Note also that this corresponds to Heuristic 3 discussed earlier, i.e., *maximize the inlet concentration of a sink to maximize material recovery.*

When inlet and outlet concentrations of the water profile are both set to their limiting (maximum) values, the latter is termed as the *limiting water profile.* As the name implies, the limiting water profile dictates that the washing operation stream is operated at its minimum flowrate for its maximum inlet and outlet concentrations. Note that since the limiting water profile represents the boundary of the operating concentrations, any alternative water supply lines below the limiting water profile are also feasible in operating the washing process (see Figure 2.4b). However, only when the process is operated at the condition of the limiting water profile (i.e., maximum inlet and outlet concentrations), the maximum water recovery potential for the overall process will be ensured.[1]

It is also worth noting that the earlier discussion is not restricted to washing operations alone. The same principle applies to all mass-exchange processes where MSA is employed, such as absorption, extraction, etc. The limiting profile can be extended to identify the minimum flowrate needed for the MSA(s) in these operations.

Example 2.3 Data extraction for extraction

Figure 2.5 shows an extraction process where an organic solvent (lean stream) is used to purify a product stream that contains an unwanted impurity (rich stream). It may be assumed that there is a negligible flowrate loss in both streams. For all the following scenarios, determine the missing parameters (F_{SK}, C_{SK}, and/or C_{SR}) for the profile of the organic solvent stream:

1. Scenario 1
   - $F_{SK} = 10$ kg/s
   - $C_{SK} = 0$ wt% (i.e., fresh solvent is used)
2. Scenario 2
   - $F_{SK} = 10$ kg/s
   - 0 wt% $\leq C_{SK} \leq 0.005$ wt%
3. Scenario 3
   - 0 wt% $\leq C_{SK} \leq 0.005$ wt%
   - 0 wt% $\leq C_{SK} \leq 0.030$ wt%

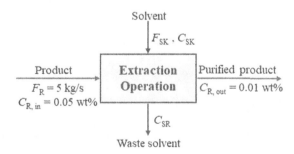

FIGURE 2.5
An extraction operation.

---

[1] This condition applies to RCN of single impurity case, and also to one of the impurity for a multiple impurity RCN, see discussion in Savelski and Bagajewicz (2000, 2003).

SOLUTION

For all scenarios, the impurity load is first determined using Equation 2.10 as follows:

$$5 \text{ kg/s } (0.0005 - 0.0001) = 0.0020 \text{ kg/s.}$$

For Scenario 1, since the inlet flowrate and concentration are fixed, we shall determine the outlet concentration of the waste solvent (C_{SR}) using Equation 2.11:

$$0.0020 \text{ kg/s} = 10 \text{ kg/s } (C_{SR} - 0,)$$

$$C_{SR} = 0.0002 = 0.02 \text{ wt\%.}$$

For Scenario 2, the earlier discussion indicates that the inlet concentration of an extraction operation should be taken at a higher value for maximizing solvent recovery. Hence, for an inlet concentration of 0.005 wt%, the outlet concentration of the waste solvent (C_{SR}) is then determined using Equation 2.11 as

$$0.0020 \text{ kg/s} = 10 \text{ kg/s } (C_{SR} - 0.00005,)$$

$$C_{SR} = 0.00025 = 0.025 \text{ wt\%}$$

In Scenario 3, the inlet and outlet concentrations of the solvent stream are set at the limiting values, and the corresponding limiting solvent flowrate can then be determined as

$$0.0020 \text{ kg/s} = F_{SK} (0.00030 - 0.00005,)$$

$$F_{SK} = 8 \text{ kg/s.}$$

The concentration versus load diagrams for all scenarios are shown in Figure 2.6. For Scenarios 1 and 2, since the solvent flowrate is fixed, we can

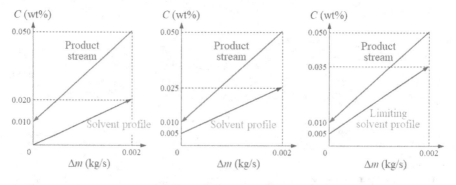

FIGURE 2.6

Concentration versus load diagrams for Example 2.3: (a) Scenario 1, (b) Scenario 2, and (c) Scenario 3—limiting solvent profile.

only alter its inlet and outlet concentrations. Comparing with Scenario 1, the outlet concentration in Scenario 2 is increased by 0.005 wt%. We will have to ensure that it is within the maximum possible value for feasible mass transfer, as discussed in Figure 2.4. For Scenario 3, we obtain the *limiting solvent profile*, where the solvent flowrate is minimized at its limiting inlet and outlet concentrations. This ensures that the solvent flowrate through the process is minimum (i.e. with the steepest solvent profile).

2.4 Data Extraction for Hydrogen-Consuming Units in Refinery

Significant amounts of hydrogen are utilized in refinery operations. Typical hydrogen-consuming units are hydrotreating and hydrocracking processes. In these processes, hydrogen gas is used to remove impurities (e.g., sulfur, nitrogen, etc.) from the crude oil stream. Since there are many hydrogen-consuming units in treating various crude products within a refinery, we may explore the opportunity for hydrogen recovery among these units. However, the extraction of data for hydrogen recovery is not as straightforward as compared with the previously discussed resource-consuming processes. In this section, we shall focus on the general operation of a typical hydrogen-consuming unit and show how data extraction could be carried out.

Figure 2.7 shows a simplified PFD for a hydrogen-consuming unit (Hallale and Liu, 2001; Alves and Towler, 2002). A liquid hydrocarbon feed is mixed with a hydrogen-rich gas stream, before being heated and sent to the hydrotreating/hydrocracking reactor. In the reactor, some hydrogen is consumed, while other light hydrocarbon (e.g., methane, ethane, etc.) and gaseous components (e.g., H_2S, NH_3) are formed. The reactor effluent is cooled and sent to a high-pressure flash separator. The bottom liquid product of the

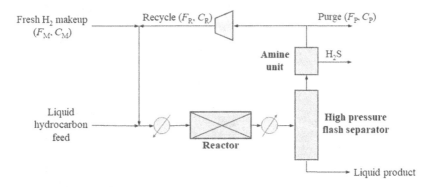

FIGURE 2.7
A simplified PFD for a hydrogen-consuming unit. (From Alves and Towler, 2002. Reproduced with permission. Copyright © 2002. American Chemical Society.)

separator is normally sent to a low-pressure separator for further purification. The top stream from the flash separator contains a significant amount of hydrogen, which is always sent for recycling. Before recycling, the gas stream is treated with an amine unit for H_2S removal. A small portion of the treated gas stream is then purged (normally to fuel or flare systems) to prevent the buildup of the contaminant, while the remainder of the gas stream is recompressed and recycled to the reactor, with some fresh hydrogen makeup.

In analyzing the hydrogen-consuming unit, we may mistakenly think that the purge stream in the recycle loop is the hydrogen source (to be sent for reuse/recycle), while the hydrogen sink corresponds to the fresh hydrogen makeup. However, this is not the case. The actual sink for a hydrogen-consuming unit is its net requirement of hydrogen that is fed to the hydrotreating/hydrocracking reactor. In other words, this corresponds to the hydrogen gas stream that is mixed with the liquid hydrocarbon feed (Figure 2.8). Since this stream is a mixture of fresh hydrogen makeup and recycle streams, we can perform simple mass balance to determine its flowrate and concentration, once the flowrates and concentrations of the makeup and recycle streams are known.

Note that the individual streams in a hydrogen-consuming unit are normally characterized in terms of its hydrogen purity (mol% or vol%). However, it is often more convenient to carry out the analysis by characterizing its impurity concentration, which is the collective impurity content in the stream. For the given flowrates and impurity concentrations of the makeup and recycle streams, the flowrate and impurity concentration of the hydrogen sink may be determined using Equations 2.12 and 2.13, respectively.

$$F_{SK} = F_M + F_R, \qquad (2.12)$$

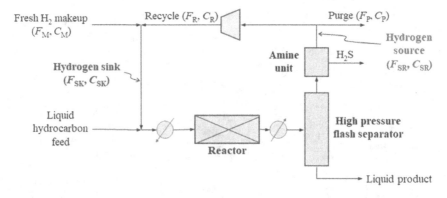

FIGURE 2.8
Identification of hydrogen sink and source. (From Hallale and Liu, 2001. Reproduced with permission. Copyright © 2001. Elsevier.)

$$C_{SK} = \frac{F_M C_M + F_R C_R}{F_M + F_R},$$ (2.13)

where F_{SK}, F_{MU}, and F_{RCY} are the respective flowrates of the hydrogen sink, makeup, and recycle streams, with their respective impurity concentrations of C_{SK}, C_{MU}, and C_{RCY}.

On the other hand, the actual source for a hydrogen-consuming unit is the stream that emits from the amine unit (i.e., after being treated for H_2S removal; see Figure 2.8). This stream is taken as the source as it contains the entire flowrate of hydrogen that may be considered for potential reuse/recycle. In other words, the summation of the recycle and purge stream flowrates gives the total flowrate of the hydrogen source (Equation 2.14), while the impurity concentrations of all these streams are essentially the same, given as in Equation 2.15:

$$F_{SK} = F_{RCY} + F_P,$$ (2.14)

$$C_{SR} = C_{RCY} = C_P.$$ (2.15)

Example 2.4 Data extraction for a refinery hydrogen network

Figure 2.9 shows a PFD for a refinery hydrogen network (Hallale and Liu, 2001), one of the motivating examples in Chapter 1. At present, two hydrogen-consuming processes, i.e., Units A and B, are fed by a fresh hydrogen source, and purge a significant amount of unused hydrogen as fuel. Note that both units have an existing internal hydrogen recycle stream.

Extract the data for this case to carry out a hydrogen recovery analysis.

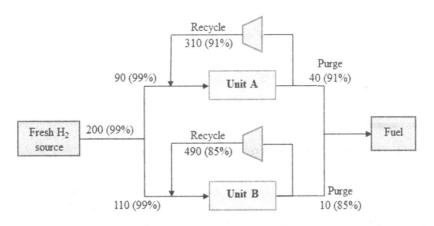

FIGURE 2.9

PFD of a refinery hydrogen network (stream flowrate given in million standard cubic feet/day—MMscfd; its hydrogen purity is given in parentheses). (From Hallale and Liu, 2001. Reproduced with permission. Copyright © 2001. Elsevier.)

TABLE 2.2

Data for Example 2.4

	Makeup Stream	Recycle Stream	Purge Stream	Hydrogen Sink	Hydrogen Source
Unit A					
Flowrate (MMscfd)	90	310	40	400	350
Concentration (%)	1	9	9	7.20	9
Unit B					
Flowrate (MMscfd)	110	490	10	600	500
Concentration (%)	1	15	15	12.43	15

SOLUTION

In the first step, all hydrogen purity in the makeup, recycle, and purge streams of both Units A and B are converted into impurity concentration, given in columns 2–4 in Table 2.2.

To identify the flowrate ($F_{SK,A}$) and concentration ($C_{SK,A}$) for the hydrogen sink of Unit A, Equations 2.12 and 2.13 are utilized, i.e.,

$$F_{SK,A} = 90 + 310 = 400 \text{ MMscfd}$$

$$C_{SK,A} = \frac{90(1) + 310(9)}{90 + 310} = 7.2\%$$

Equations 2.14 and 2.15 are then utilized to calculate the flowrate ($F_{SR,A}$) and concentration ($C_{SR,A}$) for hydrogen source of Unit A, i.e.,

$$F_{SR,A} = 310 + 40 = 350 \text{ MMscfd}$$

$$C_{SR,A} = 9\%$$

Similarly, flowrates and concentrations for the hydrogen sinks and sources for Unit B are also calculated using Equations 2.12 through 2.15. Data for both sinks are summarized in the last two columns of Table 2.2.

2.5 Data Extraction for Property Integration

Unlike water minimization and gas recovery, which are both concentration based, property integration involves properties or functionalities of streams that are not driven by their chemical constituents. Hence, to extract the correct limiting data for a property integration study, the basic background of the property mixing rule needs to be discussed. This is shown with an example in Figure 2.10, which involves the mixing of solutions A and B, with respective density ρ_A and ρ_B, mass flowrate F_A and F_B, and volumetric flowrate V_A

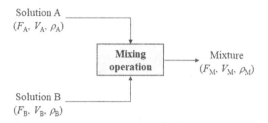

FIGURE 2.10
Mixing of solutions with different density.

and V_B. The resulting mixture consists of mass (F_M) and volumetric flowrates (V_M) and a mean density of ρ_M.

Assuming that the volumetric flowrate of the mixture is the sum of the individual solution flowrates, i.e.,

$$V_M = V_A + V_B. \tag{2.16}$$

Equation 2.16 may be written as

$$\frac{F_M}{\rho_M} = \frac{F_A}{\rho_A} + \frac{F_B}{\rho_B} \tag{2.17a}$$

and rearranged as

$$\frac{1}{\rho_M} = x_A \frac{1}{\rho_A} + x_B \frac{1}{\rho_B} \tag{2.17b}$$

where x_A and x_B are the mass fraction of solutions A and B in the mixture, respectively. A more generic form of this *mixing rule* is given as

$$\frac{1}{\rho_M} = \sum_i \frac{x_i}{\rho_i} \tag{2.17c}$$

where ρ_i is the density of solution i.

Note that Equation 2.17c is a nonlinear equation, which may lead to a complication in calculation (especially when the equation becomes highly nonlinear). A more convenient way of representation is its linearized form, given as

$$\psi_M(\rho_M) = \sum_i x_i \psi_i \left(\frac{1}{\rho_i} \right) \tag{2.18}$$

where $\psi_M(\rho_M)$ and $\psi_i(1/\rho_i)$ are the linearizing operators for the density of mixture and solution i, respectively. In other words, the nonlinear mixing rule in Equation 2.17c is transformed into a linear mixing rule given as Equation 2.18.

The mixing example in Figure 2.10 is revisited to verify that both Equations 2.19 and 2.20 are equivalent. Assume that both solutions A and B have a mass flow of 1 kg each. The density values for these solvents are given as 1000 and 1500 kg/m³, respectively. We can determine from Equation 2.17c or 2.18 that the mixture density is equivalent to 1200 kg/m³ (= 2/[1/1000 + 1/1500]). Alternatively, we can also calculate the density operator of solvents A and B as 0.001 (= 1/1000 kg/m³) and 0.00067 m³/kg (= 1/1500 kg/m³), respectively. Note that in this case, the operators become the specific gravity of the solvents. However, in some other property integration problems, the operator values may be meaningless (e.g. Examples 2.5 and 2.6). From Equation 2.19, the density operator (specific gravity) of the mixture is calculated as 0.00083 m³/kg (= 0.5(0.001) + 0.5(0.00067)) which is equivalent to the density value of 1200 kg/m³ (= 1/0.00083 kg/m³).

Similarly, we may apply the same linearizing principle for a wide range of properties P for any given mixing rule patterns. A generic form of the linearized mixing rule is given as

$$\psi_M(P_M) = \sum_i x_i \psi_i(P_i) \qquad (2.19)$$

where $\psi_M(P_M)$ and $\psi_i(P_i)$ are the linearizing operators for property P in the mixture and individual stream i, respectively.

It is interesting to note that Equation 2.19 is also applicable for concentration-based water minimization and gas recovery problems. For instance, when two water streams of equal flowrate but with respective impurity concentrations of 20 and 40 ppm are mixed, Equation 2.19 determines that the resulting mixture will have a concentration (C_M) of

$$C_M = 0.5(20) + 0.5(40) = 30 \text{ ppm.}$$

In this case, the property operator of $\psi(C)$, is equivalent to concentration. For simplicity, $\psi(P)$ will be denoted as ψ in this book. Operators for some of the common nonlinear properties are summarized in Table 2.3. Note that these operators may be determined empirically, or from first principles.

Apart from identifying the linearized mixing rule for properties, we need to define the correct limiting flowrate for the material sink in a property integration problem. This relates to the property operator of the fresh resource in the given problem. To maximize material recovery in a property integration problem, Heuristic 3 is modified as follows.

Heuristic 4—Choose the property operator of the sink that is furthest from the operator of the fresh resource.

TABLE 2.3

Operator Expressions for Nonlinear Properties

Property of Mixture, P_M	Mixing Rule	Operator
Density, ρ_M	$\dfrac{1}{\rho_M} = \sum_i \dfrac{x_i}{\rho_i}$	$\psi = \dfrac{1}{\rho_i}$
Reid vapor pressure, RVP_M	$RVP_M = \sum_i x_i RVP_i^{1.44}$	$\psi = RVP_i^{1.44}$
Electric resistivity, R_M	$\dfrac{1}{R_M} = \sum_i \dfrac{x_i}{R_i}$	$\psi = \dfrac{1}{R_i}$
Viscosity, μ_M	$\log(\mu_M) = \sum_i x_i \log(\mu_i)$	$\psi = \log(\mu_i)$
Paper reflectivity, R	$R_{\infty,M} = \sum_i x_i R_{\infty,i}^{5.92}$	$\psi = R_{\infty,i}^{5.92}$

Source: From Foo, Kazantzi, El-Halwagi, and Manan, 2006b. Reproduced with permission. Copyright © 2006. Elsevier.

The main reason behind Heuristic 4 is that there exist two distinct cases for the property integration problems with respect to the property operator value of the fresh resource. In the first case, the fresh resource has an operator value that is *superior* among all process sources. This is similar to the concentration-based water minimization and gas recovery problems, where freshwater/gas resource is always available at the lowest concentration among all sources. Hence, the limiting inlet property operator for the material sinks is set to the upper bound for a given operating range. In contrast, when a fresh resource has an operator value that is *inferior* among all sources, the limiting inlet operator for the material sink will take the lower bound of the given operating range. Doing this ensures the maximum recovery of process sources or the minimum usage of fresh resource(s).

Example 2.5 Data extraction for metal degreasing process

Figure 2.11 shows the PFD of the metal degreasing process (motivating example in Chapter 1) where fresh organic solvent is used in the degreasing and absorption units (Kazantzi and El-Halwagi, 2005; Foo et al., 2006a). The treated metal from the degreasing unit is sent to metal finishing to produce the final product. On the other hand, the used solvent from the degreasing unit is sent for reactive thermal processing for regeneration, by decomposing its grease and organic additives content. The regenerated solvent exits as the liquid product from the thermal processing unit and is then recycled to the degreasing unit, while the off-gas from the former (which contains solvent) is treated in a condenser and then an absorption unit before being sent for flare. Besides, another off-gas stream leaving the degreasing unit is also passed through a condenser before it is flared. Two generated condensate streams are observed

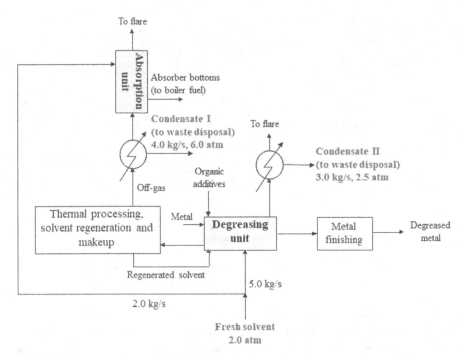

FIGURE 2.11
PFD of a metal degreasing process. (From Kazantzi and El-Halwagi, 2005. Reproduced with permission. Copyright © 2005. American Institute of Chemical Engineers (AIChE).)

to contain some solvent (source), which may be recovered to the process sinks (i.e., degreasing and absorption units) to reduce the consumption of fresh solvent.

The main property of the solvent that dictates its extent of recovery is the Reid vapor pressure (RVP), which is important in characterizing the volatility (and, indirectly, the composition) of the solvent. The general mixing rule for the RVP is given in Equation 2.20:

$$RVP_M = \sum_i x_i RVP_i^{1.44} \tag{2.20}$$

The fresh solvent has an RVP value of 2.0 atm; the RVP values for the condensate streams are given in Figure 2.11. The following process constraints on flowrate and RVP values are to be complied with for the process sinks, i.e.,

i. Degreasing unit

$$\text{Flowrate of solvent} = 5.0 \text{ kg/s} \tag{2.21}$$

$$2.0 \le \text{RVP of solvent (atm)} \le 3.0 \tag{2.22}$$

TABLE 2.4

Limiting Data for Example 2.5

Sink	F_{SKj} (kg/s)	Ψ_{SKj} (atm$^{1.44}$)
Degreasing unit	5	4.86
Absorption unit	2	7.36
Source	F_{SRi} **(kg/s)**	Ψ_{SRi} **(atm$^{1.44}$)**
Condensate I	4	13.20
Condensate II	3	3.74
Fresh solvent	To be determined	2.71

ii. Absorption unit

$$\text{Flowrate of solvent} = 2.0 \text{ kg/s} \tag{2.23}$$

$$2.0 \le \text{RVP of solvent (atm)} \le 4.0 \tag{2.24}$$

SOLUTION

Table 2.3 shows that the property operator of the RVP is given as

$$\psi = RVP_i^{1.44} \tag{2.25}$$

Equation 2.25 is first used to calculate the operator values of the fresh solvent, as well as Condensates I and II, i.e., 2.71, 13.20, and 3.74 atm$^{1.44}$, respectively.[2] It is observed that the fresh solvent has a superior operator value among all process sources. Similarly, Equation 2.25 is used to convert the operating boundary of the RVP values of the process sinks. Hence, the revised form of Equations 2.22 and 2.24 are given in their operator values as follows:

$$2.71 \le \psi \quad \text{for degreasing unit } \left(\text{atm}^{1.44}\right) \le 4.86 \tag{2.26}$$

$$2.71 \le \psi \quad \textit{for absorption unit } \left(\text{atm}^{1.44}\right) \le 7.36 \tag{2.27}$$

Following Heuristic 4, the upper limit of the property operator operating range of both sinks is taken as the limiting data for the problem, i.e., 4.86 (Condensate I) and 7.36 atm$^{1.44}$ (Condensate II), respectively. A summary of the limiting data for all process sinks and sources, along with the fresh solvent, is given in Table 2.4.

Example 2.6 Data extraction for papermaking process

Figure 2.12 shows the PFD of a Kraft papermaking process (Kazantzi and El-Halwagi, 2005; El-Halwagi, 2006; Foo et al., 2006a). Wood chips are mixed with white liquor in the Kraft digester, where they are

---

[2] This is a factitious unit without any physical meaning.

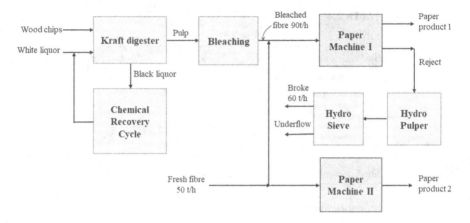

FIGURE 2.12
PFD of a papermaking process. (From Kazantzi and El-Halwagi, 2005. Reproduced with permission. Copyright © 2005. American Institute of Chemical Engineers (AIChE).)

digested and treated chemically. The black liquor from the digester is recycled, while the produced pulp is then sent to the bleaching section. Next, the bleached fiber is supplied to Paper Machines I and II to be converted into two grades of paper products. An external fresh fiber source is also fed to both paper machines to supplement its fiber need. A portion of the offspec product is rejected (termed as *reject*). This latter is sent for treatment in the hydro pulper and hydro sieve units. Two effluents are produced from these units, i.e., an underflow and an overflow waste that is termed as *broke*. The former is normally sent for burning, while the broke is sent for waste treatment. It is also observed that the broke contains some fiber content, which may be recovered to the paper machine.

A paper product property known as *reflectivity* (R_∞) is the quality of concern in determining the extent of recovery for the broke. Reflectivity is defined as the reflectance of an infinitely thick material compared with an absolute standard, i.e., magnesium oxide, and is dimensionless. The mixing rule for reflectivity R_∞ is given as

$$R_{\infty,M} = \sum_i x_i R_{\infty,i}^{5.92} \qquad (2.28)$$

Reflectivity index for the three fiber sources, i.e., fresh fiber, bleached fiber, and broke, is given as 0.95, 0.88, and 0.75, respectively. The process sources, i.e., bleached fiber and broke, are maximized before the fresh fiber is used. The present operating flowrates of these fiber sources are shown in Figure 2.12.

On the other hand, the following process constraints on flowrate and fiber reflectivity are to be complied with for process sinks in recovering fiber, i.e.,

i. Paper Machine I

$$\text{Fiber requirement} = 100 \text{ t/h} \qquad (2.29)$$

$$0.85 \leq \text{Fiber reflectivity (dimensionless)} \leq 0.95 \qquad (2.30)$$

ii. Paper Machine II

$$\text{Fiber requirement} = 40 \text{ t/h} \qquad (2.31)$$

$$0.90 \leq \text{Fiber reflectivity (dimensionless)} \leq 0.95 \qquad (2.32)$$

SOLUTION

Table 2.3 shows that the property operator of reflectivity is given as

$$\psi = R_{\infty,i}^{5.92} \qquad (2.33)$$

Operator values of the fresh and bleached fiber, as well as the broke, are determined using Equation 2.33 as 0.74, 0.47, and 0.18, respectively. It is observed that the fresh fiber has an inferior operator value among all process sources. Similarly, the operating boundary of reflectivity of the process sinks is converted into their respective operator values using Equation 2.33 as follows:

$$0.38 \leq \psi \text{ for Paper Machine I (dimensionless)} \leq 0.74 \qquad (2.34)$$

$$0.54 \leq \psi \text{ for Paper Machine II (dimensionless)} \leq 0.74 \qquad (2.35)$$

Following Heuristic 4, the lower limit of the property operator operating range of both sinks is taken as the limiting data for the problem, i.e., 0.38 (Paper Machine I) and 0.54 atm$^{1.44}$ (Paper Machine II), respectively. A summary of the limiting data for all process sinks and sources, along with the fresh solvent, is given in Table 2.5.

TABLE 2.5

Limiting Data for the Kraft Papermaking Process Example 2.6

Sink	F_{SKj} (t/h)	Ψ_{SKj}
Machine I	100	0.382
Machine II	40	0.536

Source	F_{SRi} (t/h)	Ψ_{SRi}
Bleached fiber	90	0.469
Broke	60	0.182
Fresh fiber	To be determined	0.738

Source: Kazantzi and El-Halwagi (2005).

2.6 Additional Readings

An illustration of the impact of wrongly extracted limiting data is found in Foo et al. (2006b), focusing on water network synthesis of the fixed flowrate problems.[3] In essence, the work shows that when the wrong limit of the operating range is taken as the limiting data, the synthesized RCN will experience higher consumption of fresh resource as well as higher waste discharge.

PROBLEMS
Water Minimization Problems

2.1 Figure 2.13 shows a tire-to-fuel process, where scrap tires are converted into energy (El-Halwagi, 1997; Noureldin and El-Halwagi, 1999). Scrap tires are shredded by a high-pressure water-jet before they are fed to the pyrolysis reactor. The spent water from the shredding unit is then filtered, recompressed, and recycled. Wet cake from the filtration unit is sent to the solid waste handling section. To compensate water loss in the filtered wet cake, freshwater is added to the recycled water loop at the water-jet compression station.

In the pyrolysis reactor, the shredded tires are turned into oils and gaseous fuels. The oils are sent for separation and finishing to produce liquid transportation fuels. Off-gases from the reactor are sent to a condenser to separate its light oil content from the gaseous fuel. The condensate is sent to a decanter where a two-phase liquid is produced. The organic layer from the decanter is mixed with the liquid oil from the

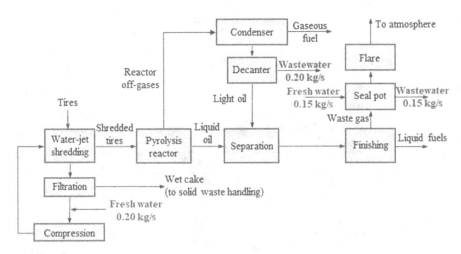

FIGURE 2.13
Tire-to-fuel process in Problem 2.1. (From El-Halwagi, 1997. With permission.).

---

[3] Reader may also solve Problems 4.1 and 4.2 to see the consequence of wrong data extraction.

reactor for separation and finishing, while the aqueous layer forms the wastewater effluent. On the other hand, a waste gas emits from the finishing section and is sent for flare. A seal pot is used to prevent the back-propagation of fire from the flare, by passing through a water stream to form a buffer zone between the fire and the source of the flare gas.

Two wastewater streams are observed in the process, i.e., effluent from the decanter and seal pot. These sources may be recovered to the process water sinks, i.e., seal pot and water-jet compression station, in order to reduce freshwater consumption. To evaluate the potential of water recovery, heavy organic content of the wastewater streams is the impurity that is of main concern. For the water sinks, the following constraints on feed flowrate and impurity concentration should be satisfied:

i. Seal pot:

$$0.10 \leq \text{Flowrate of feed water} \ (\text{kg/s}) \leq 0.20 \qquad (2.36)$$

$$0 \leq \text{Impurity concentration of feed water} \ (\text{ppm}) \leq 500 \qquad (2.37)$$

ii. Makeup to water-jet compression station:

$$0.18 \leq \text{Flowrate of feed water} \ (\text{kg/s}) \leq 0.20 \qquad (2.38)$$

$$0 \leq \text{Impurity concentration of makeup water} \ (\text{ppm}) \leq 50 \qquad (2.39)$$

On the other hand, the water sources have the following operating conditions:

i. Decanter:

$$\text{Wastewater flowrate} = 0.20 \ \text{kg/s} \qquad (2.40)$$

$$\text{Impurity concentration} = 500 \ \text{ppm} \qquad (2.41)$$

ii. Seal pot:

$$\begin{array}{c} \text{Wastewater flowrate} = \text{flowrate of its water sink} \\ \left(\text{given as in Equation 2.36}\right) \end{array} \qquad (2.42)$$

$$\text{Impurity concentration} = 200 \ \text{ppm} \qquad (2.43)$$

Note that Equation 2.42 indicates that the water sink and source of the seal pot will always have uniform flowrates. Identify the limiting data for water recovery for this process (see answers in Problem 3.1).

2.2 Figure 2.14 shows a Kraft pulping process where paper product is produced from wood chips (El-Halwagi, 1997). The wood chips (with 50% water content) are fed with steam in both chip bin and presteaming unit.

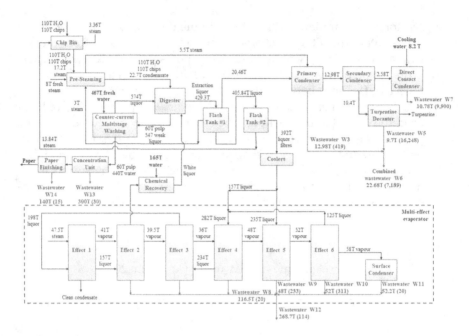

FIGURE 2.14
A Kraft pulping process in Problem 2.2 (basis: 1 h; T refers to ton; values in parenthesis indicate methanol concentration in ppm). (From Hamad et al.,, 1995. With permission.)

The steaming process will assist in the impregnation of chips with chemicals at the latter processing steps. The mixture from presteaming unit is then sent to a digester where white liquor (a chemical mixture) is used to solubilize lignin in the wood chips, which is commonly termed as the *cooking* operation. By-products from the cooking operation include methanol, turpentine, etc. The cooked chips are then sent for washing in a countercurrent multistage washer, where cooking chemicals are removed. The washed product is then sent for concentration and the finishing units to produce the paper product. In the latter units, wastewater streams (W13 and W14) are formed.

Extraction liquor from the digester is sent for two-stage flashing operations. Steams generated from these flash units are recovered to the chip bin and presteaming, respectively. The liquid portion from the second flash tank is cooled and sent to the multistage evaporation unit, where a large amount of combined wastewater stream (W12) is produced from the condensate of each effect of the evaporation unit.

Vapor stream from the first flash unit is sent to the primary condenser, along with the steam from the presteaming unit. The noncondensable material from the primary condenser is sent to the secondary condenser, where most gases are condensed. The condensate is sent to the turpentine decanter, where turpentine is recovered. The leftover noncondensable material from the secondary condenser is finally condensed by cooling water in the direct-contact condenser. The condensates from these units generate several wastewater streams, i.e., W6 and W7.

Due to the use of a significant amount of water in the Kraft pulping process, water recovery is considered to reduce the overall process freshwater and wastewater flowrates. For this process, methanol concentration is the main concern. The flowrates and methanol concentrations of the individual wastewater streams are shown in Figure 2.14. These water sources may be recovered to the water sinks, i.e., countercurrent multistage washing, chemical recovery, as well as the direct-contact condenser.

The following constraints on feed flowrate and methanol concentration should be satisfied for the water sinks while considering water recovery:

i. Countercurrent multistage washing:

$$\text{Flowrate of wash water} = 467 \ t/h \tag{2.44}$$

$$\text{Methanol concentration in wash water} \leq 20 \ ppm \tag{2.45}$$

ii. Chemical recovery

$$165 \leq \text{Flowrate of feed water} \ (t/h) \leq 180 \tag{2.46}$$

$$\text{Methanol concentration in wash water} \leq 20 \ ppm \tag{2.47}$$

iii. Direct-contact condenser

$$\text{Flowrate of feed water} = 8.2 \ t/h \tag{2.48}$$

$$\text{Methanol concentration in feed water} \leq 10 \ ppm \tag{2.49}$$

Identify the limiting data for water recovery for this process (see answer in Problem 4.1).

2.3 Figure 2.15 shows a manufacturing process where tricresyl phosphate is produced from cresol and phosphorous oxychloride (El-Halwagi, 1997). The reactants are mixed and heated before being fed to the reactor. Inert gas is added as diluent in the reactor. Effluent from the reactor is sent to a flash vessel. Liquid product from the flash is cooled and sent to a two-stage washing operation to remove the cresol content in the product stream. On the other hand, the vapor stream from the flash unit is sent to a two-stage scrubbing operation for purification, before it is sent for flare. A seal pot is used between scrubbing operation and flare to prevent backward propagation of fire.

It is observed that a significant amount of water is used in the water-using processes, i.e., washing and scrubbing operations, as well as

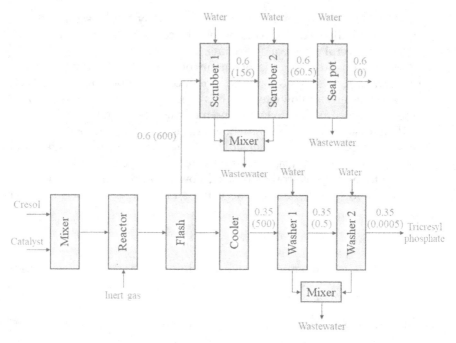

FIGURE 2.15
Tricresyl phosphate manufacturing process in Problem 2.3 (stream flowrates given in kg/s; values in parenthesis indicate cresol concentration in ppm). (From Hamad et al., 1996. With permission.)

in the seal pot. Hence, water recovery may be considered in order to reduce the freshwater consumption of the process. In considering water recovery, cresol concentration is the impurity that is of main concern for the process sinks. Since all water-using processes may be modeled as mass transfer operations, it is appropriate to take the inlet streams of these units as water sinks and their outlet streams as sources. For the water sinks, the following constraints are to be complied during water recovery:

i. Washer 1

$$\text{Flowrate of wash water} = 2.45 \text{ kg/s} \qquad (2.50)$$

$$0 \leq \text{Cresol concentration in wash water} \left(\text{ppm}\right) \leq 5 \qquad (2.51)$$

ii. Washer 2

$$\text{Flowrate of wash water} = 2.45 \text{ kg/s} \qquad (2.52)$$

$$\text{Cresol concentration in wash water} = 0 \text{ ppm} \qquad (2.53)$$

iii. Scrubber 1

$$0.7 \leq \text{Flowrate of feed water} \left(\text{kg/s}\right) \leq 0.84 \qquad (2.54)$$

$$0 \leq \text{Cresol concentration in feed water} \left(\text{ppm}\right) \leq 30 \qquad (2.55)$$

iv. Scrubber 2

$$0.5 \leq \text{Flowrate of feed water} \left(\text{kg/s}\right) \leq 0.6 \qquad (2.56)$$

$$0 \leq \text{Cresol concentration in feed water} \left(\text{ppm}\right) \leq 30 \qquad (2.57)$$

v. Seal pot

$$0.2 \leq \text{Flowrate of feed water} \left(\text{kg/s}\right) \leq 0.25 \qquad (2.58)$$

$$0 \leq \text{Cresol concentration in feed water} \left(\text{ppm}\right) \leq 100 \qquad (2.59)$$

On the other hand, the source flowrate of each unit is to be kept uniform with that of the sink. Besides, the cresol concentration of the water sources is to be determined through mass balance, i.e., based on the cresol content of the product stream/exhaust gas that is sent to the water-using processes (hint: Equations 2.10 and 2.11 are to be used here).

Identify the limiting data for water recovery for this process (see answer in Problem 4.2).

2.4 Figure 2.16 shows a water-using scheme for the bleaching section in a textile plant before carrying out water minimization study (Ujang et al., 2002). As shown, freshwater (0 mg/L) is utilized in singeing, scouring, and bleaching, as well as in mercerizing processes. These processes are treated as mass-exchange processes, as water is used as MSA to reduce the impurity load from the process streams. Hence, these processes have uniform inlet and outlet flowrates. Chemical oxygen demand (COD) concentration is taken as the main consideration for water recovery,

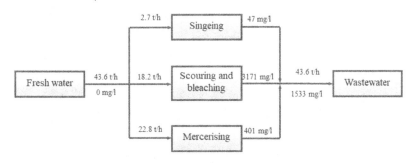

FIGURE 2.16
Water-using scheme for the bleaching section in a textile plant in Problem 2.4. (From Ujang et al., 2002. With permission.)

as it is the most significant parameter for process operations. Identify the limiting data for water recovery for this process. Note that the limiting inlet concentration for the three processes are given as 0 (singeing), 47 (scouring and bleaching) and 75 ppm (mercerizing) respectively (hint: Equations 2.10 and 2.11 may be used; see answer in Problem 3.2).

2.5 **Figure 2.17** shows a PFD for alkene production from shale gas (Foo, 2019). As shown, the shale gas stream (mainly consists of methane and other light hydrocarbons) is fed to a methanation reactor together with high-pressure steam. Methanation and water gas shift reactions occur in the reactor, where all hydrocarbon components (except methane) react with steam to form carbon monoxide and hydrogen. Effluent from the methanation reactor is then sent to an oxyhydrochlorination reactor. In the latter, hydrogen chloride (HCl) and oxygen (O_2) are fed so to convert methane into chloromethane (CH_3Cl). As methane conversion is relatively low (20%), Absorber 1 is used to recover the unconverted gases so that it may be recycled to the oxyhydrochlorination reactor. Product stream from Absorber 1 is sent to a stripper, where steam is used to separate the absorbing solvent from the hydrocarbon product. Effluent from the stripper is sent to a decanter, where solvent (to be recycled to Absorber 1) and its waste is separated; the latter is treated prior to the final disposal. Top product from the stripper is sent for purification in Distillation 1. In the latter, product is emitted from the column top stream, while its bottom stream produces wastewater. Top product stream from the distillation is sent to a chlorine-catalyzed oxidative pyrolysis (CCOP) reactor, where chloromethane is converted into ethylene, propylene, and butylene.

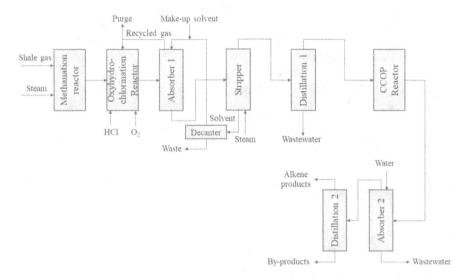

FIGURE 2.17
PFD for ethylene production from shale gas in Problem 2.5 (Foo, 2019).

The CCOP reactor effluent is fed to Absorber 2, where water is used to remove hydrogen chloride from the product steam. The absorber vapor product is sent to Distillation 2 for final purification, while its liquid bottom steam will be discarded as wastewater.

Due to large water consumption and wastewater generation, the wastewater streams from Distillation 1 and Absorber 2 are evaluated for their recovery in the water sinks. The latter include steam usage in reactor feed, as well as striping and absorbing agents. To carry out such analysis, hydrogen chloride (HCl) concentration is the main impurity in concern. The following flowrate and HCl concentration constraints are imposed for the water sinks:

i. Methanation reactor:

$$12.4 \leq \text{Steam usage in reactor } (t/h) \leq 13.6 \qquad (2.60)$$

$$\text{HCl concentration in steam} \leq 0\% \qquad (2.61)$$

ii. Stripper:

$$9.0 \leq \text{Steam as striping agent } (t/h) \leq 9.9 \qquad (2.62)$$

$$\text{HCl concentration in steam} \leq 0\% \qquad (2.63)$$

iii. Scrubber 2:

$$153.0 \leq \text{absorbing agent } (t/h) \leq 168.3 \qquad (2.64)$$

$$\text{HCl concentration in absorbing agent} \leq 5\% \qquad (2.65)$$

Identify the limiting data for water recovery for this process (see answer in Problem 3.8, and discussion in Chapter 5—Case Study).

2.6 Figure 2.18 shows a specialty chemical production process where water minimization study (Wang and Smith, 1995) is to be carried out for its process and utility sections (i.e. steam and cooling system). As shown, freshwater is fed to the reactor along with the main process feedstock. The reactor effluent is sent to a sedimentation unit where solid product is separated from its water content. The concentrated product is next sent to a cyclone unit, where water is used for product washing. The washed product is next sent to a filtration unit where its filter cake is washed with water. Several wastewater streams are generated due to the washing operations. Apart from the main operation, water is also utilized for steam generation and cooling system. In carrying out water minimization study, the presence of suspended solids is of main concern. The flowrate and suspended solids concentration for some water sinks and sources are shown in Figure 2.18, while those

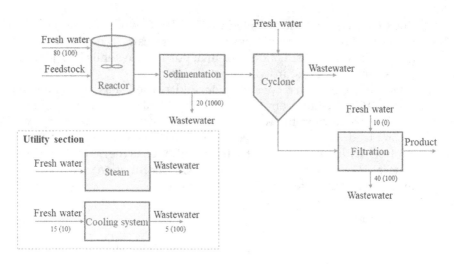

FIGURE 2.18

A specialty chemical production process in Problem 2.6 (water flowrate given in t/h; concentration given in ppm in parenthesis).

TABLE 2.6

Limiting Data for Cyclone and Cooling System in Problem 2.6

Operation	Inlet Concentration, C_{SKj} (ppm)	Outlet Concentration, C_{SRi} (ppm)	Mass Load, Δm (kg/h)
Cyclone	200	700	25
Steam system	0	10	0.1

for cyclone and cooling system are to be determined through mass balances, as they are modeled as mass-separating agent (see data given in Table 2.6). Identify the limiting data for water recovery for this process (see answer in Problem 4.6).

2.7 Figure 2.19 shows the PFD for a palm oil milling process (Chungsiriporn et al., 2006), where various processing steps are used to extract the crude palm oil from the fresh fruit bunches. When the fresh fruit bunch arrives at the mill, it is sterilized by steam to deactivate its enzymes and to loosen the fruits. Steam condensate from the sterilizer consists of oil and solid is discharged to wastewater treatment. The loosened fruits are then sent to several milling processes (e.g. rotary thresher, digester, screw press, and screener), where palm oil mixture is produced. The mixture (containing oil, water, and fine solid particles) is then fed to the settling tank where crude palm oil is produced from its overflow. Bottom sludge from the settling tank is sent to decanter and separator where residual oil is recovered. Water is added to decanter and separator for blending and balancing the oil phase for ease of oil recovery. Wastewater is produced from the separator is discarded, along with

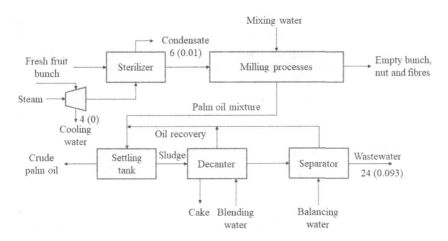

FIGURE 2.19
PFD for a palm oil milling process in Problem 2.7 (source flowrates given in t/h; NPE content is given in fraction in parenthesis).

sterilizer condensate and cooling water from the turbine. These three wastewater streams (water sources) may be considered for their recovery to the process sinks; their flowrate and concentrations are given in Figure 2.19. For water recovery consideration, the main impurity in concern is non-process elements (NPEs), which are mainly fine particles and suspended solids. The freshwater resource may be assumed to be free of NPE (0 ppm). The following flowrate and NPE concentration constraints are imposed for the water sinks:

i. Milling processes:

$$\text{Mixing water flowrate} = 14.0 \text{ t/h} \qquad (2.65)$$

$$0 \leq \text{NPE fraction} \leq 0.020 \qquad (2.66)$$

ii. Decanter:

$$\text{Blending water flowrate} = 2.5 \text{ t/h} \qquad (2.67)$$

$$0 \leq \text{NPE fraction} \leq 0.120 \qquad (2.68)$$

iii. Separator:

$$\text{Balancing water flowrate} = 7.5 \text{ t/h} \qquad (2.69)$$

$$0 \leq \text{NPE fraction} \leq 0.093 \qquad (2.70)$$

Identify the limiting data for water recovery for this process (see answer in Problem 4.8).

2.8 Figure 2.20 shows a simplified PFD of a Kraft process that is used for large-scale paper production (Parthasarathy and Krishnagopalan, 2001). As shown, wood chips are first to digester where they are treated with the white liquor. Effluent from the digester is sent to brown stock washer where it is separated from the black liquor. Next, the pulp is then sent to the bleach plant where it undergoes ClO_2 treatment (acidic) and alkali extraction (with NaOH). Both ClO_2 treatment and alkali extraction generate wastewater streams, which are treated before discharge. The black liquor from the brown stock washer is concentrated in a multi-effect evaporator before it is burned in the recovery furnace to yield the inorganic smelt. The latter is dissolved to form green liquor, which is sent to the green liquor clarifier. Effluent of the latter is sent to the causticizing process. The latter consists of several units where green liquor is converted to white liquor. The latter is sent to a white liquor clarifier, which is then fed to the digester.

As Kraft process consumes large amount of water, and generates a significant amount of wastewater, it is desired to carry out water minimization study. In doing so, it is important to pay attention to the chloride content in the water stream. It is known that the freshwater resource is available at a chloride concentration of 4.2 ppm. The identified water sinks and sources are shown in Figure 2.20. The flowrate and chloride concentration of the water sources are given in Table 2.7,

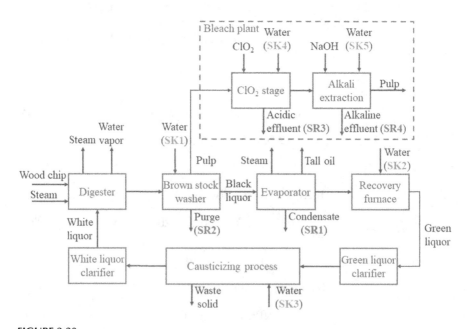

FIGURE 2.20
Simplified PFD of a Kraft process in Problem 2.8. (Adapted from Parthasarathy and Krishnagopalan, 2001.)

TABLE 2.7

Limiting Data for Kraft Process in Problem 2.8

SRi	Sources	Flowrate (t/d)	Concentration (ppm)
SR1	Evaporator condensate	7,971	0
SR2	Washer purge	945	275
SR3	ClO$_2$ stage effluent	14,520	235
SR4	Alkaline extraction effluent	13,750	504

while the following flowrate and chloride concentration constraints are imposed for water sinks (see answer in Problem 4.11):

i. Brown stock washer (SK1):

$$7,000 \leq \text{Flowrate} \left(t/d \right) \leq 10,000 \tag{2.71}$$

$$0 \leq \text{Chloride concentration} \left(ppm \right) \leq 110 \tag{2.72}$$

ii. Recovery furnace (SK2):

$$4,200 \leq \text{Flowrate} \left(t/d \right) \leq 6,900 \tag{2.73}$$

$$0 \leq \text{Chloride concentration} \left(ppm \right) \leq 5.5 \tag{2.74}$$

iii. Causticizing process (SK3):

$$2,000 \leq \text{Flowrate} \left(t/d \right) \leq 3,800 \tag{2.75}$$

$$0 \leq \text{Chloride concentration} \left(ppm \right) \leq 38.5 \tag{2.76}$$

iv. ClO$_2$ stage (SK4):

$$12,000 \leq \text{Flowrate} \left(t/d \right) \leq 14,400 \tag{2.77}$$

$$0 \leq \text{Chloride concentration} \left(ppm \right) \leq 10.12 \tag{2.78}$$

v. Alkaline extraction (SK5):

$$11,200 \leq \text{Flowrate} \left(t/d \right) \leq 13,500 \tag{2.79}$$

$$0 \leq \text{Chloride concentration} \left(ppm \right) \leq 13.2 \tag{2.80}$$

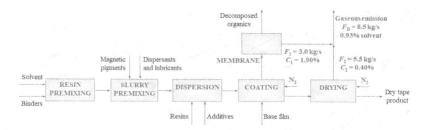

FIGURE 2.21
A magnetic tape process for Problem 2.9. (From Dunn et al., 1995. With permission.)

Utility Gas Recovery Problems

2.9 A magnetic tape manufacturing process is shown in Figure 2.21 (El-Halwagi, 1997). Coating ingredients are first dissolved in an organic solvent to form a slurry mixture. The slurry is then suspended with resin binders and special additives, which are then deposited on the base film. The coated film is then sent to the drying operation where the dry tape product is made. Note that the limiting inlet concentration for the three processes is given as o (singing), 47 (scouring and bleaching), and 75 ppm (mercerizing) respectively (hint: Equations 2.10 and 2.11 may be used; see answer in Problem 3.2).

In the coating chamber, nitrogen gas is used to induce the solvent evaporation rate that suits the deposition operation. During the operation, a small portion of the solvent is decomposed into organic species. A membrane unit is then used to purify the chamber exhaust gas. The membrane reject stream mainly consists of nitrogen gas and some leftover organic solvent. Besides, nitrogen gas is also used in the drying operation to evaporate the remaining solvent. The dryer exhaust gas is then mixed with the membrane reject stream and sent for disposal. To reduce nitrogen consumption of the process, the solvent-laden exhaust gases (source) may be recovered to the process sinks, i.e., coating and drying processes. Solvent content is the primary concern in considering nitrogen recovery. Besides, the following process constraints of the sinks are to be complied with for any nitrogen recovery scheme:

i. Coating

$$3.0 \leq \text{Flowrate of gaseous feed} \left(\text{kg/s} \right) \leq 3.2 \qquad (2.81)$$

$$0.0 \leq \text{wt\% of solvent} \leq 0.2 \qquad (2.82)$$

ii. Drying

$$5.5 \leq \text{Flowrate of gaseous feed} \left(\text{kg/s} \right) \leq 6.0 \qquad (2.83)$$

$$0.0 \leq \text{wt\% of solvent} \leq 0.1 \qquad (2.84)$$

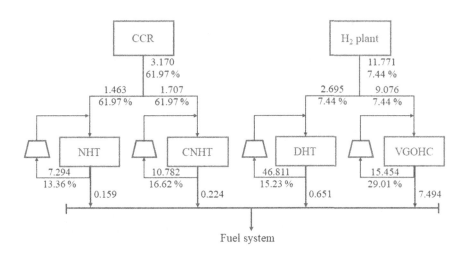

FIGURE 2.22
A hydrogen network for Problem 2.10 (numbers represent the total gas flowrate in t/h and hydrogen purity concentration in %). (From Umana et al., 2014.)

The flowrate and concentration of the nitrogen sources are shown in Figure 2.17. It may be assumed that the characteristics of these sources are independent of the feed gas. Note that the fresh nitrogen is solvent free, i.e., with solvent concentration of 0 wt%.

Identify the limiting data for nitrogen recovery in the magnetic tape manufacturing process (see answers in Problem 3.8).

2.10 A hydrogen network is shown in Figure 2.22 (Umana et al., 2014), where hydrogen recovery is considered. As shown, the network consists of an external hydrogen (H_2) plant, and an internal source, i.e. catalytic reformer (CCR). There are four hydrotreaters in the network, i.e. naphtha hydrotreater (NHT), cracked naphtha hydrotreater (CNHT), diesel hydrotreater (DHT), and vacuum gas oil hydrocracker (VGOHC).

Identify the limiting data in order to carry out a hydrogen recovery analysis (see answer in Problem 4.18).

2.11 Figure 2.23 shows a refinery hydrogen network (Alves and Towler, 2002), where hydrogen recovery is considered. Four hydrogen-consuming processes are found in the network, i.e., hydrocracking unit (HCU), naphtha hydrotreater (NHT), cracked naphtha hydrotreater (CNHT), and diesel hydrotreater (DHT). The internal recycled hydrogen stream data for these processes are given in Table 2.8.

Two in-plant hydrogen-producing facilities are also found, i.e., catalytic reforming unit (CRU) and steam reforming unit (SRU). These internal hydrogen sources are to be maximized before the purchase of fresh hydrogen. The fresh hydrogen feed has an impurity content of 5%.

Identify the limiting data in order to carry out a hydrogen recovery analysis (see answer in Problem 4.19).

TABLE 2.8

Flowrate and Concentration of Recycle Stream in Figure 2.23

Units	F_{RCY} (mol/s)	C_{RCY} (mol%)
HCU	1732.6	25.0
NHT	41.6	25.0
CNHT	415.8	30.0
DHT	277.2	27.0

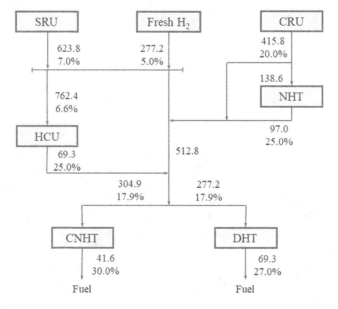

FIGURE 2.23
A refinery hydrogen network in Problem 2.11 (numbers represent the total gas flowrate in mol/s and impurity concentration in mol%). (From Alves and Towler, 2002. Reproduced with permission. Copyright ©2002. American Chemical Society.)

2.12 An existing hydrogen network of an aromatic plant (Shariati et al., 2013) is shown in Figure 2.24. The network consists of several hydrogen-consuming processes such as arofining (AF), hydrotreating (HT), hydrogenation (HG), regeneration (RG), transalkylation (TA) isomerization (IS), and disproportionation (DP). Fresh hydrogen is supplied from a catalytic reformer, which is coupled with a pressure swing adsorption (PSA) unit. Note that the outlet stream of RG is not considered for recovery due to process constraints.

   Identify the limiting data in order to carry out a hydrogen recovery analysis. Do we include/exclude hydrogen production plant and the PSA unit from the analysis (see answer in Problem 6.17)?

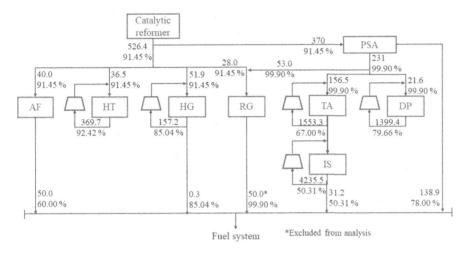

FIGURE 2.24
Existing hydrogen network for an aromatic plant in Problem 2.12 (numbers represent gas flowrate in mol/s and hydrogen purity in mol%). (From Shariati et al., 2013.)

Property Integration Problems

2.13 Figure 2.25 shows a wafer fabrication process where a large amount of ultrapure water (UPW) is consumed (Ng et al., 2009). Four UPW-using processes are the wet processing section (Wet), lithography, combined chemical and mechanical processing (CMP), and miscellaneous operations (Etc.). A total of six wastewater streams are generated from these processes, with some processes (Wet and CMP) generating two wastewater streams of different quality.

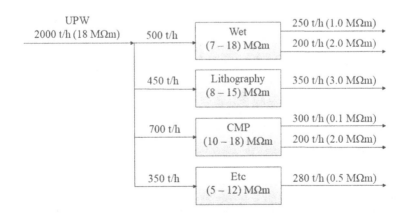

FIGURE 2.25
Usage of UPW in a wafer fabrication process in Problem 2.13 (From Ng et al., 2009. With permission.)

In order to reduce the UPW and wastewater flowrates, the wastewaters are considered for recovery to the water-using processes. The most significant water quality factor in considering water recovery is *resistivity* (R), which is an index that reflects the total ionic content in the aqueous streams, with the property mixing rule of

$$\frac{1}{R_M} = \sum_i \frac{x_i}{R_i} \tag{2.85}$$

The sink and source of the water-using processes are given as fixed flowrates, as shown in Figure 2.25. Besides, it may be assumed that the characteristics of sources are independent from those of the sinks. The sinks may accept water with a given range of resistivity values, as shown in parenthesis in Figure 2.19.

Identify the limiting data for water recovery for this process (see answer in Problem 4.19).

2.14 Recovery of acetic acid (AA) is being considered in a vinyl acetate manufacturing process, where vinyl acetate monomer (VAM) is produced from the reactants of ethylene, oxygen, and AA (El-Halwagi, 2006). Note that an equimolar of water impurity is also formed as a reaction by-product. As shown in the PFD in Figure 2.26, AA is first evaporated before it is

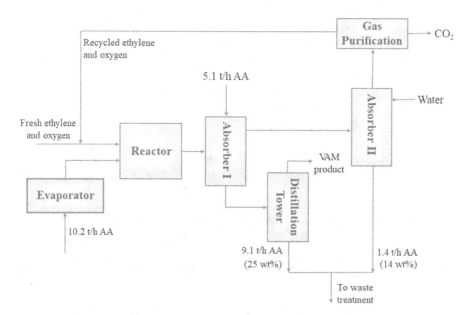

FIGURE 2.26
Vinyl acetate manufacturing process in Problem 2.14 (numbers represent AA stream flowrate; values in parenthesis represent water concentrations in AA streams). (From El-Halwagi, 2006. With permission.)

fed to the reactor. The reactor effluent is then sent to Absorber I, where AA is used to absorb VAM. The bottom stream of Absorber I is sent to the distillation tower, where VAM is recovered as top product; the latter is sent for final finishing. On the other hand, the top product stream of Absorber I, which contains unreacted ethylene and oxygen, is sent to Absorber II, where water is used as a scrubbing agent to absorb the remaining AA content. Ethylene and oxygen gases leave Absorber II in the top stream and are purified before being recycled to the reactor. The bottom product streams of Absorber II and distillation tower are sent for waste treatment. The latter consists of neutralization and biological treatment units. In considering the recovery of AA, water content is the impurity that is of main concern. Besides, the following technical constraints are to be observed in evaluating options for AA recovery.

i. Neutralization unit (in waste treatment)

$$0 \leq \text{Feed flowrate } (t/h) \leq 11.0 \tag{2.86}$$

$$0 \leq \text{AA content in feed stream } (wt\%) \leq 85 \tag{2.87}$$

ii. Evaporator

$$10.2 \leq \text{Feed flowrate } (t/h) \leq 11.2 \tag{2.88}$$

$$0.0 \leq \text{Water content in feed stream } (wt\%) \leq 10.0 \tag{2.89}$$

iii. Absorber I

$$5.1 \leq \text{Feed flowrate } (t/h) \leq 6.0 \tag{2.90}$$

$$0.0 \leq \text{Water content in feed stream } (wt\%) \leq 5.0 \tag{2.91}$$

Identify the limiting data for AA recovery for this process (see answers in Problem 3.12).

References

Alves, J. J. and Towler, G. P. 2002. Analysis of refinery hydrogen distribution systems, *Industrial and Engineering Chemistry Research*, 41, 5759–5769.
Chungsiriporn, J., Prasertsan, S., and Bunyakan, C. 2006. Minimization of water consumption and process optimization of palm oil mills, *Clean Technologies and Environmental Policy*, 8, 151–158.

Dunn, R. F., El-Halwagi, M. M., Lakin, J., and Serageldin, M. 1995. Selection of organic solvent blends for environmental compliance in the coating industries. In: Griffith, E. D., Kahn, H., and Cousins, M. C. eds., *Proceedings of the First International Plant Operations and Design Conference*, Vol. III, pp. 83–107. New York: AIChE.

El-Halwagi, M. M.1997. *Pollution Prevention Through Process Integration: Systematic Design Tools*. San Diego, CA: Academic Press.

El-Halwagi, M. M.2006. *Process Integration*. San Diego, CA: Academic Press.

Foo, D. C. Y., Kazantzi, V., El-Halwagi, M. M., and Manan, Z. A. 2006a. Cascade analysis for targeting property-based material reuse networks, *Chemical Engineering Science*, 61(8), 2626–2642.

Foo, D. C. Y., Manan, Z. A., and El-Halwagi, M. M. 2006b. Correct identification of limiting water data for water network synthesis, *Clean Technologies and Environmental Policy*, 8(2), 96–104.

Foo, D. C. Y. 2019. Optimize plant water use from your desk, *Chemical Engineering Progress*, 115, 45–51.

Hallale, N. and Liu, F. 2001. Refinery hydrogen management for clean fuels production, *Advances in Environmental Research*, 6, 81–98.

Hamad, A. A., Crabtree, E. W., Garrison, G. W., and El-Halwagi, M. M. 1996. Optimal design of hybrid separation systems for in-plant waste reduction. In: *Proceedings of the Fifth World Congress of Chemical Engineering*, San Diego, CA, Vol III, pp. 453–458.

Hamad, A. A., Varma, V., El-Halwagi, M. M., and Krishnagopalan, G. 1995. Systematic integration of source reduction and recycle reuse for cost-effective compliance with the cluster rules. In: AIChE Annual Meeting, Miami, FL.

Kazantzi, V. and El-Halwagi, M. M. 2005. Targeting material reuse via property integration, *Chemical Engineering Progress*, 101(8), 28–37.

Ng, D. K. S., Foo, D. C. Y., Tan, R. R., Pau, C. H., and Tan, Y. L. 2009. Automated targeting for conventional and bilateral property-based resource conservation network, *Chemical Engineering Journal*, 149, 87–101.

Noureldin, M. B. and El-Halwagi, M. M. 1999. Interval-based targeting for pollution prevention via mass integration, *Computers and Chemical Engineering*, 23, 1527–1543.

Parthasarathy, G. and Krishnagopalan, G. 2001. Systematic reallocation of aqueous resources using mass integration in a typical pulp mill, *Advances in Environmental Research*, 5, 61–79.

Savelski, M. and Bagajewicz, M. 2000. On the optimality conditions of water utilization systems in process plants with single contaminants, *Chemical Engineering Science*, 55, 5035–5048.

Savelski, M. and Bagajewicz, M. 2003. On the necessary conditions of optimality of water utilization systems in process plants with multiple contaminants, *Chemical Engineering Science*, 58, 5349–5362.

Shariati, M., Tahouni, N., and Panjeshahi, M. H. 2013. Investigation of different approaches for hydrogen management in petrochemical complexes, *International Journal of Hydrogen Energy*, 38, 3257–3267.

Ujang, Z., Wong, C. L., and Manan, Z. A. 2002. Industrial wastewater minimization using water pinch analysis: A case study on an old textile plant, *Water Science and Technology*, 46, 77–84.

Umana, B., Shoaib, A., Zhang, N., and Smith, R. 2014. Integrating hydroprocessors in refinery hydrogen network optimisation, *Applied Energy*, 133, 169–182.

Wang, Y. P. and Smith, R. 1994. Wastewater minimisation, *Chemical Engineering Science*, 49, 981–1006.

Wang, Y. P. and Smith, R. 1995. Wastewater minimization with flowrate constraints, *Chemical Engineering Research and Design*, 73, 889–904.

3

Graphical Targeting Techniques for Direct Reuse/Recycle

Graphical techniques are well recognized for their advantage in providing good insights for problem analysis. In this chapter, a commonly used graphical technique for the targeting of various resource conservation networks (RCNs), is introduced, i.e., *material recovery pinch diagram* (MRPD), developed independently by El-Halwagi et al. (2003) as well as Prakash and Shenoy (2005). This chapter focuses on the targeting of a direct reuse/recycle RCN for both single and multiple fresh resources. Besides, targeting for multiple fresh resources and threshold problems are also discussed.

3.1 Material Recovery Pinch Diagram

For an RCN with a single fresh resource with superior quality (highest purity or lowest impurity concentration among all sources), the steps to construct the MRPD are given as follows (El-Halwagi et al., 2003; Prakash and Shenoy, 2005):

1. Arrange the process sinks in descending order of quality levels. Do the same for process sources.

2. Calculate the maximum load acceptable by the sinks (m_{SK_j}) and the load possessed by the sources (m_{SR_i}), which are given as the product of the sink (F_{SK_j}) or source (F_{SR_i}) flowrate with its corresponding quality level (q_{SK_j} or q_{SR_i}). These load values are given by Equations 3.1 and 3.2, respectively:

$$m_{SK_j} = F_{SK_j} q_{SK_j} \tag{3.1}$$

$$m_{SR_i} = F_{SR_i} q_{SR_i} \tag{3.2}$$

3. All process sinks are plotted from the origin on a load-versus-flow-rate diagram, in descending order of their quality levels to form the *sink composite curve*. The individual sink segments are connected by linking the tail of a latter to the arrowhead of an earlier segment (Figure 3.1).

DOI: 10.1201/9781003173977-4

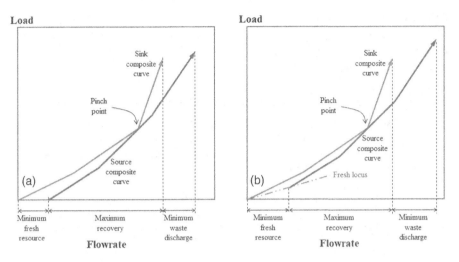

FIGURE 3.1
The MRPD for (a) pure fresh resource and (b) impure fresh resource. (From El-Halwagi et al., 2003; Prakash and Shenoy, 2005.)

4. Step 3 is repeated for all process sources to form the *source composite curve* on the same load-versus-flowrate diagram. For a feasible MRPD, the entire sink composite curve has to stay above and to the left of the source composite curve. Else, the MRPD is considered infeasible, and Step 5 is next followed.

5. For the RCN with pure (zero impurity content) fresh resources, the source composite curve is moved horizontally to the right until it just touches the sink composite curve, with the latter lying above and to the left of the source composite curve (Figure 3.1a). For an RCN with impure fresh resources, a fresh locus is to be plotted on the MRPD, with its slope corresponding to its quality level (q_R). The source composite curve is slid on the locus until it just touches the sink composite curve and to the right and below the latter (Figure 3.1b).

Several *targets* are identified from the MRPD. First, the overlap region between the two composite curves gives the maximum recovery target among all process sinks and sources of the RCN. When sources are insufficient to satisfy the requirement of the sinks, fresh resources (to be purchased) are needed for the RCN. The minimum requirement of the fresh resource is given by the horizontal distance where the sink composite curve extends to the left of the source composite curve. On the other hand, the horizontal distance where the source composite curve extends to the right of the sink composite curve represents the minimum waste generated from the RCN, which will be sent for final treatment before environment discharge.

The point where the two composite curves touch is termed as the *pinch*, which is the overall bottleneck for maximum recovery among all sinks and sources in the RCN. Next, the use of MRPD is demonstrated in several examples that follow.

Example 3.1 Water minimization for acrylonitrile (AN) production

The AN production case study in Example 2.1 is revisited here. The original limiting data for the problem are given in Table 2.1 (also found in the first three columns of Table 3.1). Determine the following targets for the problem using the MRPD:

1. Minimum fresh resource (water) for the RCN
2. Maximum water recovery between process sinks and sources
3. Minimum wastewater discharge from the RCN

SOLUTION

The five-step procedure is followed to plot the MRPD for the AN case study:

1. *Arrange the process sinks in descending order of their quality level. Do the same for process sources.*

   For the AN production case study, the quality level corresponds to the concentration of ammonia (see Example 2.1). As shown in the third column of Table 3.1, all sinks and sources are arranged in ascending order of their impurity concentrations (i.e., descending order of quality).

2. *Calculate the maximum load acceptable by the sinks (m_{SKj}) and the load possessed by the sources (m_{SRi}).*

   The load values for all sinks or sources are calculated using Equations 3.1 and 3.2, respectively, and are shown in the fourth column of Table 3.1. Note that in this case, the loads correspond

TABLE 3.1

Limiting Water Data for Plotting the MRPD for AN Case Study

Sinks, SK_j	F_{SKj} (kg/s)	C_{SKj} (ppm)	m_{SKj} (mg/s)	Cum. F_{SKj} (kg/s)	Cum. m_{SKj} (mg/s)
BFW (SK1)	1.2	0	0	1.2	0
Scrubber inlet (SK2)	5.8	10	58.0	7.0	58.0

Sources, SR_i	F_{SRi} (kg/s)	C_{SRi} (ppm)	m_{SRi} (mg/s)	Cum. F_{SRi} (kg/s)	Cum. m_{SRi} (mg/s)
Distillation bottoms (SR1)	0.8	0	0	0.8	0
Off-gas condensate (SR2)	5	14	70.0	5.8	70.0
Aqueous layer (SR3)	5.9	25	147.5	11.7	217.5
Ejector condensate (SR4)	1.4	34	47.6	13.1	265.1

to the flowrates of ammonia load in the sinks and sources, respectively.

3. *All process sinks are plotted from the origin on a load-versus-flowrate diagram, in descending order of their quality levels to form the sink composite curve. The individual sink segments are connected by linking the tail of a letter to the arrowhead of an earlier segment.*

   Since the flowrate and load values of the sink composite curve are cumulative in nature, cumulative flowrates (Cum. F_{SKj}) and cumulative loads (Cum. m_{SKj}) for all sinks are first calculated, given in the last two columns of Table 3.1. These values are then used to plot the sink composite curve for the AN case on the load-versus-flowrate diagram, as shown in Figure 3.2. Note that the slope of each segment corresponds to its quality level, i.e., ammonia concentration in this case. Hence, the sink composite curve indicates that a total flowrate of 7 kg/s is needed by sinks SK1 and SK2, and they can only accept a total impurity load of 58 mg/s.

4. *Step 3 is repeated for all process sources to form the source composite curve on the same load-versus-flowrate diagram.*

   Similar to the plotting of the sink composite curve (Step 3), the cumulative flowrates (Cum. F_{SRi}) and cumulative loads (Cum. m_{SRi}) for all sources are first calculated in the last two columns of Table 3.1. These values are used to plot the source composite curve in Figure 3.3. Similar to the sink composite curve, the slope of each segment in the source composite curve

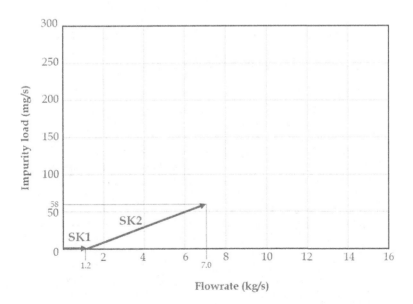

FIGURE 3.2

Plotting of sink composite curve for Example 3.1. (From El-Halwagi, 2006. Reproduced with permission. Copyright © 2006. Elsevier.)

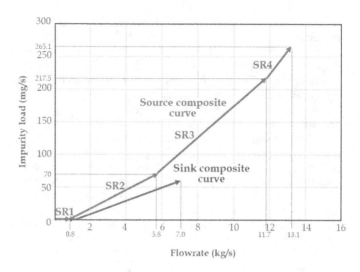

FIGURE 3.3

Plotting of source composite curve for Example 3.1 (From El-Halwagi, 2006. Reproduced with permission. Copyright © 2006. Elsevier.)

corresponds to the ammonia concentration of these sources. The source composite curve in Figure 3.3 indicates that the sources have a total water flowrate of 13.1 kg/s and possess a total impurity load of 265.1 mg/s.

Before proceeding to Step 5, the sink and source composite curves in Figure 3.3 are first examined for their feasibility. Note that there are two criteria that a feasible MRPD needs to fulfill, i.e., flowrate and load constraints. The flowrate constraint has been fulfilled in this case, as the total source flowrate (13.1 kg/s) exceeds that required by the sinks (7.0 kg/s). Nevertheless, the sinks can only tolerate a maximum total load limit of 58 mg/s impurity. However, if one were to feed the available sources to fulfill the total flowrate requirement of the sinks, the maximum total load limit of the sinks would be exceeded. This can be easily observed from Figure 3.3, where the source composite curve lies almost completely above the sink composite curve. In other words, the available sources are "too concentrated" to be fed to the sinks. In order to "dilute" the water sources for recovery, freshwater is required for the process. The minimum extent of the freshwater flowrate is to be determined in Step 5.

5. *For the RCN with pure (zero impurity content) fresh resources, the source composite curve is moved horizontally to the right until it just touches the sink composite curve, with the latter lying above and to the left of the source composite curve.*

Fresh resource for the AN case is freshwater feed, which has zero ammonia content. Hence, the source composite curve is moved horizontally to the right until it just touches the sink

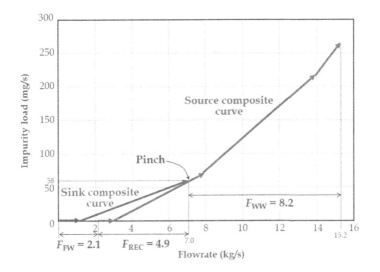

FIGURE 3.4
MRPD for Example 3.1. (From El-Halwagi, 2006. Reproduced with permission. Copyright © 2006. Elsevier.)

composite curve and lies completely below and to the right of the latter. This forms a feasible MRPD as shown in Figure 3.4, where both the flowrate and load constraints of the RCN are fulfilled.

Several network targets are readily identified from the feasible MRPD in Figure 3.4. The minimum fresh resource for the AN case, i.e., freshwater flowrate (F_{FW}), is identified as 2.1 kg/s. This represents the minimum freshwater needed by the RCN after the water sources are fully recovered to the sinks. The extent of recovery (F_{REC}), which is given by the overlapping region between the sink and source composite curves, is identified as 4.9 kg/s (=7.0–2.1 kg/s) from Figure 3.4. Besides, the excess water sources (after recovery to the sinks) emit as wastewater from the network, with its total flowrate (F_{WW}) identified as 8.2 kg/s. Figure 3.4 also shows the pinch where the two composite curves touch, which corresponds to a cumulative flowrate of 7.0 kg/s and a cumulative load of 58 mg/s. The significance of the pinch is discussed in the following section.

3.2 Significance of the Pinch and Insights from MRPD

As discussed earlier, the sink and source composite curves of a feasible MRPD touch each other at the pinch, with the latter representing the overall bottleneck for maximum material recovery in an RCN. In other words, the

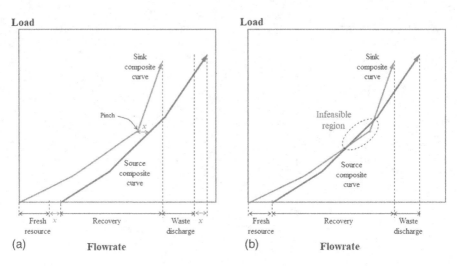

FIGURE 3.5
RCN with (a) additional and (b) insufficient resources. (From El-Halwagi, 2006. Reproduced with permission. Copyright © 2006. Elsevier.)

identified fresh resource target is the minimum requirement of a particular resource in the RCN. If one were to utilize x amount of additional resources in this RCN, the same amount of process sources (which is supposed to be recovered to the sinks) would end up being discharged as waste. This leads to an x amount of additional waste being generated from the RCN, as shown in Figure 3.5a. In contrast, if one were to utilize less than the targeted resource, the RCN would experience infeasibility in part of the network (see Figure 3.5b).

Note that the pinch always occurs at one of the source quality levels. The source where the pinch occurs is termed as the *pinch-causing source* (Manan et al., 2004; Foo et al., 2006). As shown in Figure 3.6, the pinch divides the overall network into two separate regions with respect to the pinch quality, i.e., *higher quality region* (HQR, below the pinch) and *lower quality region* (LQR, above the pinch). With respect to the pinch quality, we can easily classify to which region the individual sinks and sources belong.

Several remarks are worth mentioning for HQR and LQR. In the HQR, source composite curve is observed to have shorter horizontal distance than the sink composite curve (see Figure 3.7a), indicating that the total flowrate of process sources is less than that of the process sinks in this region. In other words, this region experiences flowrate deficit. Hence, fresh resource has to be supplied to supplement the flowrate deficit. Note that this region is the most constrained part of the RCN, as it controls the overall fresh resource flowrate for the RCN. Besides, it is noticed that the sinks in the HQR will always accept the maximum load sent from the sources (including fresh sources).

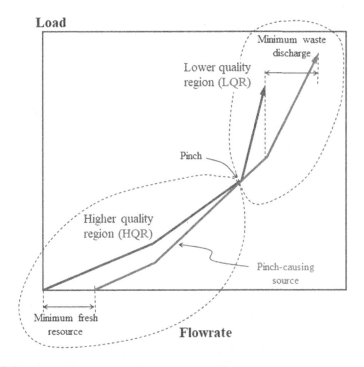

FIGURE 3.6
Pinch always occurs at the pinch-causing source, and divide the overall network into HQR and LQR.

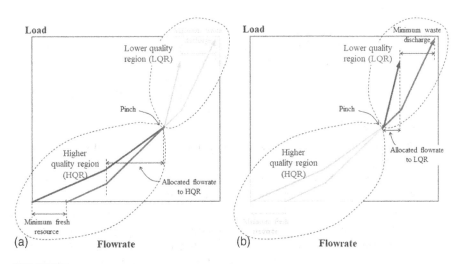

FIGURE 3.7
(a) HQR of the RCN, where flowrate deficit is experienced; (b) LQR of the RCN, where surplus source flowrate is encountered.

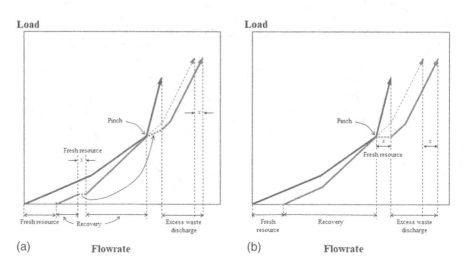

FIGURE 3.8
Cases with higher fresh resource and waste discharge: (a) the use of source in the HQR for sink in the LQR and (b) the use of fresh resource in the LQR.

On the other hand, in the LQR, the sink composite curve is observed to have a shorter horizontal distance than the source composite curve (see Figure 3.7b), indicating that the total flowrate of the process sources is larger than that of the process sinks. In other words, this region encounter a surplus of process sources; the latter will emit as waste from the RCN.

A few "golden rules" can then be derived for the HQR and LQR (Hallale, 2002; Manan et al., 2004; El-Halwagi, 2006):

1. Sources in the HQR (including fresh resource) should not be fed to sink in the LQR and may not be mixed with sources in the latter region. If a source that belongs to the HQR is being used in the LQR, such as that in Figure 3.8a, an additional x amount of fresh resource is needed in the RCN. This also results in the same amount of additional waste discharge from the network.

2. The pinch-causing source is an exception for Rule 1, as it belongs to both HQR and LQR. The horizontal distance of the pinch-causing source segment in the HQR and LQR represents its allocated flowrate in the respective regions (see Figure 3.6).

3. Fresh resource(s) shall only be used in the HQR, as this region experiences flowrate deficit. Using this resource in the LQR (where flowrate surplus is observed) will result in higher waste discharge. This is shown in Figure 3.8b. As x amount of additional fresh resource is used in the LQR, the waste discharge from the RCN also increases with the same magnitude.

Example 3.2 Water minimization for AN production (revisited)

Make use of the MRPD for the AN case study in Figure 3.4 to carry out the following tasks:

1. Identify the HQR and LQR where the process sinks and sources belong.
2. Identify the pinch-causing source and its allocated flowrates to the HQR and LQR.
3. Identify a reuse/recycle scheme for the case study based on the insights.

SOLUTION

With respect to the pinch concentration, the RCN may be divided into HQR (lower concentration) and LQR (higher concentration), as shown in Figure 3.9. Both sinks SK1 and SK2 in the sink composite curve, along with the freshwater feed (F_{FW} = 2.1 kg/s), belong to the HQR. Besides, source SR1 and a big portion of SR2 also belong to the same region. On the other hand, a small portion of SR2, as well as SR3 and SR4 belong to the LQR, while no sink is present in this region. Hence, all sources will emit as wastewater (F_{WW} = 8.2 kg/s) from the LQR.

Note that the sinks in the LQR accept the maximum impurity load sent from the sources. For this case, this corresponds to 58 mg/s ammonia content for the AN case, as shown in Figure 3.9.

It is observed from Figure 3.9 that the pinch occurs at SR2. The slope of SR2, i.e., 14 ppm, hence corresponds to the pinch concentration of the

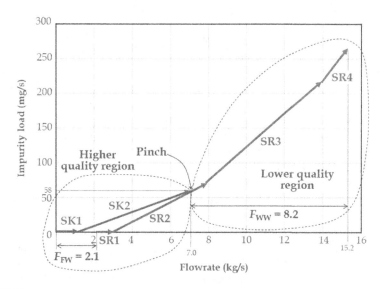

FIGURE 3.9
HQR and LQR for Example 3.2.

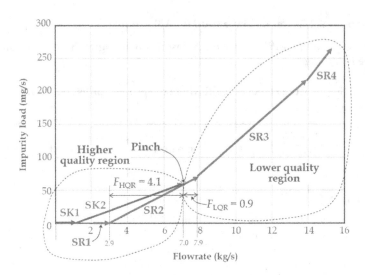

FIGURE 3.10
Pinch-causing source and its allocated flowrates for Example 3.2.

network, while SR2 (F_{SR2} = 5.0 kg/s) is identified as the pinch-causing source. As shown in Figure 3.10, the allocated flowrates of the pinch-causing source to the HQR (F_{HQR}) and LQR (F_{LQR}) are determined as 4.1 and 0.9 kg/s, respectively.

Based on these insights, a water reuse/recycle network for the AN case may be developed. In the HQR, the allocated flowrate of the pinch-causing source, i.e., SR2 to this region (F_{HQR}) will be used along with SR1 and the freshwater feed to satisfy sinks SK1 and SK2. Since SK1 (boiler feed water—BFW) does not tolerate any impurity (see Table 3.1), it is fed with freshwater. The remaining water sources, i.e., freshwater feed and SR1, and the remaining flowrate of SR2 are mixed to be fed to SK2 (scrubber inlet). As mentioned earlier, in the LQR, the unutilized sources SR3 and SR4 as well as the remaining flowrate of SR2 are discharged as wastewater. The resulting water reuse/recycle network is shown in Figure 3.10. However, note that the development of an RCN based on targeting insights will only work for simple cases.[1] For a more complicated problem, the detailed network design procedure should be used (see Chapter 8 for details).

It is also worth noting the financial saving for this case. The water network in Figure 3.11 shows that the freshwater is reduced from 7.0 kg/s (summation of sink flowrates in Table 3.1) to 2.1 kg/s, or a reduction of 70%. On the other hand, the wastewater flowrate is reduced by to 8.2 kg/s from 13.1 kg/s (summation of source flowrates in Table 3.1), corresponds to a reduction of 37.4%. Assuming the freshwater and wastewater have a unit costs of 0.001 $/kg (1 $/t), respectively, and an annual operating time of 8000 h, the cost saving

---

[1] To develop an RCN based on targeting insights, there should only be maximum one sink with nonzero quality index in the HQR.

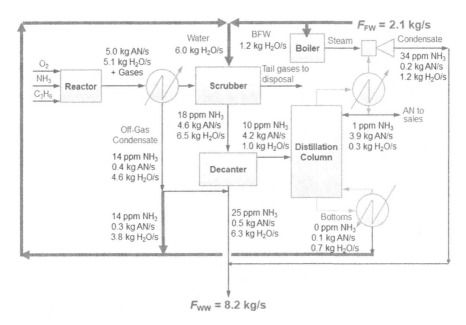

FIGURE 3.11
Water reuse/recycle network for AN case study. Note: Detailed mass balance has not been revised yet. (From El-Halwagi, 1997.)

associated with these water reduction effort may be calculated as 284,240 $/y (= (4.9 + 4.9) × 0.001 × 3600 × 8000).

On top of that, the biotreatment facility that was facing hydraulic bottleneck originally (see discussion in Example 2.1) is no longer facing its restriction. The process can now increase its production capacity as the main bottleneck has been removed. In other words, doing water recovery in this case helps to remove process debottleneck apart from achieving waste minimization objective.

Example 3.3 Hydrogen reuse/recycle in a refinery hydrogen network

The refinery hydrogen network case study in Example 2.4 (Hallale and Liu, 2001) is revisited here. The original limiting data needed for the case study are shown in Table 2.2 (also in the first three columns of Table 3.2). Determine the following targets for this RCN:

1. Minimum fresh resource (hydrogen) and purge flowrates, given that the fresh hydrogen feed has 1% impurity
2. Maximum recovery target between process sinks and sources
3. Pinch concentration that limits the maximum recovery between process sinks and sources
4. The pinch-causing source and its allocated flowrate to the HQR and LQR

TABLE 3.2

Limiting Data for Example 3.3

Sinks, SK$_j$	F_{SKj} (MMscfd)	C_{SKj} (%)	m_{SKj} (MMscfd)	Cum. F_{SKj} (MMscfd)	Cum. m_{SKj} (MMscfd)
A inlet (SK1)	400	7.20	28.8	400	28.8
B inlet (SK2)	600	12.43	74.58	1000	103.38
Sources, SR$_i$	F_{SRi} (MMscfd)	C_{SRi} (%)	m_{SRi} (MMscfd)	Cum. F_{SRi} (MMscfd)	Cum. m_{SRi} (MMscfd)
A outlet (SR1)	350	9.00	31.5	350	31.5
B outlet (SR2)	500	15.00	75	850	106.5

SOLUTION

Following Steps 1–3 of the MRPD procedure, the cumulative flowrates
and loads of the process sinks and sources are shown in the last two columns of Table 3.2. We then construct the MRPD following Steps 3 and 4,
resulting in the sink and source composite curves in Figure 3.12.

As shown in Figure 3.12, the source composite curve is found to be on
the left and above the sink composite curve. This means that all available sources have a higher concentration than the maximum acceptable
limit of the sinks. Besides, the total flowrate (850 MMscfd) of all sources
is also insufficient to be used by the sinks (1000 MMscfd). This means
that the MRPD is infeasible for both flowrate and load constraints.
Hence, fresh hydrogen is needed to restore the flowrate and load feasi-
bilities of the RCN.

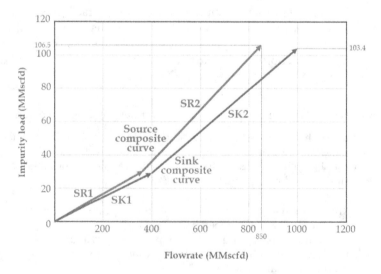

FIGURE 3.12

Plotting of composite curves for Example 3.3.

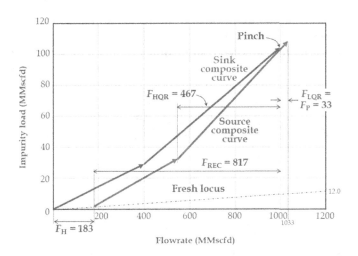

FIGURE 3.13
MRPD for Example 3.3.

Since fresh hydrogen feed has 1% impurity, a fresh locus is plotted on the MRPD with a slope that corresponds to its impurity concentration. The source composite curve is then slid on this locus until it just touches the sink composite curve and is to the right and below the latter (Figure 3.12).

From Figure 3.13, the horizontal distance of the sink composite curve that extends to the left of the source composite curve gives the minimum fresh hydrogen feed (F_H) needed by the network as 183 MMscfd. On the other hand, the horizontal distance where the source composite curve extends to the right of the sink composite curve dictates the minimum hydrogen that is purged from the network (F_P), which is identified as 33 MMscfd. The overlapping region between the two composite curves then gives the maximum recovery target (F_{REC}), which is identified as 817 MMscfd (= 1000–183 MMscfd). The pinch corresponds to the concentration of the second segment of the source composite curve, i.e., 15% of SR2. The latter is hence identified as the pinch-causing source for the problem, with 467 MMscfd (500–33 MMscfd) of its flowrate being allocated to the HQR (F_{HQR}), while the rest to the LQR (F_{LQR}). Since there is no process sink in the LQR, the allocated flowrate to this region is purged. In other words, the purge stream for this case is originated from SR2.

3.3 Targeting for Multiple Resources

In earlier sections, minimum flowrate targeting procedure is dedicated to RCN with single fresh resource of superior quality among all sources. There are also cases where RCN is supplied with multiple fresh resources of different quality levels, which call for a revised targeting procedure. Note that fresh resources should only be utilized in the HQR of an RCN, since it is the area that experiences resource deficit.

Since there is more than one fresh resource in the RCN, it is always useful to compare the relative costs among the available fresh resources. However, since these fresh resources have different impurity levels (which lead to different minimum flowrates), comparing them based on their unit cost alone is insufficient. A useful index termed as the *prioritized cost* (Shenoy and Bandyopadhyay, 2007) may be used, which is given in Equation 3.3:

$$CTP_R = \frac{CT_R}{q_R - q_P} \tag{3.3}$$

where

CT_R and CTP_R are the unit cost and prioritized costs for fresh resource

q_R and q_P are the quality indices of fresh resource and the pinch

Note that pinch quality is obtained when the RCN is supplied by a single resource. When a lower quality resource has a prioritized cost that is lower than that of a high-quality resource (i.e., $CTP_{R1} < CTP_{R2}$ and $q_{R1} > q_{R2}$), the use of the former should be maximized; otherwise, only the high-quality resource should be used.

For the case where the lower quality resource is to be maximized before the higher quality resource, a two-step targeting procedure is used (Wan Alwi and Manan, 2007). Figure 3.14 demonstrates the case where two fresh resources are available for use in an RCN, i.e., a pure primary resource that

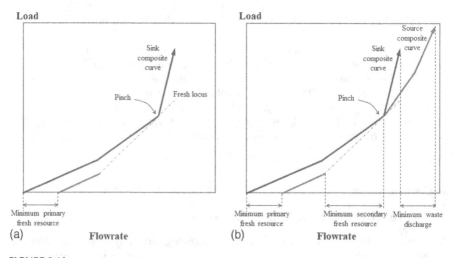

FIGURE 3.14
Minimum flowrate targeting for multiple fresh resource: (a) targeting for primary fresh resource and (b) targeting for secondary fresh resource. (From Foo, 2009. With permission.)

has superior quality among all sources and secondary resource (with inter-mediate quality level). Step 1 involves the targeting of primary resource. As shown in Figure 3.14a, a partial section of the source composite curves is first plotted, with only the source segments of better quality than of the secondary fresh resource. A locus of the secondary fresh resource (with slope corresponding to its quality level) is added at the end of the earlier source seg-ments. Next, the source segments along with the impure fresh resource locus are then slid horizontally until they stay below and to the right of the sink composite curve, and touch the latter at the pinch point (Figure 3.14a). The overhang of the sink composite curve gives the minimum flowrate of the pure fresh resource. We then proceed to Step 2 to determine the flowrate target of the secondary fresh resource. The rest of the source segments (of lower qual-ity) are connected and slide them along the locus of the impure fresh resource until they touch the sink composite curve. The horizontal distance of the fresh locus gives the minimum flowrate needed for the secondary fresh resource (Figure 3.14b). Similar to earlier cases, the overhang of the source composite curve gives the minimum flowrate of waste discharge from the RCN.

Note that apart from in-plant recovery the same targeting procedure may also be utilized for *inter-plant resource conservation network*, where resources are shared by different RCNs. For such a case, the targeting procedure may be used to set the minimum cross-plant flowrate (virtually no cost needed), which in turn leads to reduced overall fresh resource and waste flowrates (see Chapter 10 for details).

Example 3.4 Targeting for multiple freshwater sources

The AN production case study in Examples 3.1 and 3.2 is revisited. In this case, an impure freshwater source with 5 ppm ammonia content (second-ary fresh resource) is available for use besides the pure freshwater source (primary fresh resource). Unit costs for the pure (CT_{FW1}) and impure fresh-water sources (CT_{FW2}) are given as \$0.001/kg and \$0.0005/kg, respectively. Determine the minimum flowrates for both freshwater sources as well as that for wastewater. Determine also the operating costs for freshwater sources of the RCN.

SOLUTION

In Example 3.2, the targeting procedure identifies that the pinch concen-tration for this problem corresponds to 14 ppm. Hence, the prioritized costs for pure (CTP_{FW1}) and impure freshwater sources (CTP_{FW2}) are determined using Equation 3.3 as follows:

$$CTP_{FW1} = \frac{\$0.001/kg}{(14-0)ppm} = \frac{\$0.071}{gNH_3}$$

$$CTP_{FW2} = \frac{0.0005/kg}{(14-5)ppm} = \frac{\$0.056}{gNH_3}$$

Since the prioritized cost of the impure freshwater source is much lower than that of the pure freshwater source, the use of the former should be maximized prior to the latter. Following Step 1 of the targeting procedure, the source composite curve should only consist of a single segment at this stage, i.e., source SR1, which has a lower concentration (0 ppm) than that of impure freshwater source (5 ppm). Locus of the latter is then connected to the SR1 segment. Both SK1 segment and the impure freshwater locus are then slid horizontally until they stay below and to the right of the sink composite curve. Figure 3.15a shows

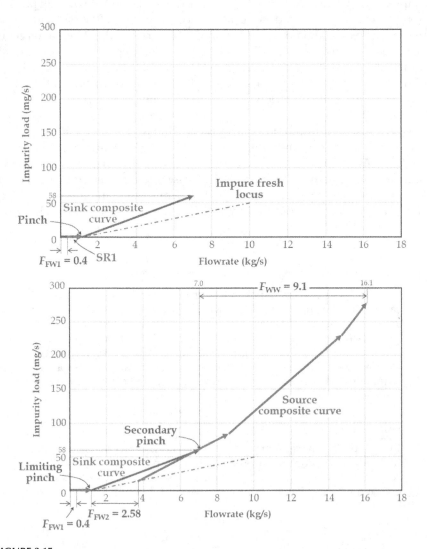

FIGURE 3.15
MRPD for Example 3.4: (a) targeting for pure freshwater source and (b) targeting for impure freshwater source.

that a pinch (5 ppm) is formed between the impure fresh locus and the sink composite curve. The overhang of the sink composite curve dictates the minimum flowrate of pure freshwater source (F_{FW1}) as 0.4 kg/s.

Next, the remaining water source segments (SR2–SR4) are then connected and slid along the locus of the impure freshwater locus until they touch the sink composite curve (Step 2). For this case, a new pinch point is formed at the end of the sink composite curve. The source segment where the pinch is formed indicates that the new pinch concentration corresponds to 14 ppm (i.e., identical to the case where a single pure freshwater source is used—Examples 3.1 and 3.2). The minimum flowrate of the impure freshwater source (F_{FW2}) is identified from the horizontal distance of the impure fresh locus, i.e., 2.58 kg/s (Figure 3.15b). Note that for this kind of *multiple pinch* cases, the highest quality pinch (i.e., 5 ppm for this case) is known as the *limiting pinch*, in which it controls the overall fresh resource flowrates (Hallale, 2002; Manan et al., 2004). Finally, the overhang of the source composite curve gives the minimum wastewater flowrate of this RCN, i.e., 9.1 kg/s.

We next determine the operating cost for freshwater sources of the RCN. With unit costs of $1/t and $0.5/t for pure ($CT_{FW1}$) and impure freshwater sources (CT_{FW2}), the freshwater cost of the RCN is determined as $0.0017/s (=0.4 kg/s × $0.001/kg + 2.58 kg/s × $0.0005/kg). On the other hand, if the impure freshwater source is not utilized, i.e., 2.1 kg/s of pure freshwater source will have to be used (see Examples 3.1 and 3.2), the freshwater cost rises to $0.0021/s (=2.1 kg/s × $0.001/kg). Nevertheless, when the unit cost of impure freshwater source is raised to $0.0007/kg, the RCN experiences higher freshwater cost, $0.0022/s (=0.4 kg/s × $0.001/kg + 2.58 kg/s × $0.0007/kg). Hence, in this latter case, it will be more sensible to make use of the freshwater source alone (with lower cost). A comparison of these scenarios is given in Table 3.3.

TABLE 3.3

Comparison of Various Cost Scenarios for Example 3.4

Scenarios	Pure Freshwater Flowrate (kg/s)	Impure Freshwater Flowrate (kg/s)	Freshwater Cost of RCN ($/s)
Pure freshwater source alone, with CT_{FW2} = $0.001/kg (Example 3.1)	2.1	0	0.0021
With pure and impure freshwater sources (CT_{FW2} = $0.0005/kg)	0.4	2.58	0.0017
With pure and impure freshwater sources (CT_{FW2} = $0.0007/kg)	0.4	2.58	0.0022

3.4 Targeting for Threshold Problems

Examples that we have discussed so far involve the use of fresh resources and also the generation of waste. There are also cases where neither fresh resource is needed nor waste is generated for an RCN. These special cases are known as the *threshold problems* (Foo, 2008). These are illustrated in the following examples.

Example 3.5 Targeting for threshold problem—zero fresh resource

Table 3.4 shows the limiting data for a water minimization case study. Target the minimum freshwater and wastewater flowrates for the water network.

SOLUTION

Following Steps 1–4 of the procedure, the MRPD for the example is plotted in Figure 3.16. However, before the source composite curve is

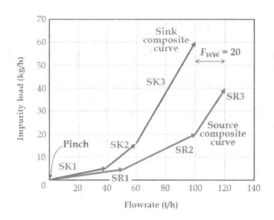

FIGURE 3.16

MRPD for Example 3.5. (From Foo, 2008. Reproduced with permission. Copyright © 2008. Elsevier.)

TABLE 3.4

Limiting Water Data for Example 3.5

SK_j	F_{SK_j} (t/h)	C_{SK_j} (ppm)	m_{SK_j} (kg/h)	Cum. F_{SK_j} (t/h)	Cum. m_{SK_j} (kg/h)
SK1	40	125	5	40	5
SK2	20	500	10	60	15
SK3	40	1125	45	100	60
SR_i	F_{SK_i} (t/h)	C_{SK_i} (ppm)	m_{SK_i} (kg/h)	Cum. F_{SK_i} (t/h)	Cum. m_{SK_i} (kg/h)
SR1	50	100	5	50	5
SR2	50	300	15	100	20
SR3	20	1000	20	120	40

being shifted (Step 5 of the MRPD procedure), it is already below and to the right of the sink composite curve. In other words, it fulfills both the flowrate and load constraints of the problem. Hence, no freshwater is needed for this case. Besides, the MRPD dictates that a wastewater flowrate (F_{WW}) of 20 t/h is generated from the network. Since the two composite curves touch each other at the lowest concentration end, the pinch for this case corresponds to the concentration of SR1, i.e., 100 ppm.

Example 3.6 Targeting for threshold problem—zero waste discharge

Table 3.5 shows the limiting water data for a water minimization case study taken from Foo (2008). Use the MRPD to target the minimum freshwater and wastewater flowrates of the network.

SOLUTION

Following the five-step procedure, the MRPD for this case study is plotted in Figure 3.16. However, as shown in Figure 3.17, even though the source composite curve lies completely below and to the right of the sink composite curve, the total available flowrate of the sources is still insufficient for the sinks. Hence, freshwater is required to supplement the flowrate requirement in the last segment of the sink composite curve. Summing the freshwater flowrates (F_{FW}) at both ends of the MRPD gives a total of 60 t/h. For better representation, the source composite curve is further shifted to the right so that the total freshwater requirement is collected on the left of the MRPD, as shown in Figure 3.18. Since the source composite curve does not extend further from the sink composite curve, this is a zero discharge RCN. Foo (2008) defined the concentration of the last segment of the source composite curve as the *threshold concentration* of the problem.

Note that the threshold problem exists in all types of RCN problems, apart from the water minimization examples shown here.

TABLE 3.5

Limiting Water Data for Example 3.6

SK_j	F_{SKj} (t/h)	C_{SKj} (ppm)	m_{SKj} (kg/h)	Cum. F_{SKj} (t/h)	Cum. m_{SKj} (kg/h)
SK1	50	20	1	50	1
SK2	20	50	1	70	2
SK3	100	400	40	170	42
SR_i	F_{SRi} (t/h)	C_{SRi} (ppm)	m_{SRi} (kg/h)	Cum. F_{SRi} (t/h)	Cum. m_{SRi} (kg/h)
SR1	20	20	0.4	20	0.4
SR2	50	100	5	70	5.4
SR3	40	250	10	110	15.4

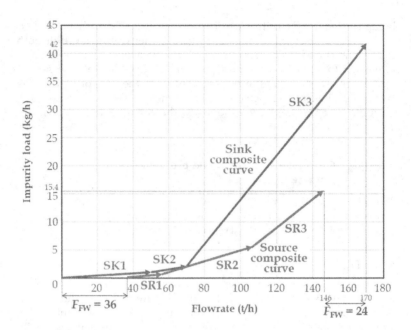

FIGURE 3.17
MRPD for Example 3.6.

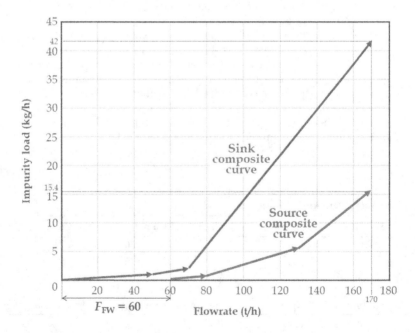

FIGURE 3.18
MRPD for Example 3.6 (with collection of total freshwater flowrates). (From Foo, 2008. Reproduced with permission. Copyright © 2008. Elsevier.)

3.5 Targeting for Property Integration

For an RCN with resource at the superior property operator level (see Section 2.5 for discussion of superior and inferior property operators), the same MRPD procedure discussed in earlier section may be used. However, for the special case where property operator of the fresh resource exists at the *inferior* level, Step 5 of the MRPD procedure is modified slightly (see modification in italics):

5. For the RCN with zero operator fresh resource of inferior value, *the sink composite curve is moved horizontally to the right until it just touches the source composite curve, with the latter lying above and to the left of the sink composite curve.* For RCN with nonzero operator fresh resource, a fresh locus is plotted on the MRPD, with its slope corresponding to the quality level of the fresh resource (q_R). The source composite curve is slid on the locus until it just touches the sink composite curve and is to the *left and above the latter.*

Two examples of property-based RCNs with fresh resources of superior and inferior property operators are next shown in the following for illustration.

Example 3.7 Solvent recovery for metal degreasing process

The metal degreasing process in Example 2.5 is revisited here. The main property for the evaluation of a reuse/recycle scheme is the Reid vapor pressure (RVP), which is converted to the corresponding operator values for the process sinks and sources in Section 2.5 and summarized in Table 2.3. Similarly, the RVP value of the fresh resource (solvent), i.e., 2.0 atm, is converted to its operator of 2.713 atm$^{1.44}$ ($= 2^{1.44}$ atm$^{1.44}$) using Equation 2.25. Identify the minimum fresh solvent and solvent waste targets for the RCN of this process.

SOLUTION

Since the fresh solvent has the lowest operator value among all sources (superior level), the earlier discussed procedure for MRPD plotting is followed to identify the RCN targets. Following Steps 1–3, the property loads as well as the cumulative flowrates and loads for the sinks or sources are determined and summarized in Table 3.6. Note that in Table 3.6, SR2 with lower operator value (higher quality) is placed before SR1, following Step 1 of the MRPD procedure. Note also that unlike the earlier examples on concentration-based RCN (e.g., water minimization, hydrogen network), the loads for these property-based composite curves (represented by the y-axis) do not possess any physical meaning. They are merely used for calculation purposes.

TABLE 3.6

Limiting Data for Metal Degreasing Process in Example 3.7

Sink, SK$_j$	F_{SKj} (kg/s)	ψ_{SKj} (atm$^{1.44}$)	m_{SKj} (kg atm$^{1.44}$/s)	Cum. F_{SKj} (kg/s)	Cum. m_{SKj} (kg atm$^{1.44}$/s)
Degreaser (SK1)	5.0	4.87	24.32	5.0	24.32
Absorber (SK2)	2.0	7.36	14.72	7.0	39.04

Source, SR$_i$	F_{SRi} (kg/s)	ψ_{SRi} (atm$^{1.44}$)	m_{SRi} (kg atm$^{1.44}$/s)	Cum. F_{SRi} (kg/s)	Cum. m_{SRi} (kg atm$^{1.44}$/s)
Condensate II (SR2)	3.0	3.74	11.22	3.0	11.22
Condensate I (SR1)	4.0	13.20	52.80	7.0	64.02

Since the fresh solvent consists of nonzero operator value, a fresh locus is added to the MRPD, with its slope being its operator value. Following Steps 3–5, the MRPD is plotted and is shown in Figure 3.19. As shown, the minimum fresh solvent (F_{FS}) needed for the problem is targeted as 2.4 kg/s. The same amount of solvent waste (F_{WS}) is generated from network. Besides, SR1 is identified as the pinch-causing source, with its property operator being the pinch of the RCN.

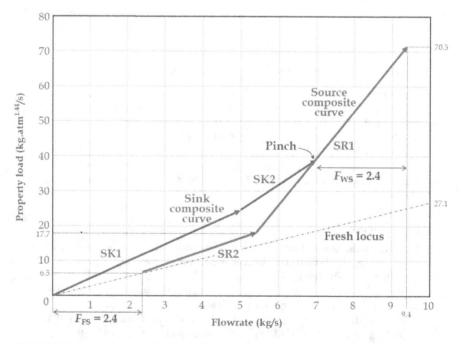

FIGURE 3.19

MRPD for Metal Degreasing Process in Example 3.7. (From Kazantzi and El-Halwagi, 2005. Reproduced with permission. Copyright © 2005. American Institute of Chemical Engineers (AIChE).)

Example 3.8 Fiber recovery for paper-making process

The paper-making process in Example 2.6 is revisited here. The property that is of concern for reuse/recycle is reflectivity, which may be converted to the corresponding operator value (see Example 2.6). The fresh resource use is fresh fiber, which has a reflectivity of 0.95 (dimensionless) or an operator of 0.738 ($=0.95^{5.92}$). Identify the minimum fresh fiber and waste targets for the RCN of this process.

SOLUTION

Since the fresh fiber has the highest operator value among all sources, it is considered to possess an inferior property operator value. Following Steps 1–2, the cumulative flowrates and loads for plotting the MRPD are calculated and summarized in Table 3.7.

Following Steps 3–4, the MRPD for the example is given in Figure 3.20. Note that the individual segments of the sink and source composite curves appear in descending slope (with descending order of quality), which is different from earlier examples. A fresh locus is also added since the fresh fiber has a nonzero operator value, with the locus slope being the property operator of the fresh fiber.

Similar to earlier cases, to ensure a feasible RCN, the two feasibility criteria on flowrate and load constraints need to be fulfilled. From Figure 3.20, it is observed that the source composite curve has a flowrate that is larger than that of the sink composite curve. This means that the flowrate constraint has been fulfilled. However, the load constraint has not been fulfilled as the two composite curves intersect each other. This can be accomplished by adding fresh resources (fiber) to the network. Following the revised Step 5 of the MRPD procedure, the source composite curve is slid along the fresh locus until it is above and to the left of the sink composite curve (Figure 3.21).

The interpretation for the MRPD is the same as in earlier examples. As shown in Figure 3.21, the minimum fresh fiber (F_{FF}) needed by the RCN is identified from the horizontal distance where the sink composite curve extends to the left of the source composite curve, i.e., 15 t/h. On the other hand, the minimum waste discharge (F_{WF}) is identified from the section where the source composite curve extends to the right of the sink composite curve, i.e., 25 t/h.

TABLE 3.7

Limiting Data for Paper-Making Process in Example 3.8

Sink, SK_j	F_{SKj} (t/h)	Ψ_{SKj} (Dimensionless)	m_{SKj} (t/h)	Cum. F_{SKj} (t/h)	Cum. m_{SKj} (t/h)
Machine II (SK2)	40	0.536	21.44	40	21.44
Machine I (SK1)	100	0.382	38.20	140	59.64

Source, SR_i	F_{SRi} (t/h)	Ψ_{SRi} (Dimensionless)	m_{SRi} (t/h)	Cum. F_{SRi} (t/h)	Cum. m_{SRi} (t/h)
Bleached Fiber (SR1)	90	0.469	42.23	90	42.23
Broke (SR2)	60	0.182	10.93	150	53.16

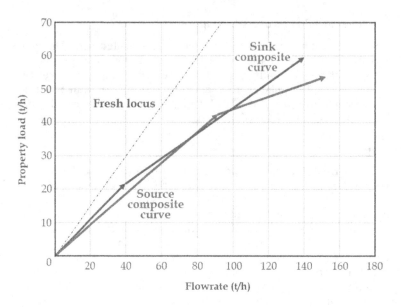

FIGURE 3.20
Plotting of sink and source composite curve for Example 3.8. (From Kazantzi and El-Halwagi, 2005. Reproduced with permission. Copyright © 2005. American Institute of Chemical Engineers (AIChE).)

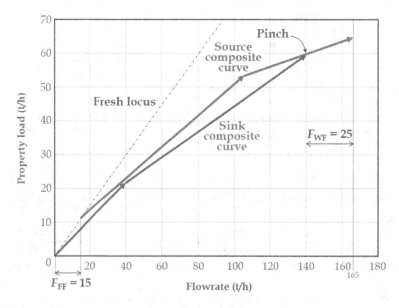

FIGURE 3.21
MRPD for Example 3.8. (From Kazantzi and El-Halwagi, 2005. Reproduced with permission. Copyright © 2005. American Institute of Chemical Engineers (AIChE).)

3.6 Additional Readings

Apart from the MRPD, many different graphical techniques have also been developed for the targeting of material reuse/recycle in the RCN in the past decade. These include the *limiting composite curve* (Wang and Smith, 1994, 1995), *surplus diagram* (Alves and Towler, 2002; Hallale, 2002), source composite curve (Bandyopadhyay, 2006; Bandyopadhyay et al., 2006), etc. Each of these methods has its own strength and limitations. A review that compares these techniques may be found in Foo (2009).

PROBLEMS
Water Minimization Problems

3.1 For the tire-to-fuel process in Problem 2.1, determine the minimum flowrates of freshwater (0 ppm) and wastewater. Use the limiting data in Table 3.8 (see answer in Problem 7.3).

3.2 For the textile plant in Problem 2.4, determine the minimum flowrates of freshwater (0 ppm) and wastewater for the process. The limiting data is given in Table 3.9.

3.3 Table 3.10 summarizes the limiting data for the classical water minimization example from Wang and Smith (1994). Identify the minimum freshwater and wastewater flowrates when the RCN is fed with single pure freshwater resource (see answers in Problem 8.16).

3.4 Table 3.11 summarizes the limiting data for a water minimization example taken from Polley and Polley (2000). Identify the minimum

TABLE 3.8

Limiting Data for Tire-To-Fuel Process in Problem 3.1

j	Sinks, SK_j	F_{SKj} (kg/s)	C_{SKj} (ppm)	i	Sources, SR_i	F_{SRi} (kg/s)	C_{SRi} (ppm)
1	Seal pot	0.1	500	1	Decanter	0.2	500
2	Water jet makeup	0.18	50	2	Seal pot	0.1	200

TABLE 3.9

Limiting Data for Textile Plant in Problem 3.2

j	SK_j	F_{SKi} (t/h)	C_{SKi} (ppm)	i	SR_i	F_{SRi} (t/h)	C_{SRi} (ppm)
1	Singeing	2.67	0	1	Singeing	2.67	47
2	Mercerizing	22.77	75	2	Mercerizing	22.77	476
3	Scouring and bleaching	18.17	47	3	Scouring and bleaching	18.17	3218

TABLE 3.10

Limiting Data for Classical Water Minimization Example (Wang and Smith, 1994) in Problem 3.3

Sinks, SK_j	F_{SKj} (t/h)	C_{SKj} (ppm)	Sources, SR_i	F_{SRi} (t/h)	C_{SRi} (ppm)
SK1	20	0	SR1	20	100
SK2	100	50	SR2	100	100
SK3	40	50	SR3	40	800
SK4	10	400	SR4	10	800

TABLE 3.11

Limiting Data for Water Minimization Example (Polley and Polley, 2000) in Problem 3.4

j	Sinks, SK_j	F_{SKj} (t/h)	C_{SKj} (ppm)	i	Sources, SR_i	F_{SRi} (t/h)	C_{SRi} (ppm)
1	SK1	50	20	1	SR1	50	50
2	SK2	100	50	2	SR2	100	100
3	SK3	80	100	3	SR3	70	150
4	SK4	70	200	4	SR4	60	250

freshwater and wastewater flowrates for this case for the following scenarios:

a. Single pure freshwater feed, i.e., zero impurity content (solution is found in Example 12.3).

b. Single impure freshwater feed, with 10 ppm impurity content.

c. Two freshwater feeds, i.e., pure (0 ppm, $1/ton) and impure freshwater feeds (80 ppm, $0.2/ton). Determine also the operating cost for freshwater feed.

3.5 Figure 3.22 shows a process flow diagram for a bulk chemical production (Foo, 2012). Two gas phase reactants are fed to the reactor to be converted into the desired product with the presence of water (as reaction carrier). In the separation unit, the unconverted reactant is recovered to the reactor, while the separated products are sent to the downstream purification units, which consist of decanter and dryer. Wastewater is generated from the decanter due to the huge amount of water being used as reaction carrier as well as the steam condensate used in the separation unit. Besides, a significant amount of water is also utilized in the utility section of the plant. This includes the makeup water for boiler and cooling tower (with a large evaporation loss) as well as water for general plant and vessel cleaning. Note that due to piping configuration, water for vessel cleaning also contributes to the decanter effluent. Apart from the decanter effluent, two other water sources are also

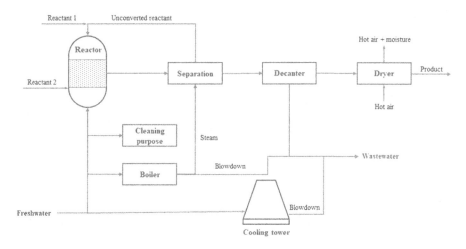

FIGURE 3.22
Process flow diagram for a bulk chemical production in Problem 3.5. (From Foo, 2011.)

found in the utility section, i.e., blowdown streams from the cooling tower and boiler.

Limiting data for water minimization are summarized in Table 3.12. Suspended solid is taken as the primary impurity for water recovery. Determine the minimum flowrates of freshwater and wastewater for the process. Freshwater feed is assumed to have zero impurity (see answers in Problem 7.6).

3.6 Figure 3.23 shows an organic chemical production (Hall, 1997; Foo, 2008). A large amount of water is consumed as reactor feed, as well as for utility and washing water (in washers and for hosepipe). Besides, a large amount of wastewater is also produced from the various processing units. The limiting flowrates and concentrations for the water sinks and sources are given in Figure 3.23. Determine the minimum flowrates of freshwater and wastewater for the process. Freshwater feed is assumed to have zero impurity (see answers in Problem 7.11).

TABLE 3.12

Limiting Data for Bulk Chemical Production in Problem 3.5

j	SK_j	F_{SKj} (t/h)	C_{SKj} (ppm)	i	SR_i	F_{SRi} (t/h)	C_{SRi} (ppm)
1	Reactor	10.5	0	1	Decanter	13.6	20
2	Boiler makeup	0.6	0	2	Cooling tower blowdown	1.0	400
3	Cooling tower makeup	9.3	50	3	Boiler blowdown	0.2	400
4	Cleaning	3.5	100	4			

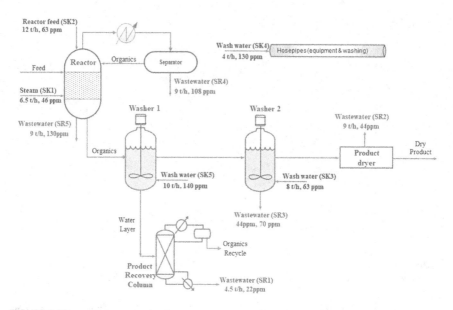

FIGURE 3.23
An organic chemical production in Problem 3.6. (From Foo, 2008. Reproduced with permission. Copyright ©2008. Elsevier.)

3.7 Use the insights from the MRPD to develop the reuse/recycle scheme for the threshold problems in Examples 3.5 and 3.6. Show the RCN in a superstructural representation similar to that in Figure 1.17.

3.8 For the alkene production case (Foo, 2019) in Problem 2.5, identify the minimum fresh and wastewater flowrate targets for its RCN. The limiting data (extracted using principles in Chapter 2) are given in Table 3.13. Fresh water supply is assumed to be free from impurity (see answers and discussion in Chapter 5—Industrial Case Study).

3.9 Figure 3.24 shows the water circuit of a brick manufacturing process (Skouteris et al., 2018). To reduce water footprint of this process, water minimization study is to be carried out. Table 3.14 shows the limiting water data for the process. Use MRPD to determine the minimum freshwater and wastewater flowrates for its water network, assuming two freshwater feeds are available, with impurity (COD) content of 0 and 55 ppm respectively (see answers in Problem 8.19).

TABLE 3.13

Limiting Data for Alkene Production Case in Problem 3.8 (Foo, 2019)

j	SK_j	F_{SKi} (t/h)	C_{SKi} (%)	i	SR_i	F_{SRi} (t/h)	C_{SRi} (%)
1	Steam usage in reactor	12.4	0	1	Distillation effluent	35.8	1.8
2	Steam usage in stripper	9.0	0	2	Absorber effluent	180	15
3	Absorbing agent	153	5				

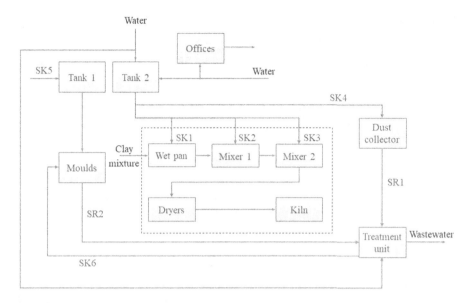

FIGURE 3.24
Water network for a brick manufacturing process (Skouteris et al., 2018) in Problem 3.9.

TABLE 3.14

Limiting Water Data for a Brick Manufacturing Process (Skouteris et al., 2018) in Problem 3.9

j	Sinks, SK_j	F_{SKj} (t/d)	C_{SKj} (ppm)	i	Sources, SR_i	F_{SKi} (t/d)	F_{SKi} (ppm)
1	Wet pan	16.2	0	1	Spent water from dust extractor	41.8	90
2	Mixer 1	14.9	0	2	Spent water used for cleaning	693.7	180
3	Mixer 2	38.3	0				
4	Dust collection system	41.8	0				
5	Molds cleaning	210.3	65				
6	Gully cleaning	483.4	75				

Utility Gas Recovery Problems

3.10 For the magnetic tape process in Problem 2.9 (El-Halwagi, 1997), identify the minimum fresh nitrogen feed and purge gas flowrate targets for its RCN. The limiting data are given in Table 3.15 (note: beware of the impurity concentration when plotting the source composite curve; see answer in Problem 8.22).

3.11 Figure 3.25 shows a process where a large amount of oxygen gas is utilized (Foo and Manan, 2006). As shown, oxygen is used in both the oxidation processes as well as to enhance the combustion system and

TABLE 3.15

Limiting Data for Magnetic Tape Process (El-Halwagi, 1997) in Problem 3.10

	Sinks, SK$_j$	F_{SKj} (kg/s)	C_{SKj} (wt%)	i	Sources, SR$_i$	F_{SRi} (kg/s)	C_{SRi} (wt%)
1	Drying	5.5	0.1	1	Membrane retentate	3	1.9
2	Coating	3	0.2	2	Drying	5.5	0.4

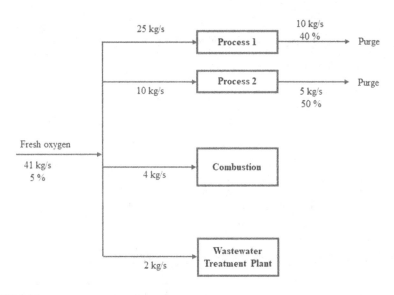

FIGURE 3.25

An oxygen recovery problem (Foo and Manan, 2006) in Problem 3.11. (From Foo, 2006. Reproduced with permission. Copyright © 2006. American Chemical Society.)

the aerobic section of the wastewater treatment plant. The purge oxygen streams from the oxidation processes are still of good quality and may be considered for reuse/recycling. Limiting data for the process sinks and sources are given in Table 3.16. The fresh oxygen supply has a supply of 5% impurity. Determine the minimum fresh oxygen gas and purge flowrate targets for the RCN (see answer in Problem 8.23).

TABLE 3.16

Limiting Data for Oxygen Recovery Problem in Problem 3.11

i	Sinks	F_{SKj} (kg/s)	C_{SKj} (wt%)	j	Sources	F_{SRi} (kg/s)	C_{SRi} (wt%)
1	Process 1	25	10	1	Process 1	10	40
2	Process 2	10	10	2	Process 2	5	50
3	Combustion	4	65				
4	Wastewater treatment	2	65				

Property Integration Problems

3.12 For the vinyl acetate manufacturing process in Problem 2.14 (El-Halwagi, 2006)., identify the minimum fresh and waste acetic acid flowrate targets for its RCN. The limiting data are given in Table 3.17. Assume that the fresh acetic acid does not poses any water content.

3.13 Figure 3.26 shows a microelectronics manufacturing facility where resource conservation is considered (Gabriel et al., 2003; El-Halwagi, 2006). Wafer fabrication (Wafer Fab) and combined chemical and mechanical processing (CMP) are identified as both sinks and sources for the reuse/recycle problem. Both units accept the ultrapure water (UPW) as feed and produce some rejects that may be recovered, i.e., 50% and 100% spent rinse streams. The property that is of concern for water recovery is resistivity (R), with the mixing rule given in Equation 2.64. The limiting data for the process are shown in Table 3.18. Determine the minimum flowrate targets for UPW as well as the reject streams for this process. The UPW has a resistivity of 18,000 kΩ/cm (see answer in Problem 8.30).

3.14 Table 3.19 shows the limiting data for a paper recovery study (Kit et al., 2011). The property that is of concern for paper reuse/recycle is the water fraction in the paper material. Fresh fiber with zero water fraction may be used to supplement the process sources. Use the MRPD to determine the minimum amount of fresh and waste fiber for the study.

TABLE 3.17

Limiting Data for Vinyl Acetate Manufacturing Process in Problem 3.12

j	Sinks, SK$_j$	F_{SK_j} (t/h)	ψ_{SK_j} (wt%)	i	Sources, SR$_i$	F_{SR_i} (t/h)	ψ_{SR_i} (wt%)
1	Evaporator	10.2	10	1	Absorber II bottom	1.4	14
2	Absorber I	5.1	5	2	Distillation bottom	9.1	25

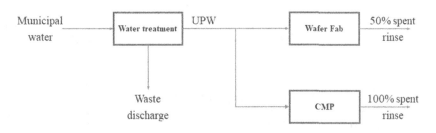

FIGURE 3.26
A microelectronics manufacturing facility in Problem 3.13. (From El-Halwagi, 2006. Reproduced with permission. Copyright © 2006 Elsevier.)

TABLE 3.18

Limiting Data for Microelectronics Manufacturing Facility in Problem 3.13

j	Sink	F_{SKj} (gal/min)	R_{SKj} (kΩ/cm)	ψ_{SKj} (cm/kΩ)
1	Wafer Fab	800	16,000	0.0000625
2	CMP	700	10,000	0.0001000
i	Source	F_{SRi} (gal/min)	R_{SRi} (kΩ/cm)	ψ_{SRi} (cm/kΩ)
1	50% spent rinse	1,000	8,000	0.000125
2	100% spent rinse	1,000	2,000	0.000500

TABLE 3.19

Limiting Data for Paper Recovery Study in Problem 3.14

j	Sinks	F_{SKj} (t/year)	Ψ_{SKj} (Water Fraction)	i	Sources	F_{SRi} (t/year)	ψ_{SRi} (Water Fraction)
1	Magazine	400	0.038	1	Magazine	336	0.135
2	Manila card	400	0.049	2	Manila card	2.8	0.144

References

Alves, J. J. and Towler, G. P. 2002. Analysis of refinery hydrogen distribution systems, *Industrial and Engineering Chemistry Research*, 41, 5759–5769.

Bandyopadhyay, S. 2006. Source composite curve for waste reduction, *Chemical Engineering Journal*, 125, 99–110.

Bandyopadhyay, S., Ghanekar, M. D., and Pillai, H. K. 2006. Process water management, *Industrial and Engineering Chemistry Research*, 45, 5287–5297.

El-Halwagi, M. M.1997. *Pollution Prevention Through Process Integration: Systematic Design Tools*. San Diego, CA: Academic Press.

El-Halwagi, M. M.2006. *Process Integration*. San Diego, CA: Elsevier Inc.

El-Halwagi, M. M., Gabriel, F., and Harell, D. 2003. Rigorous graphical targeting for resource conservation via material recycle/reuse networks, *Industrial and Engineering Chemistry Research*, 42, 4319–4328.

Foo, D. C. Y. 2008. Flowrate targeting for threshold problems and plant-wide integration for water network synthesis, *Journal of Environmental Management*, 88(2), 253–274.

Foo, D. C. Y. 2009. A state-of-the-art review of pinch analysis techniques for water network synthesis, *Industrial and Engineering Chemistry Research*, 48(11), 5125–5159.

Foo, D. C. Y. 2012. Resource conservation through pinch analysis. In: Foo, D. C. Y., El-Halwagi, M. M. and Tan, R. R. eds., *Recent Advances in Sustainable Process Design and Optimisation*. World Scientific.

Foo, D. C. Y. and Manan, Z. A. 2006. Setting the minimum flowrate targets for utility gases using cascade analysis technique, *Industrial and Engineering Chemistry Research*, 45(17), 5986–5995.

Foo, D. C. Y., Manan, Z. A., and Tan, Y. L. 2006. Use cascade analysis to optimize water networks, *Chemical Engineering Progress*, 102(7), 45–52.

Foo, D. C. Y. 2019. Optimize plant water use from your desk, *Chemical Engineering Progress*, 115, 45–51 (February 2019).

Gabriel, F. B., Harell, D. A., Dozal, E., and El-Halwagi, M. M. 2003. Pollution targeting for functionality tracking. *AIChE Spring Meeting*, New Orleans, LA, March 31–April 4, 2003.

Hall, S. G. 1997. Water and effluent minimisation. Institution of Chemical Engineers, North Western Branch Papers, No. 4, November 12, 1997.

Hallale, N. 2002. A new graphical targeting method for water minimisation, *Advances in Environmental Research*, 6(3), 377–390.

Hallale, N. and Liu, F. 2001. Refinery hydrogen management for clean fuels production, *Advances in Environmental Research*, 6, 81–98.

Kazantzi, V. and El-Halwagi, M. M. 2005. Targeting material reuse via property integration, *Chemical Engineering Progress*, 101(8), 28–37.

Kit, S. G., Wan Alwi, S. R., and Manan, Z. A. 2011. A new graphical approach for simultaneous targeting and design of a paper recycling network, *Asia-Pacific Journal of Chemical Engineering*, 6(5), 689–812.

Manan, Z. A., Tan, Y. L., and Foo, D. C. Y. 2004. Targeting the minimum water flowrate using water cascade analysis technique, *AIChE Journal*, 50(12), 3169–3183.

Polley, G. T. and Polley, H. L. 2000. Design better water networks, *Chemical Engineering Progress*, 96(2), 47–52.

Prakash, R. and Shenoy, U. V. 2005. Targeting and design of water networks for fixed flowrate and fixed contaminant load operations, *Chemical Engineering Science*, 60(1), 255–268.

Shenoy, U. V. and Bandyopadhyay, S. 2007. Targeting for multiple resources, *Industrial and Engineering Chemistry Research*, 46, 3698–3708.

Skouteris, G., Ouki, S., Foo, D. C. Y., Saroj, D., Altini, M., Melidis, P., Cowley, B., Ells, G., Palmer, S., and O'Dell, S. 2018. water footprint assessment and water pinch analysis techniques for sustainable water management in the brick-manufacturing industry, *Journal of Cleaner Production*, 172, 786–794.

Wan Alwi, S. R. and Manan, Z. A. 2007. Targeting multiple water utilities using composite curves, *Industrial and Engineering Chemistry Research*, 46, 5968–5976.

Wang, Y. P. and Smith, R. 1994. Wastewater minimisation, *Chemical Engineering Science*, 49, 981–1006.

Wang, Y. P. and Smith, R. 1995. Wastewater minimization with flowrate constraints, *Chemical Engineering Research and Design*, 73, 889–904.

4

Algebraic Targeting Techniques for Direct Reuse/Recycle*

In Chapter 3, graphical technique that provides various useful insights for a resource conservation network (RCN) prior to detailed network design has been demonstrated. However, all graphical techniques are associated with several common pitfalls, such as low accuracy, as well as cumbersome drawing exercises. In this chapter, a well-established algebraic technique *known as material cascade analysis* (MCA) is introduced. The technique can be used to determine various RCN targets for direct reuse/recycle scheme, for both single and multiple fresh resources. Targeting for threshold problems is also discussed. Depending on the application, the MCA technique may be further categorized as follows: *water cascade analysis* (WCA; see Manan et al., 2004; Foo et al., 2006b; Foo, 2007, 2008), *gas cascade analysis* (GCA; Foo and Manan, 2006), and *property cascade analysis* (PCA; Foo et al., 2006a). However, all of these possess similar structure.

4.1 Generic Procedure for Material Cascade Analysis Technique

Steps to construct the MCA are given as follows:

1. Quality levels (q_k) of all process sinks and sources are arranged in descending order. Quality level for the fresh resource is added, if it does not already exist for the process sinks and sources. An arbitrary value is also added for the lowest quality level.
2. Flowrates of the process sinks (F_{SKj}) and sources (F_{SRi}) are located at their quality levels in columns 2 and 3 of Table 4.1, respectively. When there is more than one sink and source at the same quality level, their individual flowrates should be added.

---

* All examples of this chapter are available in spreadsheet, downloadable from book website.

TABLE 4.1

MCA for Minimum Resource Targeting

q_k	$\sum_j F_{SK_j}$	$\sum_i F_{SR_i}$	$\sum_i F_{SR_i} - \sum_j F_{SK_j}$	$F_{C,k}$	Δm_k	$Cum.\Delta m_k$	$F_{R,k}$
				F_R			
q_k	$\left(\sum_j F_{SK_j}\right)_1$	$\left(\sum_i F_{SR_i}\right)_1$	$\left(\sum_i F_{SR_i} - \sum_j F_{SK_j}\right)_1$	$\Downarrow$			
				$F_{C,k}$	Δm_k		
q_{k+1}	$\left(\sum_j F_{SK_j}\right)_{k+1}$	$\left(\sum_i F_{SR_i}\right)_{k+1}$	$\left(\sum_i F_{SR_i} - \sum_j F_{SK_j}\right)_{k+1}$	$\Downarrow$		$Cum.\Delta m_{k+1}$	$F_{R,k+1}$
				$F_{C,k+1}$	Δm_{k+1} $\Downarrow$		
$\vdots$ $\vdots$	$\vdots$	$\vdots$	$\vdots$	$\vdots$	$\vdots$	$\vdots$	$\vdots$
$\vdots$ $\vdots$	$\vdots$	$\vdots$	$\vdots$	$\vdots$	$\vdots$	$\vdots$	$\vdots$
$\vdots$ $\vdots$	$\vdots$	$\vdots$	$\vdots$	$\vdots$	$\vdots$	$\vdots$	$\vdots$
q_{n-2}	$\left(\sum_j F_{SK_j}\right)_{n-2}$	$\left(\sum_i F_{SR_i}\right)_{n-2}$	$\left(\sum_i F_{SR_i} - \sum_j F_{SK_j}\right)_{n-2}$	$\Downarrow$		$Cum.\Delta m_{n-2}$	
				$F_{C,n-2}$	Δm_{n-2} $\Downarrow$		
q_{n-1}	$\left(\sum_j F_{SK_j}\right)_{n-1}$	$\left(\sum_i F_{SR_i}\right)_{n-1}$	$\left(\sum_i F_{SR_i} - \sum_j F_{SK_j}\right)_{n-1}$	$\Downarrow$		$Cum.\Delta m_{n-1}$	$F_{R,n-1}$
				$F_{C,k-1} = F_D$	Δm_{n-1} $\Downarrow$		
q_n						$Cum.\Delta m_n$	$F_{R,n}$

Source: From Foo, 2007. Reproduced with permission. Copyright 2007 Elsevier.

3. At each quality level k, the total flowrate of the process sink(s) is deducted from that of the process source(s), with the net flowrate given (see column 4, Table 4.1). For each level, a positive net flowrate indicates the presence of excess process source(s), while a negative value indicates insufficient process source(s) in fulfilling the requirement of the process sink(s).

4. The net material flowrate is cascaded down the quality levels to yield the cumulative flowrate ($F_{C,k}$) in column 5. The first entry of this column corresponds to the fresh resource consumption for the RCN (F_R). However, at this stage, it is assumed that no fresh resource is used, i.e., $F_R = 0$.

5. In column 6, the impurity/property load in each quality interval (Δm_k) is calculated. The load value is given by the product of the cumulative flowrate ($F_{C,k}$, column 5), combined with the difference across two quality levels ($q_{k+1} - q_k$), where $F_{C,k}$ is located.

6. The load values are cascaded down the quality levels to yield the cumulative load (Cum. Δm_k) in column 7. If negative Cum. Δm_k values are observed, and Steps 7 and 8 are followed.

7. The *interval fresh resource flowrate* $(F_{R,k})$ is calculated for each quality level in column 8. This flowrate value is determined by dividing the cumulative loads (Cum. Δm_k, column 7) by the difference between the quality levels of interest (q_k) and of the fresh resource (q_R). This is given by

$$F_{R,k} = \frac{\text{Cum.}\Delta m_k}{\left(q_k - q_R\right)} \tag{4.1}$$

8. The absolute value of the largest negative $F_{R,k}$ in column 8 is identified as the *minimum fresh resource consumption* (F_R) of the network. Steps 4–6 are repeated by using the newly obtained F_R value in the first entry of column 5.

For convenience, it is always better to show the results of the MCA in tabulated form. When negative Cum. Δm_k values are observed in Step 7 (column 7), the result is known as the *infeasible cascade table*. After all negative Cum. Δm_k values in column 7 are removed (by carrying out Step 8), a *feasible cascade table* is created.

Several RCN targets are readily obtained from the feasible cascade table, similar to the graphical techniques described in Chapter 3. Apart from the fresh resource target (F_R) identified in Step 8, the final entry of column 5 in the feasible cascade table indicates the minimum waste discharge (F_D) from the RCN. Besides, the material recovery pinch is found at the quality level where zero Cum. Δm_k value (column 7) is observed. Process source that exists at the pinch is known as the *pinch-causing source*, where a portion of this source is allocated to the higher quality region (HQR) above the pinch, and the rest to the lower quality region (LQR) below the pinch. These flowrate allocation targets $(F_{HQR}$ and F_{LQR} for HQR and LQR, respectively) are observed in column 5 (among the $F_{C,k}$ values) in the intervals just above and just below the pinch quality level (Manan et al., 2004; Foo et al., 2006a,b).

Example 4.1 Water minimization for acrylonitrile (AN) production

Water minimization case study of the AN process given in Examples 2.1 and 3.1 is revisited here. The limiting water data for the problem are given in Table 4.2 (reproduced from Table 2.1). Assuming pure freshwater (i.e., no impurity) is used, perform the following tasks using the WCA technique:

1. Determine the minimum freshwater and wastewater flowrates, as well as the pinch concentration of the RCN.

TABLE 4.2

Limiting Water Data for AN Production

Water Sinks, SK_j			
j	Stream	Flowrate F_{SKj} (kg/s)	NH_3 Concentration C_{SKj} (ppm)
1	BFW	1.2	0
2	Scrubber inlet	5.8	10

Water Sources, SR_i			
i	Stream	Flowrate F_{SRi} (kg/s)	NH_3 Concentration C_{SRi} (ppm)
1	Distillation bottoms	0.8	0
2	Off-gas condensate	5	14
3	Aqueous layer	5.9	25
4	Ejector condensate	1.4	34

2. Perform the WCA using MS Excel. Identify the flowrate allocation targets of the pinch-causing source to the HQR and LQR.
3. Develop the RCN structure based on insights gained from WCA.

SOLUTION

1. Determine the minimum freshwater and wastewater flowrates, as well as the pinch concentration of the RCN.

The eight-step procedure is followed to carry out WCA for the AN case study:

   i. *Quality levels (q_k) of all process sinks and sources are arranged in descending order. Quality level for the fresh resource is also added, if it does not already exist for the process sinks and sources. An arbitrary value is also added for the highest quality level.*

   As mentioned in Examples 2.1 and 3.1, the quality level in this case corresponds to the concentration of ammonia (NH_3) in the water sinks and sources. We then arrange the concentrations in the final column of Table 4.2 in ascending order, i.e., 0, 10, 14, 25, and 34 ppm (i.e., descending order of quality). Note that the freshwater concentration (0 ppm) is already present among the concentration levels of the water sinks and sources. For the highest quality level, it is appropriate to set the highest concentration of the unit used in this case, i.e., 1,000,000 ppm. All quality levels are shown in the first column (C_k) of the cascade table in Table 4.3.

   ii. *Flowrates of the process sinks (F_{SKj}) and sources (F_{SRi}) are located at their quality levels in columns 2 and 3 of the cascade table.*

   iii. *At each quality level k, the total flowrate of the process sink(s) is deducted from that of the process source(s), with the net flowrate given in column 4.*

TABLE 4.3

Preliminary Cascade Table for Steps 2–4

C_k (ppm)	$\sum_j F_{SK_j}$ (kg/s)	$\sum_i F_{SK_i}$ (kg/s)	$\sum_i F_{SK_i}$ (kg/s) $-$ $\sum_j F_{SK_j}$ (kg/s)	$F_{C,k}$ (kg/s)
				$F_{FW} = 0$
0	1.2	0.8	−0.4	
				−0.4
10	5.8		−5.8	
				−6.2
14		5	5	
				−1.2
25		5.9	5.9	
				4.7
34		1.4	1.4	
				6.1
1,000,000				

    iv. *The net material flowrate is cascaded down the quality levels to yield the cumulative flowrate (F_{ck}) in column 5. The first entry of this column corresponds to the fresh resource consumption for the RCN (F_R). However, at this stage, it is assumed that no fresh resource is used, i.e., $F_R = 0$.*

        Following Steps 2–4, a preliminary cascade table is constructed in Table 4.3. In the final column, fresh resource usage (corresponds to freshwater feed, F_{FW} in this case) is taken as zero.

    v. *In column 6, the impurity/property load in each quality interval (Δm_k) is calculated.*

    vi. *The load values are cascaded down the quality levels to yield the cumulative load (Cum. Δm_k) in column 7. If negative Cum. Δm_k values are observed, and Steps 7 and 8 are carried out.*

    vii. *The interval fresh resource flowrate ($F_{R,k}$) is calculated for each quality level in column 8.*

        Following Steps 5–7, an infeasible cascade table is constructed (Table 4.4). In column 7, negative Cum. Δm_k values are observed in the concentration levels of 10, 14, and 25 ppm (Step 6), which means that the RCN experiences load infeasibility. Hence, the *interval freshwater flowrate ($F_{FW,k}$)* for each concentration level is calculated and shown in column 8 of Table 4.4.

    viii. *The absolute value of the largest negative $F_{R,k}$ in column 8 is identified as the minimum fresh resource consumption (F_R) of the network. Steps 4–6 are repeated by using the newly obtained F_R value in the first entry of column 5.*

        In column 8 of Table 4.4, the largest negative $F_{FW,k}$ value is identified as −2.06 kg/s. Next, its absolute value is identified as

TABLE 4.4

Infeasible Cascade Table

C_k (ppm)	$\sum_j F_{SKj}$ (kg/s)	$\sum_i F_{SKi}$ (kg/s)	$\sum_i F_{SKi} - \sum_j F_{SKj}$ (kg/s)	$F_{C,k}$ (kg/s)	Δm_k (mg/s)	Cum. Δm_k (mg/s)	$F_{FW,k}$ (kg/s)
				$F_{FW} = 0$			
0	1.2	0.8	−0.4				
				−0.4	−4		
10	5.8		−5.8			−4	−0.40
				−6.2	−24.8		
14		5	5			−28.8	−2.06
				−1.2	−13.2		
25		5.9	5.9			−42	−1.68
				4.7	42.3		
34		1.4	1.4			0.3	0.01
				6.1	6,099,792.6		
1,000,000						6,099,792.9	6.10

the minimum freshwater consumption (F_{FW}) of the network. Repeating Steps 4–6 with $F_{FW} = 2.06$ kg/s results in the feasible cascade table in Table 4.5. In Table 4.5, the final entry of column 5 indicates the minimum waste (wastewater) discharge from the RCN, with a flowrate (F_{WW}) of 8.16 kg/s. The material recovery pinch is found at the concentration level where zero Cum. Δm_k value is observed in column 7, i.e., 14 ppm.

TABLE 4.5

Feasible Cascade Table for AN Case

C_k (ppm)	$\sum_j F_{SKj}$ (kg/s)	$\sum_i F_{SRi}$ (kg/s)	$\sum_i F_{SRi} - \sum_j F_{SKj}$ (kg/s)	$F_{C,k}$ (kg/s)	Δm_k (mg/s)	Cum.Δm_k (mg/s)
				$F_{FW} = 2.06$		
0	1.2	0.8	−0.4			
				1.66	16.57	
10	5.8		−5.8			16.57
				−4.14	−16.57	
14		5	5			**0.00**
				0.86	9.43	**(PINCH)**
25		5.9	5.9			9.43
				6.76	60.81	
34		1.4	1.4			70.24
				$F_{WW} = 8.16$	8,156,865.51	
100,0000						8,156,935.76

2. Perform the WCA using MS Excel. Identify the flowrate allocation targets of the pinch-causing source to the HQR (F_{HQR}) and
LQR (F_{LQR}).

The WCA is re-performed using MS Excel. Steps 1–3 are
done in columns B–F of the spreadsheet (see Figure 4.1). Next,
Steps 4–8 are done in columns G–J, which forms the infeasible
cascade table. In Step 8, the largest negative value is identified
among the entries in column J, and is returned as the minimum
freshwater flowrate (F_{FW}) as the first entry of column K (cell K4).
Steps 4–6 are repeated to construct the feasible cascade table in
columns K–M. As shown in Figure 4.1, the freshwater (F_{FW}) and
wastewater flowrates (F_{WW}) are identified as 2.06 and 8.16 kg/s,
identical to those reported in Table 4.5.

Figure 4.1 also shows the pinch concentration is identified at 14
ppm. As the pinch concentration divides the network into HQR
and LQR, process sinks and sources with concentration lower
than 14 ppm, i.e., SK1, SK2, and SR1 (identified from Table 4.2),
belong to the HQR, while those with concentration higher than
14 ppm, i.e., SR3 and SR4, belong to the LQR. However, process
source at 14 ppm, i.e., the pinch-causing source (SR2), belongs
to both regions. As shown in column K of Figure 4.1, 4.14 kg/s
of this pinch-causing source is sent to the HQR (F_{HQR}, cell K8,
the negative sign indicates water is sent from lower to higher quality
levels), while 0.86 kg/s to the LQR (F_{LQR}, cell K10).

3. Develop the RCN structure based on insights gained from WCA.

For a simple case such as this, we can develop the RCN structure based on the flowrate allocation insights of the cascade table.
As shown in column K of Figure 4.1, 2.06 kg/s of freshwater is fed
to the RCN. A portion of this freshwater flowrate (0.4 kg/s) is consumed at 0 ppm level, where a negative net flowrate is observed.

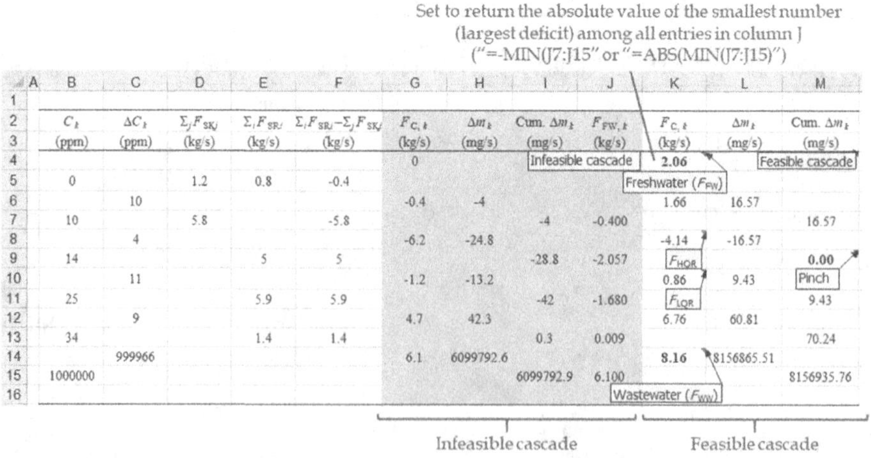

Set to return the absolute value of the smallest number
(largest deficit) among all entries in column J
("=-MIN(J7:J15" or "=ABS(MIN(J7:J15)")

A	B	C	D	E	F	G	H	I	J	K	L	M
1												
2	C_i	ΔC_i	$\Sigma_j F_{SKj}$	$\Sigma_i F_{SRi}$	$\Sigma_i F_{SRi} - \Sigma_j F_{SKj}$	$F_{C,i}$	Δm_i	Cum. Δm_i	$F_{FW,i}$	$F_{C,i}$	Δm_i	Cum. Δm_i
3	(ppm)	(ppm)	(kg/s)	(kg/s)	(kg/s)	(kg/s)	(mg/s)	(mg/s)	(kg/s)	(kg/s)	(mg/s)	(mg/s)
4						0		Infeasible cascade		2.06		Feasible cascade
5	0		1.2	0.8	-0.4					Freshwater (F_{FW})		
6		10				-0.4	-4			1.66	16.57	
7	10		5.8		-5.8			-4	-0.400			16.57
8		4				-6.2	-24.8			-4.14	-16.57	
9	14		5	5				-28.8	-2.057	F_{HQR}		0.00
10		11				-1.2	-13.2			0.86	9.43	Pinch
11	25		5.9	5.9				-42	-1.680	F_{LQR}		9.43
12		9				4.7	42.3			6.76	60.81	
13	34		1.4	1.4				0.3	0.009			70.24
14		999966				6.1	6099792.6			8.16	8156865.51	
15	1000000							6099792.9	6.100			8156935.76
16										Wastewater (F_{WW})		

Infeasible cascade   Feasible cascade

FIGURE 4.1
WCA with MS Excel; and identification of F_{HQR} and F_{LQR} for AN case.

Since this level only consists of SK1 (boiler feed water—BFW) and SR1 (distillation bottoms), both freshwater feed and SR1 are used to fulfill the flowrate requirement of SK1.

At 10 ppm level, where SK2 (scrubber inlet) is found, the remaining freshwater flowrate (1.66 kg/s) is consumed by this sink. However, SK2 has a flowrate requirement of 5.8 kg/s. Hence, 4.14 kg/s of water flowrate is supplied by the allocation flowrate of the pinch-causing source, SR2 (off-gas condensate, 14 ppm), to the HQR (F_{HQR}). With the flowrate allocation to SK2, the pinch-causing source SR2 has its flowrate reduced to 0.86 kg/s, which then leaves as allocation flowrate to the LQR (F_{LQR}). This allocation flowrate is then cascaded to lower concentration levels. In the absence of process sink at 25 and 34 ppm levels, the allocation flowrate of SR2 is mixed with other unutilized sources (SR3 and SR4) to be discharged as wastewater stream from the RCN.

With the earlier-discussed insights, we can develop the structure for the RCN in Figure 4.2, which is essentially the same as that in Figure 3.11. However, note that the development of an RCN based on targeting insights will only work for simple cases such as this example.[1] For more complicated problems, a detailed network design procedure should be used (see Chapter 8 for details).

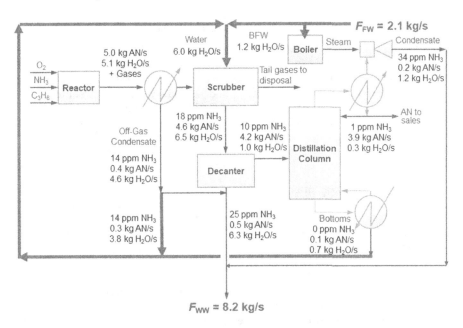

$F_{WW} = 8.2$ kg/s

FIGURE 4.2
RCN structure for AN case study (Example 4.1). Note: Detailed mass balance has not been revised yet.

---

[1] There should be maximum one sink with non-zero quality index in the HQR.

4.2 Targeting for Multiple Fresh Resources

When an RCN is served by several fresh resources of different quality, the MCA procedure is to be revised in order to determine the minimum resource flowrates. In principle, the revised procedure consists of three steps (Foo, 2007):

1. Flowrate targeting for lower quality resource—this step determines the minimum flowrate for lower quality resource in the absence of higher quality resource.

2. Flowrate targeting for higher quality resource—minimum higher quality resource flowrate needed to supplement the lower quality resource is determined.

3. Flowrate adjustment of lower quality resource—the lower quality resource flowrate is reduced, as higher quality resource is used for the RCN.

Note that flowrate targeting/adjustment for each step should follow the MCA procedure as described in the previous section.

Example 4.2 Targeting for multiple freshwater sources

The AN production case study in Example 4.1 is revisited. Similar to Example 3.4, a secondary impure freshwater source with 5 ppm ammonia content is available for use apart from the primary pure freshwater source. Unit costs for the primary (CT_{FW1}) and secondary freshwater sources (CT_{FW2}) are given as \$0.001/kg and \$0.0005/kg, respectively. Determine the minimum flowrates for both freshwater sources and wastewater stream, as well as the operating cost for freshwater sources of the RCN.

SOLUTION

Since the unit costs of both freshwater sources are identical to those in Example 3.4, their prioritized costs remain identical. Hence the use of the secondary impure freshwater is maximized before the primary freshwater source is utilized. A three-step procedure is followed to carry out the targeting task:

1. *Flowrate targeting for lower quality resource.*

   Following the previously described MCA procedure, an infeasible cascade for targeting the minimum flowrate of the secondary freshwater source (5 ppm) is given in Table 4.6. As shown in the last column of Table 4.6, the minimum flowrate for this water source is determined as 3.2 kg/s.

2. *Flowrate targeting for higher quality resource.*

   We then include the targeted flowrate of secondary freshwater source (F_{FW2} = 3.2 kg/s) into the column for water sources in

TABLE 4.6

Targeting for Secondary Freshwater Source: Infeasible Cascade

C_k (ppm)	$\sum_j F_{SKj}$ (kg/s)	$\sum_i F_{SRi}$ (kg/s)	$\sum_i F_{SRi} - \sum_j F_{SRj}$ (kg/s)	$F_{C,k}$ (kg/s)	Δm_k (mg/s)	Cum. Δm_k (mg/s)	$F_{FW2,k}$ (kg/s)
0	1.2	0.8	−0.4				
				−0.4	−2		
5			0			−2.00	
				−0.4	−2		
10	5.8		−5.8			−4.00	−0.80
				−6.2	−24.8		
14		5	5			−28.80	−3.20
				−1.2	−13.2		
25		5.9	5.9			−42.00	−2.10
				4.7	42.3		
34		1.4	1.4			0.30	0.01
				6.1	6,099,792.6		
1,000,000						6,099,792.90	6.10

the cascade table. A pinch concentration is observed at 14 ppm, where zero cumulative load is found (see Table 4.7). However, it is also observed that a negative cumulative load value is found at 5 ppm. In other words, the RCN experiences load infeasibility at this level. Hence, we next proceed to determine the minimum

TABLE 4.7

Targeting for Primary Freshwater Source: Infeasible Cascade

C_k (ppm)	$\sum_j F_{SKj}$ (kg/s)	$\sum_i F_{SRi}$ (kg/s)	$\sum_i F_{SRi} - \sum_j F_{SRj}$ (kg/s)	$F_{C,k}$ (kg/s)	Δm_k (mg/s)	Cum. Δm_k (mg/s)	$F_{FW1,k}$ (kg/s)
0	1.2	0.8	−0.4				
				−0.40	−2.00		
5		$F_{FW2} = 3.2$	3.2			−2.00	−0.40
				2.80	14.00		
10	5.8		−5.8			12.00	1.20
				−3.00	−12.00		
14		5	5			**0.00**	0.00
				2.00	22.00	**(PINCH)**	
25		5.9	5.9			22.00	0.88
				7.90	71.10		
34		1.4	1.4			93.10	2.74
				9.30	9,299,683.80		
1,000,000						9,299,776.90	9.30

flowrate of primary freshwater source (0 ppm) needed to restore the feasibility of the RCN. The interval flowrates for the primary freshwater source are then determined for each concentration level. As shown in the final column of Table 4.7, the minimum flowrate for this water source is determined as 0.4 kg/s.

3. *Flowrate adjustment of lower quality resource.*

The targeted flowrate for the primary water source is included in the first level of column 5 in the cascade table (Table 4.8). However, we then observe that a new pinch concentration has emerged at 5 ppm and that the original pinch concentration at 14 ppm (due to the use of secondary freshwater source) has disappeared. It means that the secondary freshwater source has a larger flowrate than its minimum requirement, due to the use of the primary freshwater source. Hence, we proceed to reduce the flowrate of the secondary freshwater source. The last column in Table 4.8 tabulates the *excessive interval flowrates* supplied by the secondary freshwater source to each concentration level, which is determined using Equation 4.1. For instance, the excessive interval flowrate at 10 ppm is determined as follows:

$$\frac{16 \text{ mg/s}}{(10-5)\text{ppm}} = 3.2 \text{ kg/s}$$

From the last column of Table 4.8, it is observed that the minimum excessive interval flowrate is found at 14 ppm,

TABLE 4.8

Flowrate Adjustment for Secondary Freshwater Source

C_k (ppm)	$\sum_j F_{SKj}$ (kg/s)	$\sum_i F_{SRi}$ (kg/s)	$\sum_i F_{SRi} - \sum_j F_{SRj}$ (kg/s)	$F_{C,k}$ (kg/s)	Δm_k (mg/s)	Cum. Δm_k (mg/s)	$F_{FW2,k}$ (kg/s)
				$F_{FW1} = 0.40$			
0	1.2	0.8	−0.4				
					0.00	0.00	
5		$F_{FW2} = 3.2$	3.2			0.00	
					3.20	16.00	(PINCH)
10	5.8		−5.8			16.00	3.20
					−2.60	−10.40	
14		5	5			5.60	0.62
					2.40	26.40	
25		5.9	5.9			32.00	1.60
					8.30	74.70	
34		1.4	1.4			106.70	3.68
				$F_{WW} = 9.70$	9,699,670.20		
1,000,000						9,699,776.90	9.70

TABLE 4.9

Minimum Flowrates for Primary and Secondary Freshwater Sources

C_k (ppm)	$\sum_j F_{SRj}$ (kg/s)	$\sum_i F_{SRi}$ (kg/s)	$\sum_i F_{SRi} - \sum_j F_{SRj}$ (kg/s)	$F_{C,k}$ (kg/s)	Δm_k (mg/s)	Cum. Δm_k (mg/s)
				$F_{FW1} = 0.40$		
0	1.2	0.8	−0.4			
				0.00	0.00	
5		$F_{FW2} = 2.58$	3.2			0.00
				2.58	12.89	(PINCH)
10	5.8		−5.8			12.89
				−3.22	−12.89	
14		5	5			0.00
				1.78	19.56	(PINCH)
25		5.9	5.9			19.56
				7.68	69.10	
34		1.4	1.4			88.66
				$F_{WW} = 9.08$	9,077,469.13	
1,000,000						9,077,557.79

i.e., 0.62 kg/s. In other words, 0.62 kg/s excessive flowrate of the secondary freshwater source has been supplied to the RCN. Hence, this excessive flowrate is to be deducted from the earlier targeted secondary freshwater source flowrate (in Step 1), i.e., 3.2–0.62 = 2.58 kg/s. Including this reduced flowrate in the cascade table restores the original pinch concentration at 14 ppm (see Table 4.9). As discussed in Example 3.4, for a *multiple-pinch* RCN, the higher quality pinch (5 ppm) is termed as the *limiting pinch*, which controls the overall fresh resource flowrates of the RCN. Note that since both freshwater sources have unit costs identical to those given in Example 3.4, the operating cost of freshwater is essentially equal to that given in Example 3.4, i.e., $0.0017/s (= 0.4 kg/s × $0.001/kg + 2.58 kg/s × $0.0005/kg).

4.3 Targeting for Threshold Problems

As discussed in Chapter 3, there exist two distinct cases of threshold problems, i.e., RCN with zero fresh resource and/or zero waste discharge. For RCN with zero fresh resource, the procedure described earlier for MCA may be followed without modification. Note, however, that only Steps 1–6 of the procedure are needed to establish the RCN targets, while Steps 7 and 8 are omitted.

TABLE 4.10

Limiting Water Data for Example 4.3

SK_j	F_{SKj} (t/h)	C_{SKj} (ppm)	SR_i	F_{SRi} (t/h)	C_{SRi} (ppm)
SK1	40	125	SR1	50	100
SK2	20	500	SR2	50	300
SK3	40	1125	SR3	20	1000

On the other hand, for RCN with zero waste discharge, performing Steps 1–8 of the MCA procedure will result in negative waste discharge. Hence, Step 9 is added to establish the rigorous targets for this kind of RCN, which is described as follows:

9. The absolute value of the negative waste discharge flowrate is added to the fresh resource consumption determined in Step 8 to obtain the rigorous fresh resource consumption (F_R) for the threshold case. Steps 4–6 are repeated by using the newly obtained F_R value in the first entry of column 5.

Example 4.3 Targeting for threshold problem—zero fresh resource

Table 4.10 shows the limiting data for a water minimization case study described in Example 3.5. Determine the following RCN targets using the WCA technique:

1. Minimum freshwater flowrate
2. Minimum wastewater flowrate
3. Pinch concentration for the RCN

SOLUTION

Following Steps 1–6 of the procedure, the cascade table for the example is constructed in Table 4.11. Since none of the Cum. Δm_k values in column 7 are negative, the WCA procedure is completed. The first and final entries in column 5 indicate that the RCN requires no freshwater feed ($F_{FW} = 0$) but will generate a wastewater flowrate (F_{WW}) of 20 t/h. The pinch for this problem is located at the highest concentration level, i.e., 0 ppm. This means that all water sinks and sources are located in the LQR.

Example 4.4 Targeting for threshold problem—zero waste discharge

Table 4.12 shows the limiting data for a water minimization case study described in Example 3.6. Determine the minimum freshwater and wastewater flowrates using the WCA technique.

SOLUTION

Following Steps 1–6 of the procedure, the infeasible cascade table for the example is constructed in Table 4.13. Since negative Cum. Δm_k values are observed in column 7, the interval freshwater flowrates are determined for

TABLE 4.11

Cascade Table for Example 4.3

C_k (ppm)	$\sum_j F_{SKj}$ (t/h)	$\sum_i F_{SKi}$ (t/h)	$\sum_i F_{SKi} - \sum_j F_{SKj}$ (t/h)	$F_{C,k}$ (t/h)	Δm_k (kg/h)	Cum. Δm_k (kg/h)
				$F_{FW} = 0$		
0			0			(PINCH)
				0	0	
100		50	50			0
				50	1.25	
125	40		−40			1.25
				10	1.75	
300		50	50			3
				60	12	
500	20		−20			15
				40	20	
1000		20	20			35
				60	7.5	
1125	40		−40			42.5
				$F_{WW} = 20$	19,977.5	
1,000,000			0			20,020

each concentration level (Step 7), with the results shown in the final column of Table 4.13. The largest negative value for this column is observed at the lowest concentration level (1,000,000 ppm), i.e., 59.97 t/h. This means that 59.97 t/h of freshwater is needed at this concentration level.

Following Step 8 of the procedure, 59.97 t/h freshwater is added at the first entry of column 5, with the cascade result shown in Table 4.14. However, the wastewater flowrate is observed to have a negative value at the last entry of column 5. Obviously, this is an infeasible solution, which means that the freshwater flowrate is insufficient to supplement the requirement of the water sink for the network.

Following Step 9 of the procedure, absolute value of the negative wastewater discharge is added to the freshwater flowrate, which results in a total feed of 60 t/h. A revised cascade table with 60 t/h freshwater feed is constructed in Table 4.15. It is observed that the RCN has zero

TABLE 4.12

Limiting Water Data for Example 4.4

SK_j	F_{SKj} (t/h)	C_{SKj} (ppm)	SR_i	F_{SKi} (t/h)	C_{SRi} (ppm)
SK1	50	20	SR1	20	20
SK2	20	50	SR2	50	100
SK3	100	400	SR3	40	250

TABLE 4.13

Infeasible Cascade Table for Example 4.4

C_k (PPm)	$\sum_j F_{SKj}$ (t/h)	$\sum_i F_{SKi}$ (t/h)	$\sum_i F_{SKi} - \sum_j F_{SKj}$ (t/h)	$F_{C,k}$ (t/h)	Δm_k (kg/h)	Cum. Δm_k (kg/h)	$F_{FW,k}$ (t/h)
				0.00			
0							
				0.00	0.00		
20	50	20	−30			0.00	0.00
				−30.00	−0.90		
50	20		−20			−0.90	−18.00
				−50.00	−2.50		
100		50	50			−3.40	−34.00
				0.00	0.00		
250		40	40			−3.40	−13.60
				40.00	6.00		
400	100		−100			2.60	6.50
				−60.00	−59,976.00		
1,000,000						−59,973.40	−59.97

TABLE 4.14

Cascade Table with Infeasible Wastewater Flowrate

C_k (ppm)	$\sum_j F_{SKj}$ (t/h)	$\sum_i F_{SKi}$ (t/h)	$\sum_i F_{SKi} - \sum_j F_{SKj}$ (t/h)	$F_{C,k}$ (t/h)	Δm_k (kg/h)	Cum. Δm_k (kg/h)
				$F_{FW} = 59.97$		
0						
				59.97	1.20	
20	50	20	−30			1.20
				29.97	0.90	
50	20		−20			2.10
				9.97	0.50	
100		50	50			2.60
				59.97	9.00	
250		40	40			11.59
				99.97	15.00	
400	100		−100			26.59
				$F_{WW} = -0.03$	−26.59	
1,000,000						0.00 (PINCH)

TABLE 4.15

Feasible Cascade Table for Example 4.4

C_k (ppm)	$\sum_j F_{SKj}$ (t/h)	$\sum_i F_{SKi}$ (t/h)	$\sum_i F_{SKi} - \sum_j F_{SKj}$ (t/h)	$F_{C,k}$ (t/h)	Δm_k (kg/h)	Cum. Δm_k (kg/h)
				$F_{FW} = 60.00$		
0						
				60.00	1.20	
20	50	20	−30			1.20
				30.00	0.90	
50	20		−20			2.10
				10.00	0.50	
100		50	50			2.60
				60.00	9.00	
250		40	40			11.60
				100.00	15.00	
400	100		−100			26.60
				$F_{WW} = 0.00$	0.00	
1,000,000						26.60
						(THRESHOLD)

wastewater discharge ($F_{WW} = 0$). The flowrate targets are identical to those identified using graphical method (see Figure 3.18). The concentration where the pinch was formed earlier (Table 4.14) is now termed as the *threshold point* of the RCN (Foo, 2008).

Note again that apart from water minimization examples presented here, threshold problems also exist in other RCN problems such as hydrogen network and property integration systems.

4.4 Targeting for Property Integration with Inferior Property Operator Level

Procedure for carrying out MCA for most property integration problems (i.e., PCA) remains the same as described earlier. However, the cumulative load in column 7 is now termed as the *property load* (ΔL_k). Besides, in special case where the property operator of the fresh resource exists at the inferior level, modified Steps 1 and 7 are required (see modification in italic case):

1. *Property operator levels* (ψ_k) *of all process sinks and sources are arranged in descending order. Operator level for the fresh resource is added, if it does not already exist for the process sinks and sources. Zero value is also added for the lowest quality level.*

2. The interval fresh resource flowrate ($F_{R,k}$) is calculated for each *operator* level in column 8. This flowrate value is determined by dividing the cumulative loads (Cum. ΔL_k, column 7) by the difference between the *quality level of the fresh resource* (q_R) *and the quality level of interest* (q_k). This is given by

$$F_{R,k} = \frac{\Delta L_k}{\left(q_R - q_k\right)} \tag{4.2}$$

Example 4.5 Targeting for property integration—fresh resource with inferior operator value

Table 4.16 shows the limiting data for the fiber recovery case study (Kazantzi and El-Halwagi, 2005) described in Examples 2.6 and 3.8. Use MS Excel to determine the minimum fresh and waste fiber targets using the PCA technique.

SOLUTION

Following the revised Step 1, the property operator levels are arranged in descending order, as shown in column B of Figure 4.3. A zero-quality level

TABLE 4.16

Limiting Data for Fiber Recovery Case Study in Example 4.5

Sink, SK$_j$	F_{SKj} (t/h)	Ψ_{SKj} (Dimensionless)	Source, SR$_i$	F_{SRi} (t/h)	Ψ_{SRi} (Dimensionless)
Machine I (SK1)	100	0.382	Bleached fiber (SR1)	90	0.469
Machine II (SK2)	40	0.536	Broke (SR2)	60	0.182
			Fresh fiber (FF)	To be determined	0.738

A	B	C	D	E	F	G	H	I	J	K	L	M
1												
2	Ψ	$\Delta\Psi$	$\Sigma_j F_{SKj}$ (t/h)	$\Sigma_i F_{SRi}$ (t/h)	$\Sigma_i F_{SRi} - \Sigma_j F_{SKj}$ (t/h)	$F_{C,t}$ (t/h)	ΔL_t (t/h)	Cum. ΔL_t (t/h)	$F_{F,t}$ (t/h)	$F_{C,t}$ (t/h)	ΔL_t (t/h)	Cum. ΔL_t (t/h)
3												
4								Infeasible		14.98		Feasible
5	0.738				0					Fresh fiber		
6		0.202				0	0			14.98	3.03	
7	0.536		40		-40			0	0			3.03
8		0.067				-40	-2.68			-25.02	-1.68	
9	0.469			90	90			-2.68	-9.9628253			1.35
10		0.087				50	4.35			64.98	5.65	
11	0.382		100		-100			1.67	4.6910112			7.00
12		0.200				-50	-10			-35.02	-7.00	
13	0.182			60	60			-8.33	-14.982014			0.00
14		0.182				10	1.82			24.98	4.55	Pinch
15	0.000				0			-6.51	-8.8211382			4.55
16										Waste fiber (F_{WF})		

FIGURE 4.3
PCA for Example 4.5 (fiber recovery case study).

is also added. Next, following Steps 2–6 of the MCA procedure, an infeasible cascade table for the example is constructed in Figure 4.3. Since negative Cum. ΔL_k values are observed in column I, and the interval fresh fiber flowrates ($F_{FF,k}$) are calculated for each operator level in column J (Step 7).

Finally, feasible cascade is constructed in columns K–M of Figure 4.3, following Step 8 of the MCA procedure. The minimum fresh fiber requirement (F_{FF}) is targeted at 14.98 t/h, while the waste fiber target (F_{WF}) is identified at 24.98 t/h. These targets match those obtained via the graphical targeting technique described in Chapter 3 (Figure 3.21).

PROBLEMS
Water Minimization Problems

4.1 For the Kraft pulping process described in Problem 2.2, determine the minimum flowrates of freshwater and wastewater for the RCN using WCA for the following cases. Limiting data for the RCN are given in Table 4.17. Assume pure freshwater is used.

  a. Correctly extracted limiting data—sources are segregated (use data in Table 4.17a) (answers are found in Problem 8.15).
  b. Incorrectly extracted limiting data—sources are not segregated (use data in Table 4.17b).[2]

TABLE 4.17

Limiting Water Data for Kraft Pulping Process in Problem 4.1

j	Sinks, SK$_j$	F_{SKj} (t/h)	C_{SKj} (ppm)	i	Sources, SRi	F_{SRi} (t/h)	C_{SRi} (ppm)
(a) Correctly extracted limiting data							
1	Pulp washing	467	20	1	W3	12.98	419
2	Chemical recovery	165	20	2	W5	9.7	16,248
3	Condenser	8.2	10	3	W7	10.78	9,900
				4	W8	116.5	20
				5	W9	48	233
				6	W10	52	311
				7	W11	52.2	20
				8	W13	300	30
				9	W14	140	15
(b) Incorrectly extracted limiting data							
1	Pulp washing	467	20	1	W6	22.68	7189
2	Chemical recovery	165	20	2	W7	10.78	9900
3	Condenser	8.2	10	3	W12	268.7	114
				4	W13	300	30
				5	W14	140	15

---

[2] See Section 2.1 for discussion on segregation of material sources.

TABLE 4.18

Limiting Water Data for the Tricresyl Phosphate Manufacturing Process in Problem 4.2

j	Sinks, SK$_j$	F_{SKj} (kg/s)	C_{SKj} (ppm)	*i*	Sources, SR$_i$	F_{SKi} (kg/s)	C_{SKi} (ppm)
(a) Incorrectly extracted limiting data							
1	Washing I	2.45	5	1	Washing I	2.45	76.36
2	Washing II	2.45	0	2	Washing II	2.45	0.07
3	Scrubber I	0.84	30	3	Scrubber I	0.84	410.57
4	Scrubber II	0.60	30	4	Scrubber II	0.6	144
5	Flare seal pot	0.25	100	5	Flare seal pot	0.25	281.5
(b) Correctly extracted limiting data							
1	Washing I	2.45	5	1	Washing I	2.45	76.36
2	Washing II	2.45	0	2	Washing II	2.45	0.07
3	Scrubber I	0.70	30	3	Scrubber I	0.70	410.57
4	Scrubber II	0.50	30	4	Scrubber II	0.50	144
5	Flare seal pot	0.20	100	5	Flare seal pot	0.20	281.5

    c. Determine the additional operating cost incurred with the use of incorrect limiting data in (b) as compared to the case with correct limiting data in (a). Assume an annual operating time of 8000 h and unit costs for freshwater and wastewater as \$1.5/ton and \$10/ton, respectively.

4.2 For the tricresyl phosphate manufacturing process in Problem 2.3, determine the minimum flowrates of freshwater (0 ppm) and wastewater for the RCN for the following cases.

    a. Incorrectly extracted limiting data—flowrates for sources SR3, SR4, and SR5 are taken as the largest values (use data in Table 4.18a).

    b. Correctly extracted limiting data—flowrates for sources SR3, SR4, and SR5 are taken as the smallest values (use data in Table 4.18b).[3]

4.3 Table 4.19 shows the limiting data for a water minimization case study (Jacob et al., 2002). Determine the minimum flowrates of freshwater (0 ppm) and wastewater for the RCN (see answer in Problem 8.10).

TABLE 4.19

Limiting Water Data for Problem 4.3

SK$_j$	F_{SKj} (t/h)	C_{SKj} (ppm)	SR$_i$	F_{SKi} (t/h)	C_{SKi} (ppm)
SK1	1200	120	SR1	500	100
SK2	800	105	SR2	2000	110
SK3	500	80	SR3	400	110
			SR4	300	60

Source: Jacob et al. (2002).

---

[3] See Section 2.2 for discussion on data extraction principles. Refer to Equations 2.54, 2.56, and 2.58 for operating flowrates of process sources.

TABLE 4.20

Limiting Water Data for Problem 4.4

Sinks, SK_j	F_{SK_j} (t/h)	C_{SK_j} (ppm)	Sources, SR_i	F_{SK_i} (t/h)	C_{SK_i} (ppm)
1	36.364	25	1	36.364	80
2	44.308	25	2	44.308	90
3	22.857	25	3	22.857	200
4	60.000	50	4	60.000	100
5	40.000	50	5	40.000	800
6	12.500	400	6	12.500	800
7	5.000	200	7	5.000	600
8	10.000	0	8	10.000	100
9	80.000	50	9	80.000	300
10	43.333	150	10	43.333	300

Source: Savelski and Bagajewicz (2001).

4.4 Limiting data for a water network with 10 pairs of water sinks and sources are summarized in Table 4.20 (Savelski and Bagajewicz, 2001). Use WCA to identify the minimum freshwater and wastewater flowrates for this RCN. Assume pure freshwater is used (see answer in Problem 8.17).

4.5 An existing water network for a paper milling process is shown in Figure 4.4 (Foo et al., 2006b). The feedstock for the paper mill (mainly consisting of old newspapers and magazines) is first blended with dilution water and chemicals to form a pulp slurry called *stock*. The stock is sent to the forming section of the paper machine to form paper sheets. In the paper machine, freshwater is fed to the forming and pressing sections

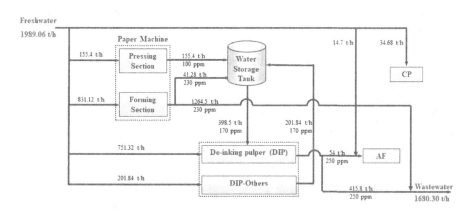

FIGURE 4.4
Water network of a paper mill in Problem 4.5. (From Foo, 2006b. Reproduced with permission. Copyright 2006 American Institute of Chemical Engineers [AIChE].)

TABLE 4.21

Limiting Water Data for Paper Milling Process in Problem 4.5

j	Sinks, SK_j	F_{SKj} (t/h)	C_{SKj} (ppm)	i	Sources, SR_i	F_{SKi} (t/h)	C_{SKi} (ppm)
1	Pressing section	155.40	20	1	Pressing section	155.40	100
2	Forming section	831.12	80	2	Forming section	1305.78	230
3	DIP-Others	201.84	100	3	DIP-Others	201.84	170
4	DIP	1149.84	200	4	DIP	469.80	250
5	Chemical preparation (CP)	34.68	20				
6	Approach flow (AF)	68.70	200				

Source: Foo et al., 2006b.

to remove debris. At the same time, wastewater is removed from the stock during paper sheet formation. A big portion of this wastewater stream is sent for waste treatment, while the rest to the water storage tank for reuse.

In the de-inking pulper (DIP) and its associated processes ("DIP-Others"), a mixture of freshwater and spent water (from the water storage) is used to remove printing ink from the main stock. A big portion of the DIP effluent is sent for waste treatment, whereas others are mixed with freshwater to be reused in the approach flow (AF) system. Besides, the effluent from "DIP-Others" is sent to the water storage tank for reuse/recycle. Freshwater is also consumed in the chemical preparation (CP) unit to dilute the de-inking chemicals.

It is believed that further water minimization is possible with this process. In considering water recovery, total suspended solid (TSS) is the most significant water quality factor that is of concern. Table 4.24 summarizes the limiting water data for the case study. Use WCA to identify the minimum freshwater and wastewater flowrate targets for the process. Assume pure freshwater is used.

4.6 Limiting data for water minimization for a specialty chemical production process (Wang and Smith, 1995) are shown in Table 4.22. Freshwater

TABLE 4.22

Limiting Water Data for Specialty Chemical Production Process in Problem 4.6

Sinks	F_{SKj} (t/h)	C_{SKj} (ppm)	Sources	F_{SKi} (t/h)	C_{SKi} (ppm)
Reactor	80	100	Reactor	20	1000
Cyclone	50	200	Cyclone	50	700
Filtration	10	0	Filtration	40	100
Steam system	10	0	Steam system	10	10
Cooling system	15	10	Cooling system	5	100

Source: Wang and Smith, 1995.

TABLE 4.23

Limiting Water Data for the Palm Oil Mill Case Study in Problem 4.8

j	Sinks, SK$_j$	F_{SKj} (t/h)	C_{SKj} (wt/wt)	*i*	Sources, SR$_i$	F_{SKi} (t/h)	C_{SKi} (wt/wt)
1	Mixing water	14.0	0.02	1	Separator outlet	24	0.093
2	Blending water	2.5	0.12	2	Sterilizer condensate	6	0.010
3	Balancing water	7.5	0.093	3	Cooling water	4	0

Source: Chungsiriporn et al., 2006.

in this case has zero impurity. Determine the minimum flowrates of freshwater and wastewater for the process (see answer in Problem 8.7).

4.7 Develop the RCN structure for the water minimization case studies in Examples 4.3 and 4.4, using the targeting insights from WCA.

4.8 A palm oil mill (Chungsiriporn et al., 2006) has undergone a water minimization study. The limiting water data are shown in Table 4.23. Freshwater in this case is free from impurity. Determine the minimum flowrates of freshwater and wastewater for the water network of the mill (see answer in Problem 8.4). Develop the RCN structure using the targeting insights.

4.9 A water minimization study was carried out in a steel plant (Tian et al., 2008). The limiting water data are shown in Table 4.24. Determine the minimum freshwater and wastewater flowrates for the water network. Assume freshwater has an impurity (chlorine) content of 20 mg/L (see answer in Problem 8.5).

4.10 Table 4.25 shows the limiting water data for a water minimization study in a bioethanol production plant (Liu et al., 2019). Determine the minimum freshwater and wastewater flowrates for the water network, assuming the freshwater feed has an impurity (COD) content of 20 mg/L.

4.11 For the Kraft pulping process in Problem 2.8, water minimization study is carried out to analyze its water-saving potential (Parthasarathy and Krishnagopalan, 2001). Impurity that is of concern for water recovery is the chloride content in the water sources. The freshwater source available for service has a chloride content of 4.2 ppm. Determine the minimum freshwater and wastewater flowrates for the RCN with the correctly extracted limiting data given in Table 4.26a. Determine also their suboptimum values when the water sink are incorrectly extracted

TABLE 4.24

Limiting Water Data for the Steel Plant in Problem 4.9

j	Sinks, SK$_j$	F_{SKj} (t/day)	C_{SKj} (mg/L)	*i*	Sources, SR$_i$	F_{SKi} (t/day)	F_{SKi} (mg/L)
1	Hot air furnace	1680	90	1	Hot air furnace	960	400
2	Blast furnace	1920	75	2	Blast furnace	1440	100
3	Power plant	1680	40	3	Power plant	1320	45

Source: Tian et al., 2008.

TABLE 4.25

Limiting Water Data for the Bioethanol Production Plant in Problem 4.10

j	Sinks, SK$_j$	F_{SKj} (t/h)	C_{SKj} (ppm)	i	Sources, SR$_i$	F_{SKi} (t/h)	F_{SKi} (ppm)
1	Circulating cooling	2.42	5	1	Circulating cooling	2.42	100
2	Workshops cleaning	1.09	60	2	Workshops cleaning	1.09	1500
3	Steam condensate	2.70	0	3	Steam condensate	2.70	20
4	Facilities cleaning	2.58	50	4	Facilities cleaning	2.58	1000
5	Sterilization	0.50	50	5	Sterilization	0.50	400

Source: Liu et al., 2019.

TABLE 4.26

Limiting Water Data for Kraft Pulping Process in Problem 4.11

j	Sinks, SK$_j$	F_{SKj} (t/d)	C_{SKj} (ppm)	i	Sources, SR$_i$	F_{SKi} (t/d)	C_{SKi} (ppm)
(a) Correctly extracted limiting data							
1	Washer	7,000	110	1	Evaporation condensate	7,971	0
2	Recovery furnace	4,200	5.5	2	Washer purge	945	275
3	Washers filters	2,000	38.5	3	Acid bleach effluent	14,520	235
4	ClO$_2$ stage	12,000	10.12	4	Alkali bleach effluent	13,750	504
5	Alkaline extraction	11,200	13.2				
(b) Incorrectly extracted limiting data							
1	Washer	10,000	110	1	Evaporation condensate	7,971	0
2	Recovery furnace	6,900	5.5	2	Washer purge	945	275
3	Washers filters	3,800	38.5	3	Acid bleach effluent	14,520	235
4	ClO$_2$ stage	14,400	10.12	4	Alkali bleach effluent	13,750	504
5	Alkaline extraction	13,500	13.2				

Source: Parthasarathy and Krishnagopalan (2001).

at their maximum flowrates, with data in Table 4.26b (see answers in Problem 8.20).

4.12 Rework Scenarios (a)–(c) in Problem 3.4 by using WCA. Identify the minimum freshwater and wastewater flowrates for the RCN.[4] In addition, determine also the minimum flowrates for the following scenarios. In all cases, assume that the lower quality freshwater source(s) has lower prioritized cost as compared to the higher quality source(s) (solutions are found in Examples 6.1, 7.1, and Problem 7.1).

a. Three freshwater sources are available for use: one pure (0 ppm) and two impure freshwater sources (20 and 50 ppm, respectively; the solution is found in Problem 7.1).

b. Same freshwater sources are available for use as in (d). However, the 20 ppm freshwater source has a limited flowrate of 40 t/h.

[4] Answer for Scenario (a)—single pure fresh water resource is available in Example 6.1.

TABLE 4.27

Limiting Water Data for Literature Example in Problem 4.13

Sinks, SK_j	F_{SKj} (t/h)	C_{SKj} (ppm)	Sources, SR_i	F_{SKi} (t/h)	C_{SKi} (ppm)
1	20	0	1	20	100
2	100	50	2	100	100
3	40	50	3	40	800
4	10	80			

Source: Wang and Smith (1995).

4.13 Limiting data for literature example (Wang and Smith, 1995) are shown in Table 4.27. Determine the minimum flowrates of freshwater and wastewater for the following scenarios. In all cases, assume that the lower quality freshwater source(s) has a lower prioritized cost as compared to the higher quality source(s).

a. Two freshwater sources are available for use: pure (0 ppm) and impure freshwater sources (25 ppm).

b. Two freshwater sources are available for use: pure (0 ppm) and impure freshwater sources (40 ppm).

c. Three freshwater sources are available for use: one pure (0 ppm) and two impure freshwater sources (25 and 50 ppm).

What conclusions can we make by observing the earlier-discussed scenarios?

4.14 For the alkene production process in Problems 2.5 and 3.8 (Foo, 2019), determine the minimum fresh water and wastewater flowrates, assuming that the fresh water is free from impurity. Limiting water data for water minimization is shown in Table 3.13 (see answers and discussion in Chapter 5—Industrial Case Study).

4.15 Figure 4.5 shows a chlor-alkali process where water minimization is carried out (Ng et al., 2014). The feed material consists of industrial salt and water, which are mixed to produce brine. The latter is fed along with de-ionized water to an electrolyzer. In the latter, the diluted brine undergoes electrolysis to form sodium hydroxide (NaOH), hydrogen (H_2), and chlorine (Cl_2) gases. The gaseous products are sent to a reactor where hydrogen chloride (HCl) gas is produced, while some Cl_2 gas also is reacted with NaOH in another reactor to form sodium hypochlorite (NaOCl). The HCl gas is sent to an absorption column where it is absorbed by de-ionized water to produce hydrochloric acid (HCl aqueous). Several products are formed in the downstream process by reacting hydrochloric acid with methanol (CH_3OH), calcium carbonate ($CaCO_3$), and dissolved scrap iron (Fe), i.e., methyl chloride (CH_3Cl), calcium chloride ($CaCl_2$), and ferric chloride ($FeCl_3$).

As the process consists of acid and base sections, water minimization study will be carried out separately in these sections, with limiting

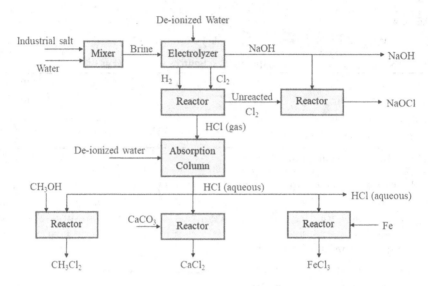

FIGURE 4.5
A chlor-alkali process for Problem 4.15. (From Ng et al., 2014.)

TABLE 4.28

Limiting Water Data for Chlor-Alkali Process in Problem 4.15 (Acid Section)

j	Sinks, SK_j	F_{SKj} (t/h)	C_{SKj} (ppm)	i	Sources, SR_i	F_{SKi} (t/h)	C_{SKi} (ppm)
1	Regen unit-acid	38	2	1	Weak acid water	50	200
2	Cl Blower	62	3000	2	$CaCl_2$ (scrubbing)	18	5000
3	$CaCl_2$ (scrubbing)	18	5000	3	$FeCl_3$ plant	10	3000
4	$FeCl_3$ (dilution)	4	3000	4	Cl blower	62	2500
5	HCl scrubber	360	2	5	HCl scrubber	124	2

water data shown in Tables 4.28 (acid section) and 4.29 (base section), respectively. Note that ion concentration is the major concern for water minimization in this process. Fresh resource to be utilized in this case is de-ionized water, with is free of ion content. Determine the minimum freshwater and wastewater flowrates for the water network of this process.

Utility Gas Recovery Problems

4.16 For the refinery hydrogen network in Examples 2.4 and 3.3, use GCA to identify the minimum flowrates of the fresh hydrogen feed and purge stream for the RCN (for fresh hydrogen feed with 1% impurity). See limiting data in Table 2.2 (see answers in Problem 8.24). Identify also which hydrogen source(s) is being purged from the RCN.

TABLE 4.29

Limiting Water Data for Chlor-Alkali Process in Problem 4.15 (Base Section)

j	Sinks, SK_j	F_{SKj} (t/h)	C_{SKj} (ppm)	i	Sources, SR_i	F_{SKi} (t/h)	C_{SKi} (ppm)
1	Cooling tower	200	2700	1	Cooling tower	144	3500
2	Boiler	120	2	2	Boiler	70	160
3	Regen unit	150	2	3	Pump Seals	100	3000
4	Regen unit-base	30	2	4	Weak alkaline water	168	200
5	DI unit	30	2500	5	Vapor condensate	180	2
6	Pump seals	100	3000	6	Purge brine	144	3000
7	Electrolysis	260	2	7	Methanol spray	8	3000
8	Methanol spray	12	5000	8	DI water regen	30	3500
9	NaOCl (dilution)	50	200	9	Hypo decomp	8	1000
10	Plant washing	374	3500	10	Wash water	400	4500
11	Soda ash prep	16	2700	11	Domestic use	66	2500
12	Floc prep	3	2700				
13	Na_2SO_3 prep	4	2700				
14	Saturator	40	2000				
15	Domestic use	90	2500				

4.17 Use GCA to identify the minimum flowrate targets of fresh nitrogen feed and purge gas stream for the magnetic tape process in Problems 2.9 and 3.10 (use limiting data in Table 3.15). Fresh nitrogen is free of impurity (see answer in Problem 8.22).

4.18 For the refinery hydrogen network (Umana et al., 2014) in Problem 2.10, identify the minimum flowrates of fresh hydrogen feed (7.44% impurity content) and purge stream for the RCN using the GCA technique. Limiting data are given in Table 4.30a (see answer in Problem 8.27). Determine also the wrong flowrate targets with incorrectly extracted limiting data in Table 4.30b, where hydrogen makeup flowrates are taken as sinks; while purge streams as sources (see Figure 2.22).

4.19 For the refinery hydrogen network (Alves and Towler, 2002) in Problem 2.11, identify the minimum flowrates of fresh hydrogen feed (5% impurity content) and purge stream for the RCN using the GCA technique. Limiting data are given in Table 4.31. The fresh hydrogen feed has an impurity content of 5%. Identify also which hydrogen source(s) is being purged from the RCN (see answer in Problem 8.25).

4.20 Data for a refinery hydrogen network are shown in Table 4.32 (Foo and Manan, 2006); determine the minimum flowrates of the fresh hydrogen feed and purge stream from the RCN using the GCA technique. The fresh hydrogen feed has an impurity of 0.1% (note that this is a multiple-pinch problem; see answer in Problem 8.26).

TABLE 4.30

Limiting Data for Hydrogen Network in Problem 4.18

j	Sinks, SK_j	F_{SKj} (t/h)	C_{SKi} (%)	i	Sources, SR_i	F_{SKj} (t/h)	C_{SKi} (%)
(a) Correctly extracted limiting data							
1	NHT	8.757	21.48	1	CCR	3.170	61.97
2	CNHT	12.489	22.82	2	NHT	7.453	13.36
3	DHT	49.506	14.81	3	CNHT	11.006	16.62
4	VGOHC	24.530	21.03	4	DHT	47.462	15.23
				5	VGOHC	22.948	29.01
(b) Incorrectly extracted limiting data							
1	NHT	8.757	21.48	1	CCR	3.170	61.97
2	CNHT	12.489	22.82	2	NHT	7.453	13.36
3	DHT	49.506	14.81	3	CNHT	11.006	16.62
4	VGOHC	24.530	21.03	4	DHT	47.462	15.23
				5	VGOHC	22.948	29.01

Source: Umana et al. (2014).

TABLE 4.31

Limiting Data for Hydrogen Network in Problem 4.19

j	Sinks, SK_j	F_{SKj} (mol/s)	C_{SKi} (mol%)	i	Sources, SR_i	F_{SKj} (mol/s)	C_{SKi} (mol%)
1	HCU	2495.0	19.39	1	HCU	1801.9	25.0
2	NHT	180.2	21.15	2	NHT	138.6	25.0
3	CNHT	720.7	24.86	3	CNHT	457.4	30.0
4	DHT	554.4	22.43	4	DHT	346.5	27.0
				5	SRU	623.8	7.0
				6	CRU	415.8	20.0

Source: Alves and Towler, 2002

TABLE 4.32

Limiting Data for Hydrogen Network in Problem 4.20

Sinks, SK_j	F_{SKj} (mol/s)	C_{SKi} (mol%)	Sources, SR_i	F_{SKj} (mol/s)	C_{SKi} (mol%)
1	120.00	0.10	1	80.00	1.70
2	27.80	1.40	2	75.00	15.00
3	80.00	2.50	3	28.55	4.00
4	60.00	2.50	4	80.00	5.00
5	100.00	3.00	5	120.00	10.00
6	150.00	10.00	6	40.00	1.70
			7	80.00	2.50

Source: Foo and Manan, 2006.

Property Integration Problems

4.21 Use PCA to identify the minimum fresh solvent and solvent waste targets for the metal degreasing process in Examples 2.5 and 3.7. Verify the targets with those identified using the graphical targeting technique described in Example 3.7. Also, identify the source that is discharged as waste from the RCN.

4.22 Identify the minimum fresh and solvent waste targets for the vinyl acetate-manufacturing process in Problems 2.14 and 3.12. Limiting data for the problem are given in Table 3.17. Verify the targets with those identified using the graphical targeting technique in Problem 3.12. Identify also the sources that are discharged as waste from the RCN.

4.23 For the wafer fabrication problem in Problem 2.13. Carry out flowrate targeting for the following scenarios (answers are available in Problem 8.31):

a. Identify the minimum ultrapure water (UPW—fresh resource) and wastewater flowrates for direct reuse/recycle scheme in the wafer fabrication process (FAB), utilizing the limiting data in Table 4.33a. Identify also the sources that are discharged as waste from the RCN.

TABLE 4.33

Limiting Data for Wafer Fabrication Process in Problem 4.23

j	Sinks, SK_j	F_{SKj} (t/h)	ψ_{SKj} ([MΩ m]$^{-1}$)	i	Sources, SR_i	F_{SKi} (t/h)	ψ_{SRi} ([MΩ m]$^{-1}$)
(a) FAB section							
1	Wet	500	0.143	1	Wet I	250	1
2	Lithography	450	0.125	2	Wet II	200	0.5
3	CMP	700	0.100	3	Lithography	350	0.333
4	Etc	350	0.200	4	CMP I	300	10
				5	CMP II	200	0.5
				6	Etc	280	2
(b) Entire plant							
1	Wet	500	0.143	1	Wet I	250	1
2	Lithography	450	0.125	2	Wet II	200	0.5
3	CMP	700	0.100	3	Lithography	350	0.333
4	Etc	350	0.200	4	CMP I	300	10
5	Cleaning	200	125.000	5	CMP II	200	0.5
6	Cooling tower makeup	450	50.000	6	Etc	280	2
7	Scrubber	300	100.000	7	Cleaning	180	500
				8	Scrubber	300	200
				9	UF reject	30% of UF inlet	100
				10	RO reject	30% of RO inlet	200

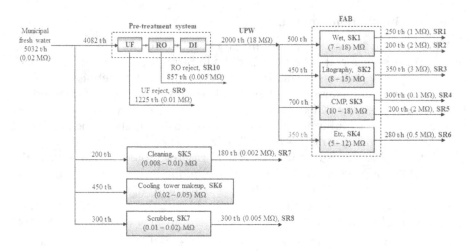

FIGURE 4.6

A wafer fabrication process in Problem 4.23. (From Ng,, 2009. Reproduced with permission. Copyright 2009 Elsevier.)

b. The overall plant for wafer fabrication is given in Figure 4.6. Apart from the FAB section, three other process units are also consuming large amount of municipal freshwater. Identify the minimum flowrates for the fresh resources (UPW and municipal freshwater) and wastewater for direct reuse/recycle scheme for the entire plant. The limiting data for this scenario are given in Table 4.33b. The municipal freshwater has a resistivity value of 0.02 MΩ m. Note that the reject streams of the reverse osmosis (RO) and ultra-filtration (UF) units will have to be adjusted based on the targeted UPW flowrate.

4.24 Table 4.34 shows the limiting data for a paper recovery study (Kit et al., 2011). The property that is of concern for paper reuse/recycle is the water fraction in the paper material. Fresh fiber with zero water fraction may be used to supplement the process sources. Determine the minimum amount of fresh and waste fiber for the study.

TABLE 4.34

Limiting Data for Paper Recovery Study (Kit et al., 2011) in Problem 4.24

Sinks	F_{SKj} (t/year)	ψ_{SKj} (Water Fraction)	Sources	F_{SKi} (t/year)	ψ_{SRi} (Water Fraction)
Newspaper	400	0.004	Newspaper	383.36	0.104
Tissue	400	0.006	Tissue	60.74	0.106
A4 paper	400	0.061	A4 paper	221.02	0.155
Coagulated box	600	0.161	Coagulated box	160.50	0.245

References

Alves, J. J. and Towler, G. P. 2002. Analysis of refinery hydrogen distribution systems, *Industrial and Engineering Chemistry Research*, 41, 5759–5769.

Chungsiriporn, J., Prasertsan, S., and Bunyakan, C. 2006. Minimization of water consumption and process optimization of palm oil mills, *Clean Technologies and Environmental Policy*, 8, 151–158.

Foo, D. C. Y. 2007. Water cascade analysis for single and multiple impure fresh water feed, *Chemical Engineering Research and Design*, 85(A8), 1169–1177.

Foo, D. C. Y. 2008. Flowrate targeting for threshold problems and plant-wide integration for water network synthesis, *Journal of Environmental Management*, 88(2), 253–274.

Foo, D. C. Y., Kazantzi, V., El-Halwagi, M. M., and Manan, Z. A. 2006a. Cascade Analysis for targeting property-based material reuse networks, *Chemical Engineering Science*, 61(8), 2626–2642.

Foo, D. C. Y. and Manan, Z. A. 2006. Setting the minimum flowrate targets for utility gases using cascade analysis technique, *Industrial and Engineering Chemistry Research*, 45(17), 5986–5995.

Foo, D. C. Y., Manan, Z. A., and Tan, Y. L. 2006b. Use cascade analysis to optimize water networks, *Chemical Engineering Progress*, 102(7), 45–52.

Foo, D. C. Y. 2019. Optimize plant water use from your desk, *Chemical Engineering Progress*, 115, 45–51.

Jacob, J., Kaipe, H., Couderc, F., and Paris, J. 2002. Water network analysis in pulp and paper processes by pinch and linear programming techniques, *Chemical Engineering Communications*, 189(2), 184–206.

Kazantzi, V. and El-Halwagi, M. M. 2005. Targeting material reuse via property integration, *Chemical Engineering Progress*, 101(8), 28–37.

Kit, S. G., Wan Alwi, S. R., and Manan, Z. A. 2011. A new graphical approach for simultaneous targeting and design of a paper recycling network, *Asia-Pacific Journal of Chemical Engineering*, 6(5), 689–812.

Liu, H., Lijun Ren, L., Zhuo, H., and Fu, S. 2019. water footprint and Water pinch analysis in ethanol industrial production for water management, *Water*, 11, 518.

Manan, Z. A., Tan, Y. L., and Foo, D. C. Y. 2004. Targeting the minimum water flowrate using water cascade analysis technique, *AIChE Journal*, 50(12), 3169–3183.

Ng, D. K. S., Foo, D. C. Y., Tan, R. R., Pau, C. H., and Tan, Y. L. 2009. Automated targeting for conventional and bilateral property-based resource conservation network, *Chemical Engineering Journal*, 149, 87–101.

Ng, D. K. S., Chew, I. M. L., Tan, R. R., Foo, D. C. Y., Ooi, M. B. L., and El-Halwagi, M. M. 2014. RCNet: An optimisation software for the synthesis of resource conservation networks, *Process Safety and Environmental Protection*, 92(6), 917–928.

Parthasarathy, G. and Krishnagopalan, G. 2001. Systematic reallocation of aqueous resources using mass integration in a typical pulp mill, *Advances in Environmental Research*, 5, 61–79.

Savelski, M. J. and Bagajewicz, M. J. 2001. Algorithmic procedure to design water utilization systems featuring a single contaminant in process plants, *Chemical Engineering Science*, 56, 1897–1911.

Tian, J. R., Zhou, P. J., and Lv, B. 2008. A process integration approach to industrial water conservation: A case study for a Chinese steel plant, *Journal of Environmental Management*, 86, 682–687.

Umana, B., Shoaib, A., Zhang, N., and Smith, R. 2014. Integrating hydroprocessors in refinery hydrogen network optimisation, *Applied Energy*, 133, 169–182.

Wang, Y. P. and Smith, R. 1995. Wastewater minimization with flowrate constraints, *Chemical Engineering Research and Design*, 73, 889–904.

5

*Automated Targeting Model for Direct Reuse/Recycle Networks**

In this chapter, the basic framework for *automated targeting model* (ATM) is introduced. The basic ATM framework is based on the *material cascade analysis* (MCA) technique described in Chapter 4. In this chapter, the ATM is demonstrated for direct material reuse/recycle network. A new variant of the direct reuse/recycle scheme, i.e., *bilateral problem*, is also introduced in the chapter. Both concentration- and property-based resource conservation networks (RCNs) are considered. The ATM is demonstrated using MS Excel and LINGO.

5.1 Basic Framework and Mathematical Formulation of ATM

The basic framework of the ATM takes the form of a cascade diagram, as shown in Figure 5.1. The six-step procedure to formulate the ATM is given as follows (Ng et al., 2009a,b,c):

1. Arrange quality levels (q_k) of all process sinks and sources in descending order. When the quality levels for fresh resources do not coincide with any of the process sinks and sources that are present, additional levels are added. An arbitrary value is also added at the final level q_n (lowest among all quality levels) to allow the calculation of the residual impurity/property load.

2. Locate flowrates of all process sinks ($F_{SKj,k}$) and sources ($F_{SRi,k}$) at their respective quality level k. When there are more than one sink/source at the same quality levels, their flowrates are added. For RCNs with multiple fresh resources, the fresh resources may also be located at the quality levels like other process sources.

3. Construct the *material cascade*. This is done following Equations 5.1 and 5.2. At every level k, the *residual material flowrate* (δ_k) is determined from the summation of the residual material flowrate cascaded from an earlier quality level (δ_{k-1}) with the *net material flowrate* at level k, i.e., $F_{Net,k}$. The latter is calculated using Equation 5.2, where

---

* All examples of this chapter are available in electronic files, downloadable from book website.

DOI: 10.1201/9781003173977-6   *131*

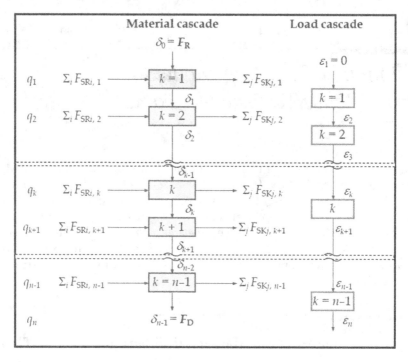

FIGURE 5.1
Basic framework of the ATM. (From Ng et al., 2009a. With permission.)

total sink flowrate ($F_{SKj,k}$) is deducted from the total source flowrate ($F_{SRi,k}$) at level k.

$$\delta_k = \delta_{k-1} + F_{\text{Net},k} \quad k = 1, 2, \ldots, n-1 \tag{5.1}$$

$$F_{\text{Net},k} = \sum_i F_{SRi,k} - \sum_j F_{SKj,k} \quad k = 1, 2, \ldots, n-1 \tag{5.2}$$

4. Construct the *load cascade*. Equation 5.3 indicates that the *residual impurity/property load* (ε_k) at the first quality level has zero value, i.e., $\varepsilon_1 = 0$, while residual loads in other quality levels are contributed by the residual load cascaded from previous level (ε_{k-1}), and load within each quality interval. The latter is given by the product of the net material flowrate from previous level k (δ_{k-1}) and the difference between two adjacent quality levels ($q_k - q_{k-1}$), given as in Equation 5.4:

$$\varepsilon_k = \begin{cases} 0 & k = 1 \\ \varepsilon_{k-1} + \Delta m_k & k = 2, 3, \ldots, n \end{cases} \tag{5.3}$$

$$\Delta m_k = \delta_{k-1}\left(q_k - q_{k-1}\right) \tag{5.4}$$

5. Define the inequality constraints. Equation 5.5 next indicates that the net material flowrate that enters the first (δ_0) and last levels (δ_{n-1}) should take non-negative values. Apart from the residual load at level 1 (which is set to zero), all residual loads must also take non-negative values, as indicated by Equation 5.6.

$$\delta_{k-1} \geq 0 \ k = 1, n \tag{5.5}$$

$$\varepsilon_k \geq 0 \ k = 2, 3, \ldots, n \tag{5.6}$$

6. Formulate the optimization objective. The latter can be set to determine the minimum fresh resource flowrate (F_R; Equations 5.7), or the minimum operating cost (OC; Equations 5.8). OC of the RCN is usually contributed by fresh resource and/or waste stream flowrates (F_D) of the RCN, with unit costs of CT_R and CT_D, respectively.

$$\text{Minimize } F_R = \delta_0 \tag{5.7}$$

$$\text{Minimize } OC = F_R CT_R + F_D CT_D \tag{5.8}$$

Note that the flowrates of fresh resource (F_R), net material ($F_{Net,k}$), residual *material* (δ_k), and residual load at each level k (ε_k) are process variables, which will be determined when the ATM is solved. On the other hand, the quality level (q_k) as well as flowrates of sink ($F_{SRi,k}$) and source ($F_{SKj,k}$) at each level k are parameters. Hence, the ATM is a linear program (LP) that can be solved to obtain a global optimum solution.

Note that the ATM has similar characteristics as the MCA technique in Chapter 4. Upon solving the ATM, the first and last entries of the material cascade represent the minimum flowrates of fresh resource ($\delta_0 = F_R$) and waste discharge ($\delta_{n-1} = F_D$) of the RCN, respectively. On the other hand, the residual load that vanishes ($\varepsilon_k = 0$) indicates the pinch point of the RCN.

Example 5.1 ATM with MS Excel[1] – acrylonitrile (AN) production case study

The water minimization case of AN process (El-Halwagi, 1997) in Examples 2.1, 3.1, and 4.1 is revisited, with limiting water data given in Table 5.1 (reproduced from Table 4.2). Assuming that pure freshwater is used, solve the following tasks:

1. Determine the minimum freshwater flowrate with the objective in Equations 5.7. Determine also its corresponding minimum wastewater flowrate, pinch concentration, pinch-causing source, as well as the flowrate allocation targets of the pinch-causing source to the higher (HQR) and lower quality (LQR) regions.
2. Determine the minimum OC with the objective in Equations 5.8, given that the unit costs of freshwater (CT_{FW}) and wastewater (CT_{WW}) are given as 0.001 \$/kg, respectively (or 1 \$/t).

---

[1] Solution files of Excel and LINGO are made available on book support website, while details for LINGO solution are found in Edition 1.

TABLE 5.1

Limiting Water Data for AN Production Case Study

	Water Sinks, SK_j		
j	Stream	Flowrate F_{SKj} (kg/s)	NH_3 Concentration C_{SKj} (ppm)
1	BFW	1.2	0
2	Scrubber inlet	5.8	10
	Water Sources, SR_i		
i	Stream	Flowrate F_{SRi} (kg/s)	NH_3 Concentration C_{SRi} (ppm)
1	Distillation bottoms	0.8	0
2	Off-gas condensate	5	14
3	Aqueous layer	5.9	25
4	Ejector condensate	1.4	34

Source: El-Halwagi, 1997.

SOLUTION

To solve the water minimization problem with MS Excel, the ATM will have a structure as shown in Figure 5.2. Note that the latter is similar to the MCA in Chapter 4. In Figure 5.2, quality levels in column B are expressed in terms of impurity concentration (C_k). Column C is used to calculate the difference between two adjacent levels. In columns D and E, the flowrate of process water sinks and sources are located. Material

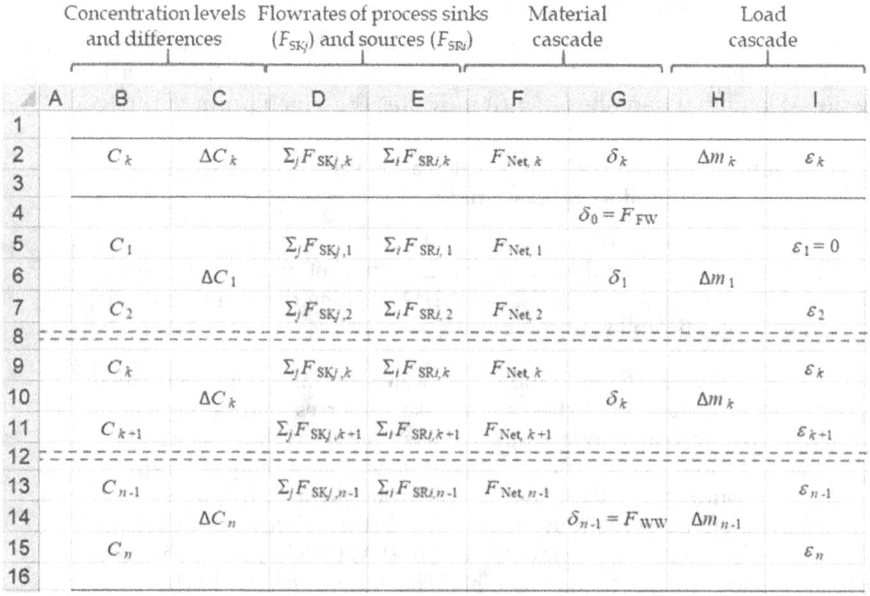

FIGURE 5.2
Basic structure of ATM in MS Excel.

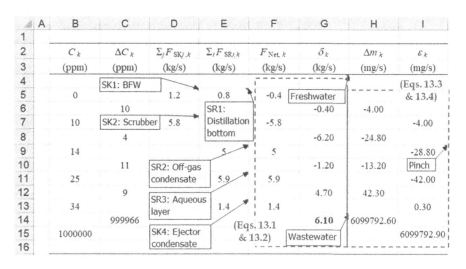

FIGURE 5.3
Cascade diagram for AN production case study (Example 5.1).

and load cascades are to be carried out in columns F–I. The six-step procedure is followed to solve the ATM for the AN production case study.

1. *Arrange quality levels (q_k) of all process sinks and sources in descending order.*
2. *Locate flowrates of the process sinks and sources at their respective quality level k.*

   Following Steps 1 and 2, the ATM for the AN case will have the structure in Figure 5.3. Concentration levels are arranged in column B in ascending order (i.e., descending order of quality), with highest possible concentration (i.e., 1,000,000 ppm) added as the final level. In column C, the difference between two adjacent levels are calculated. Next, flowrates of water sinks and sources are added in columns D and E, at their respective concentration levels.
3. *Construct the material cascade.*
4. *Construct the load cascade.*

   At every concentration level k, residual material flowrate (δ_k) and net material flowrate ($F_{Net,k}$) are calculated using Equations 5.1 and 5.2; these form the material cascade in columns F and G (see Figure 5.3). Equations 5.3 and 5.4 are next used to determine the residual impurity load at every level k (ε_k), which form the load cascade in columns H and I (see Figure 5.3).
5. *Define the inequality constraints.*
6. *Formulate the optimization objective.*

   This step is carried out using the Solver add-in of MS Excel.[2] Figure 5.4 shows the setup in Solver where δ_0 and δ_{n-1} values are set to be non-negative following Equation 5.5. Similarly,

---

[2] Solver add-in may be activated with the following path: File/Options/Add-ins/Excel Add-ins.

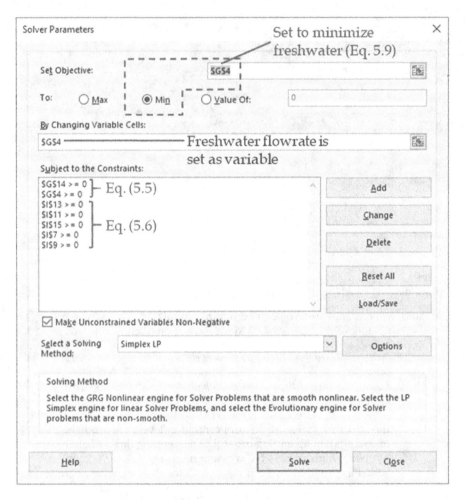

FIGURE 5.4
Solver setting for AN production case study (Example 5.1).

all residual loads are also set to non-negative values following Equation 5.6.

To determine the minimum freshwater flowrate (F_{FW}) of the water network, a revised form of Equation 3.7 is used (see setting of Solver is shown in Figure 5.4):

$$\text{Minimize } F_{FW} = \delta_0 \tag{5.9}$$

By executing Solver, the result indicates that a solution is found, with all constraints and optimality conditions met. As shown in Figure 5.5, the minimum freshwater flowrate is determined as 2.06 kg/s (δ_0; cell G4). Since the ATM has similar

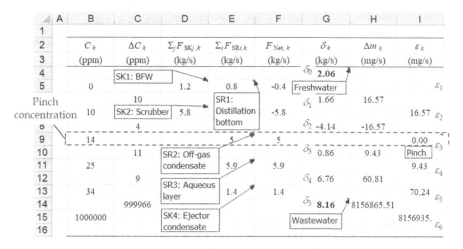

FIGURE 5.5
Results of ATM for AN production case study—minimum fresh water flowrate (Example 5.1).

characteristics as the MCA, insights for MCA may be extended here for the ATM. From Figure 5.5, the minimum wastewater flowrate for the water network is determined as 8.16 kg/s (δ_5; cell G14). The pinch corresponds to the concentration where the residual load vanishes ($\varepsilon_3 = 0$), i.e., 14 ppm. This indicates that SR2 is the pinch-causing source of the problem. Among its total flowrate of 5 kg/s (see Table 5.1), its allocated flowrates to the HQR and LQR are identified from the net material flowrates that are just lower and higher than the pinch, i.e., δ_2 (4.14 kg/s; cell G8) and δ_3 (0.86 kg/s; cell G10), respectively (see Sections 3.2 and 4.1 for more details on *flowrate allocation targets*).

Finally, note that the results of ATM in Figure 5.5 are similar to those obtained using the MCA technique (e.g., see Figure 4.1). The water network that achieves the established RCN targets is shown in Figure 4.2.

To determine the minimum *OC* of the water network, a revised form of Equation 5.8 is used:

$$\text{Minimize } OC = F_{FW}CT_{FW} + F_{WW}CT_{WW} \qquad (5.10)$$

Setting of MS Excel spreadsheet and Solver are shown in Figure 5.6. By executing Solver, the result indicates that the minimum *OC* for this case is determined as 0.10 $/s (or 36 $/h)., with all constraints and optimality conditions met. As shown in Figure 5.6, the minimum freshwater flowrate, along with the other RCN targets (wastewater flowrate, pinch concentration, etc.) are identical to those obtained with the objective of minimizing fresh water flowrate (Figure 5.6). In other words, both objectives of minimum fresh water and *OC* yield identical solution for this case study.

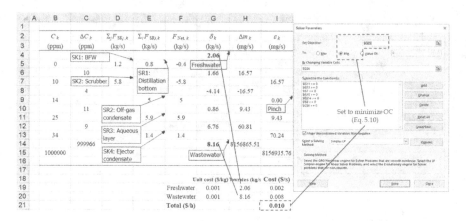

FIGURE 5.6
Results of ATM for AN production case study—minimum OC (Example 5.1).

5.2 Incorporation of Process Constraints into ATM

Since the ATM is constructed on an optimization framework, additional process constraints (e.g., flowrate requirement of certain process sinks, etc.) may be included in the model. Note however that adding certain constraints may sometimes entail a nonlinear program (NLP) for the ATM. For an NLP model, commercial optimization software (e.g., LINGO, GAMS) may be used as they are capable of solving NLP problems in reaching global solution.

Example 5.2 Flowrate constraints for AN production case study

Water minimization of the AN process in Example 5.1 is revisited. Instead of using the limiting water data in Figure 5.1, one may also make use of the original process constraints given as in Example 2.2. One of the relevant constraints indicates the flowrate requirement of process sink SK2, i.e., scrubber, as follows:

$$5.8 \le \text{flowrate of wash feed}\ (\text{kg/s}) \le 6.2 \tag{5.11}$$

Use the ATM to determine the optimum flowrate for the scrubber wash water (SK2) as well as the minimum freshwater and wastewater flowrate targets for the case study.

SOLUTION

Since only an additional flowrate constraint is added to the problem, the same ATM model in Example 5.1 may be extended for use here. An additional constraint is to be added to MS Excel Solver, so that the flowrate of SK2 (F_{SK2}) is now treated as a variable instead of a fixed flowrate value. Both Excel spreadsheet and Solver setting are shown in Figure 5.7.

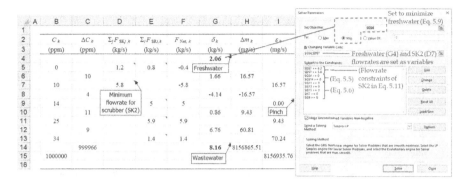

FIGURE 5.7
Solver setting for AN production case study, with flowrate constraints added for SK2 (Example 5.2).

Figure 5.7 indicates that the freshwater flowrate of the water network remains identical to that in Example 5.1. The model also determines that SK2 should be operated at the optimum flowrate of 5.8 kg/s. This justifies the use of Heuristic 2 in Chapter 2 in identifying limiting data for an RCN (see detailed discussion in Section 2.2).

Industrial Case Study

Figure 5.8 shows the process flow diagram (PFD) for an alkene production process, where shale gas is used to produce ethylene and other alkene products (Foo, 2019; see other process details in Problem 2.5).

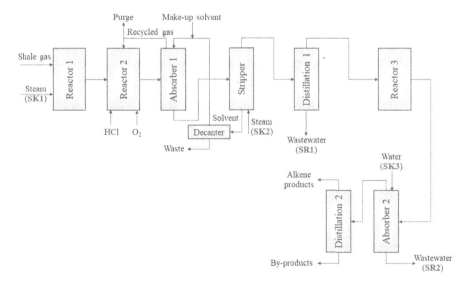

FIGURE 5.8
PFD for an alkene production process (Foo et al., 2019).

TABLE 5.2

Limiting Water Data for Alkene Production Case Study

	Water Sinks, SK$_j$		
j	Stream	Flowrate F_{SKj} (t/h)	HCl Concentration C_{SKj} (%)
1	Steam use in Reactor 1	$12.4 \leq F_{SK1} \leq 13.6$	0
2	Stripping agent for Stripper	$9.0 \leq F_{SK2} \leq 9.9$	0
3	Absorbing agent in Absorber 2	$153.0 \leq F_{SK3} \leq 168.3$	5
	Water Sources, SR$_i$		
i	Stream	Flowrate F_{SRi} (t/h)	HCl Concentration C_{SRi} (%)
1	Distillation 1 effluent	35.8	1.8
2	Absorber 2 effluent	180	15

Source: Foo et al., 2019.

The process involves huge water consumption as water is used as steam in reactor feed, as well as striping and absorbing agents (water sinks). Water minimization potential is evaluated for the recovery of its wastewater streams from Distillation 1 and Absorber 2 (water sources), with limiting water data given in Table 5.2. As shown, all water sinks have their respective flowrate operating ranges.

Water minimization study is carried out for the following scenarios:

- Scenario 1: minimum freshwater flowrate.
- Scenario 2: minimum *OC* solution, with unit cost of freshwater set to 1 $/t, while that of wastewater is set to 3 $/t.

As the water sinks have their respective flowrate operating ranges, additional constraints are added in MS Excel Solver (see Figure 5.9). When the objective is set to minimize the freshwater flowrate (Scenario 1),

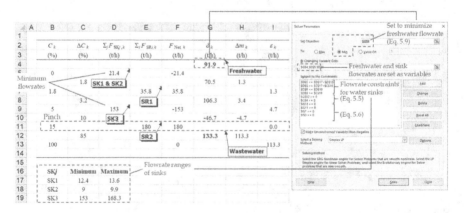

FIGURE 5.9
Results of ATM for alkene production case study—minimum flowrate solution.

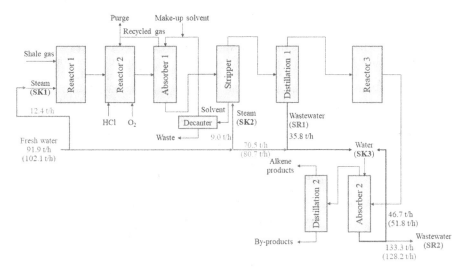

FIGURE 5.10
Revised PFD for alkene production case study, with water recovery scheme (flowrates for Scenario 2 are given in parenthesis).

the ATM determines that a total of 91.9 t/h of fresh water is needed for the water network, while the corresponding wastewater flowrate is determined as 133.3 t/h, as shown in Figure 5.9. The latter also shows that the pinch concentration is identified at 15%, and all water sinks are operated at their minimum flowrates. A revised PFD with water recovery scheme is shown in Figure 5.10.

For Scenario 2, the objective is set to minimize the OC, which is contributed by cost of freshwater and wastewater. Calculation of the OC is carried out in the main spreadsheet (cells G16:I19; see Figure 5.11). By executing Solver, the ATM determines that the OC is reported as

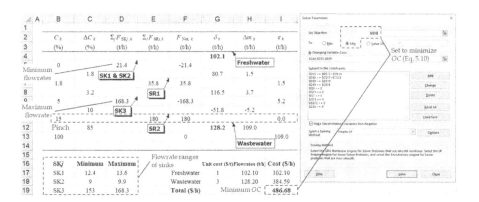

FIGURE 5.11
Results of ATM for alkene production case study—minimum OC solution.

486.7 \$/h, with freshwater and wastewater flowrates corresponding to 102.1 and 128.2 t/h, respectively (Figure 5.11). Two observations are worth mentioning. First, the fresh water flowrate is determined as 11% higher than that obtained in Scenario 1, while its wastewater flowrate is 3.8% lower. This is because as the objective is set to minimize the *OC*, the ATM tends to minimize the cost of wastewater (3 \$/t) as it is three times higher than that of freshwater (1 \$/t). Besides, it is also observed that the ATM determines that the flowrates of SK1 and SK2 are kept to the minimum (12.4 + 9.0 = 21.4 t/h), while that of SK3 are maximized, i.e., 168.3 t/h. This contradicts with Heuristic 2 as discussed in Chapter 2, where water sink flowrate should be kept to its minimum value in order to minimize the overall fresh resource.[3] The revised PFD for this scenario is similar to that given in Figure 5.10 (with different water flowrates indicated by values in parenthesis). As shown, the maximized flowrate of SK3 allows more water recovery (from SR2) to take place, which leads to reduced wastewater flowrate eventually.

A comparison between both scenarios is given in Figure 5.12. It is observed that when the ATM is solved to minimize its freshwater flowrate (Scenario 1), higher wastewater flowrate will be generated. Due to the high treatment cost of wastewater, the minimum cost solution (Scenario 2) determines that SK3 is to be operated at its maximum flowrate, so that more water may be recovered to this sink; doing this leads to lower wastewater flowrate, but with higher freshwater. The selection of solution for implementation lies with the plant authority, with consideration of other factors, e.g., freshwater availability (Scenario 1 is favored), wastewater discharge consideration, etc.

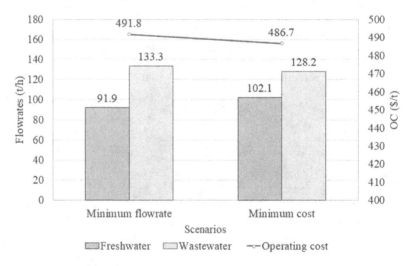

FIGURE 5.12
Scenarios comparison for alkene production case study.

---

[3] See Section 2.2 for detailed discussion.

5.3 ATM for Property Integration with Inferior Operator Level

Since the ATM is based on the MCA technique, modification is hence needed when it is used for cases when fresh resource has an inferior a property operator value among all material sources (see details in Section 2.5). Steps 1 and 4 to formulate the ATM is modified as follows:

1. Arrange quality levels (q_k) of all process sinks and sources in *ascending* order. When the quality levels for fresh resources do not coincide with any of the process sinks and sources that are present, additional levels are added. An arbitrary value is also added at the final level q_n (*highest* among all quality levels) to allow the calculation of the residual property load.

2. Equation 5.4 of the ATM that is used to determine the load within each quality interval is modified as follows:

$$\Delta m_k = \delta_{k-1}\left(q_k - q_{k-1}\right) \tag{5.12}$$

Example 5.3 Targeting for property integration— Kraft paper making case study[4]

Table 5.3 shows the limiting data for the Kraft paper making process in Examples 2.6, 3.8, and 4.5. Determine the minimum fresh and waste fiber targets using the ATM.

SOLUTION

The six-step procedure is followed to formulate the ATM for the AN case study, with revised steps 1 and 4 (with Equation 5.12). Following steps 1–5, the ATM may be constructed in MS Excel similar to Example 5.1. Its Solver setup is shown in Figure 5.13, where the objective in Equation 5.7 is revised to minimize the use of fresh fiber (F_{FF}), as given in Equation 5.13:

$$\text{Minimize } F_{FF} \tag{5.13}$$

TABLE 5.3

Limiting Data for Kraft papermaking process (Example 5.3)

Sink, SK$_j$	F_{SKj} (t/h)	ψ_{SKj} (Dimensionless)	Source, SRi	F_{SRi} (t/h)	ψ_{SRi} (Dimensionless)
Machine I (SK1)	100	0.382	Bleached fiber (SR1)	90	0.469
Machine II (SK2)	40	0.536	Broke (SR2)	60	0.182
			Fresh fiber (FF)	To be determined	0.738

---

[4] Solution with LINGO is made available on book support website, while details are found in Edition 1.

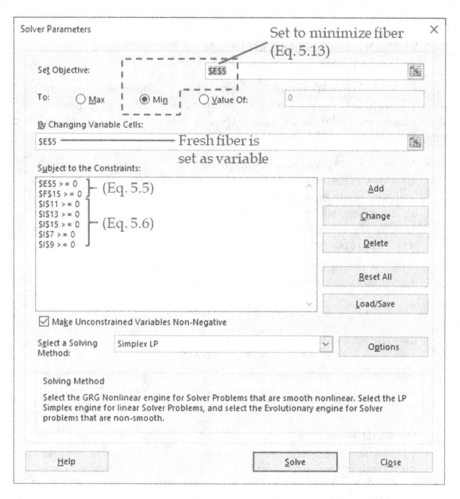

FIGURE 5.13
Solver setting for Kraft papermaking process (Example 5.3).

Upon execution, the Solver result indicates that a solution is found, with all constraints and optimality conditions met. The minimum fresh fiber is determined as 14.98 t/h, with a cascade diagram shown in Figure 5.14. From the latter, it is also observed that the minimum waste fiber for the RCN is determined as 24.98 t/h. The pinch operator is identified at the operator of 0.182 (dimensionless), corresponding to the residual load that vanishes ($\varepsilon_5 = 0$). The pinch-causing source is hence identified for the source at the operator level of 0.182. The allocated flowrates to the LQR and HQR are identified from the net material flowrates at δ_4 (35.02 t/h) and δ_5 (24.98 t/h), respectively. Note that Figure 5.14 corresponds to the MCA results in Figure 4.3.

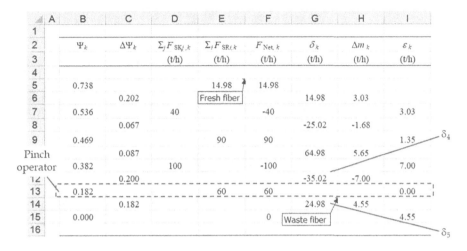

	A	B	C	D	E	F	G	H	I
1									
2		Ψ_k	$\Delta\Psi_k$	$\Sigma_j F_{SKj,k}$	$\Sigma_i F_{SRi,k}$	$F_{Net,k}$	δ_k	Δm_k	ε_k
3				(t/h)	(t/h)	(t/h)	(t/h)	(t/h)	(t/h)
4									
5		0.738			14.98	14.98			
6			0.202		Fresh fiber		14.98	3.03	
7		0.536		40		-40			3.03
8			0.067				-25.02	-1.68	
9		0.469			90	90			1.35
Pinch			0.087				64.98	5.65	
operator		0.382		100		-100			7.00
12			0.200				-35.02	-7.00	
13		0.182			60	60			0.00
14			0.182				24.98	4.55	
15		0.000				0	Waste fiber		4.55
16									

FIGURE 5.14
Results of ATM for Kraft papermaking process (Example 5.3).

5.4 ATM for Bilateral Problems

Another special case in property integration is *bilateral problem*, where multiple fresh resources are used, and the operator values of these fresh resources are at the superior and inferior levels among all process sources, respectively. Hence, we will have to formulate the ATM to handle both superior and inferior cases simultaneously. This is shown with the following example.

Example 5.4 Material reuse/recycle in palm oil milling process

Figure 5.15 shows the clay bath operation of a palm oil milling process (Ng et al., 2009c), which is used to separate the palm kernel from its shell and the uncracked nuts (from the cracked mixture stream) based on flotation principle. Due to density difference, the clay bath suspension is operated with a density of 1120 kg/m³. This allows the kernel to emit as the overflow stream, while shell fractions and the uncracked nuts with higher density are withdrawn as the bottom stream. Freshwater and clay (fresh resources) are fed to the premixing tank to achieve the desired density of the clay suspension, before it is fed to the clay bath. Sieving is used in both top and bottom streams to separate the solids (kernel, shell, and uncracked nuts) from the liquid portion.

To reduce fresh resource usage, the liquid portion from the overflow stream (SR1) is considered for recovery to the premixing tank (SK1). However, the recovery of wastewater from the bottom flow stream is not considered due to its high impurity content. Freshwater and clay may be used to supplement the recovery of overflow stream in order to meet its flowrate and density requirement of SK1. The limiting data of the sinks

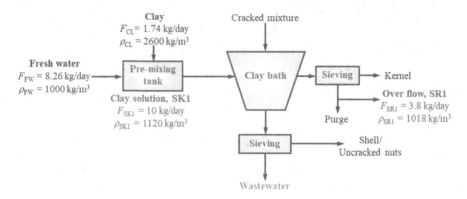

FIGURE 5.15
Clay bath operation in palm oil milling process for Example 5.4. (From Ng, et al. 2009c. Reproduced with permission. Copyright 2009 Elsevier.)

and sources (including fresh resources) are given in Table 5.4. Note that the density values and the corresponding property operators of the sinks and sources are given in columns 3 and 4, respectively.

Complete the following tasks:

1. Determine the minimum OC of the RCN, assuming that freshwater (CT_{FW}) and clay (CT_{CL}) have unit costs of $0.001/kg and $0.25/kg, respectively.
2. Develop the material recovery scheme for this RCN based on targeting insights.

SOLUTION

From Table 5.4, it is observed that the fresh resources available for this case have operator values at the superior (clay) and inferior (freshwater) levels, respectively. In other words, one of the two fresh resources will be placed at the lowest level of the material cascade, for either case where the superior or inferior cascade is formulated with the ATM. Hence, in order to optimize the use of both resources simultaneously, both superior and inferior cascades are to be formulated within the same ATM.

TABLE 5.4

Limiting Data for Palm Oil Milling Case Study (Example 5.4)

Sink, SK_j	F_{SKj} (kg/day)	Density, ρ_{SKj} (kg/m³)	$\psi_{SKj} \times 10^{-4}$ (m³/kg)
Clay solution (SK1)	10.0	1120	8.93
Source, SR_i	**F_{SRi} (kg/day)**	**Density, ρ_{SRi} (kg/m³)**	**$\psi_{SRi} \times 10^{-4}$ (m³/kg)**
Overflow (SR1)	3.8	1018	9.82
Freshwater (FW)	To be determined	1000	10
Clay (CL)	To be determined	2600	3.85

The six-step procedure may be followed to formulate the superior or inferior cascade for the ATM.

1. *Arrange quality levels (q_k) of all process sinks and sources in descending (superior cascade) and ascending order (inferior cascade).*

   The operator levels for the superior cascade are arranged in ascending order (Figure 5.16a), while the reverse is applied for operator levels of the inferior cascade (Figure 5.16b). An arbitrary large value of 100 is added at the lowest operator level in the superior cascade, while zero property operator is added at the final level of the inferior cascade. Note that for both cascades, one of the fresh resources is placed at the highest level,

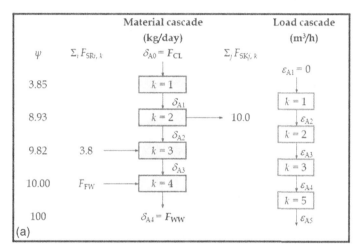

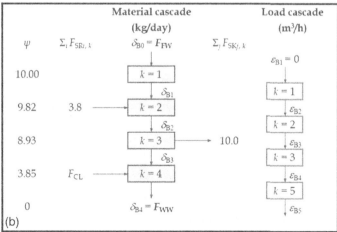

FIGURE 5.16
Cascade diagram for Example 5.5: (a) superior cascade and (b) inferior cascade. (From Ng, et al., 2009c. Reproduced with permission. Copyright 2009 Elsevier.)

while the other fresh resource is treated as a process source at its corresponding operator level (the same strategy applies to all cases with multiple fresh resources). Note also that to avoid duplication of variables, all variables for the superior cascade are represented with subscript A, while those of the inferior cascade are with subscript B. Note however that the flowrate variables for both resources, i.e., freshwater (F_{FW}) and clay (F_{CL}) remain identical in both cascades.

2. *Locate flowrates of the process sinks and sources at their respective quality level k.*
3. *Construct the material cascade.*

Equations 5.1 and 5.2 describe the material cascades across all quality levels:

$$\delta_{A1} = \delta_{A0} \qquad \delta_{B1} = \delta_{B0}$$

$$\delta_{A2} = \delta_{A1} + (0 - 10.0) \qquad \delta_{B2} = \delta_{B1} + (3.8 - 0)$$

$$\delta_{A3} = \delta_{A2} + (3.8 - 0) \qquad \delta_{B3} = \delta_{B2} + (0 - 10)$$

$$\delta_{A4} = \delta_{A3} + (F_{FW} - 0) \qquad \delta_{B4} = \delta_{B3} + (F_{CL} - 0)$$

The following equations are added to relate the net material flowrates that enter the first and the last operator levels to the minimum fresh resource and wastewater (F_{WW}) flowrates, respectively. Note that the fresh resource at the first operator level is different for both cascades, i.e., clay (with flowrate F_{CL}), for superior cascade, while freshwater (with flowrate F_{FW}) for inferior cascade. Note however that both cascades share the same flowrate for wastewater (F_{WW}) at the final operator level:

$$\delta_{A0} = F_{CL} \qquad \delta_{B0} = F_{FW}$$

$$\delta_{A4} = F_{WW} \qquad \delta_{B4} = F_{WW}$$

4. *Construct the load cascade.*

Equation 5.3 next gives the property load cascades across all operator levels in the superior (left) and inferior (right) cascades, respectively:

$$\varepsilon_{A1} = 0 \qquad \varepsilon_{B1} = 0$$

$$\varepsilon_{A2} = \varepsilon_{A1} + \Delta m_{A1} \qquad \varepsilon_{B2} = \varepsilon_{B1} + \Delta m_{B1}$$

$$\varepsilon_{A3} = \varepsilon_{A2} + \Delta m_{A2} \qquad \varepsilon_{B3} = \varepsilon_{B2} + \Delta m_{B2}$$

$$\varepsilon_{A4} = \varepsilon_{A3} + \Delta m_{A3} \qquad \varepsilon_{B4} = \varepsilon_{B3} + \Delta m_{B3}$$

$$\varepsilon_{A5} = \varepsilon_{A4} + \Delta m_{A4} \qquad \varepsilon_{B5} = \varepsilon_{B4} + \Delta m_{B4}$$

Property load within each quality interval is given by Equation 5.4, for superior (left) and inferior (right) cascades, respectively:

$$\Delta m_{A1} = \delta_{A1}(8.93 - 3.85) \qquad \Delta m_{B1} = \delta_{B1}(10 - 9.82)$$

$$\Delta m_{A2} = \delta_{A2}(9.82 - 8.93) \qquad \Delta m_{B2} = \delta_{B2}(9.82 - 8.93)$$

$$\Delta m_{A3} = \delta_{A3}(10 - 9.82) \qquad \Delta m_{B3} = \delta_{B3}(8.93 - 3.85)$$

$$\Delta m_{A4} = \delta_{A4}(100 - 10) \qquad \Delta m_{B4} = \delta_{B4}(3.85 - 0)$$

Equation 5.5 indicates that net material flowrate that enters the first and last levels should take non-negative values:

$$\delta_{A0} \geq 0 \qquad \delta_{B0} \geq 0$$

$$\delta_{A4} \geq 0 \qquad \delta_{B4} \geq 0$$

Equation 5.6 indicates that all residual loads (except that at level 1) should take non-negative values:

$$\varepsilon_{A2} \geq 0 \qquad \varepsilon_{B2} \geq 0$$

$$\varepsilon_{A3} \geq 0 \qquad \varepsilon_{B3} \geq 0$$

$$\varepsilon_{A4} \geq 0 \qquad \varepsilon_{B4} \geq 0$$

$$\varepsilon_{A5} \geq 0 \qquad \varepsilon_{B5} \geq 0$$

5. *Formulate the optimization objective.*

The minimum OC of the RCN is given as the product of the flowrates of freshwater (F_{FW}) and clay (F_{CL}) with their individual unit cost, respectively. For this example, the latter are given as \$0.001/kg (CT_{FW}) and \$0.25/kg (CT_{CL}). Hence, the optimization objective in Equation 5.8 takes the form as in Equation 5.12:

$$\text{Minimize } OC = F_{FW}CT_{FW} + F_{CL}CT_{CL} \qquad (5.12)$$

The LINGO formulation for this model can be written as follows:

```
Model:
! Optimization objective: to minimize annual operating cost
(AOC);
min = OC;
OC = (FW*0.001 + CL*0.25);
```

```
! Material cascade for superior (left) and inferior (right);

DA1 = DA0 ;                          DB1 = DB0 ;
DA2 = DA1 - 10 ;           DB2 = DB1 + 3.8 ;
DA3 = DA2 + 3.8 ;          DB3 = DB2 - 10;
DA4 = DA3 + FW ;           DB4 = DB3 + CL ;

! The net material flowrates entering the first level
correspond to flowrates of fresh clay (CL, superior) and
fresh water (FW, inferior):

! The net material flowrates entering the last level
correspond to wastewater (WW) flowrates for both cascades;

DA0 = CL ;          DB0 = FW ;
DA4 = WW ;          DB4 = WW ;

! Property load cascade for superior (left) and inferior
cascades (right);

EA1 = 0 ;           EB1 = 0 ;
EA2 = EA1 + MA1;    EB2 = EB1 + MB1 ;
EA3 = EA2 + MA2;    EB3 = EB2 + MB2 ;
EA4 = EA3 + MA3;    EB4 = EB3 + MB3 ;
EA5 = EA4 + MA4;    EB5 = EB4 + MB4 ;

! Property load within each operator interval for superior
(left) and inferior cascades (right);

MA1 = DA1*(8.93 - 3.85);   MB1 = DB1*(10 - 9.82) ;
MA2 = DA2*(9.82 - 8.93);   MB2 = DB2*(9.82 - 8.93) ;
MA3 = DA3*(10 - 9.82);     MB3 = DB3*(8.93 - 3.85) ;
MA4 = DA4*(100 - 10);      MB4 = DB4*(3.85 - 0) ;

! Net material flowrates entering the first last levels
should take non-negative values ;

DA0 >= 0 ;          DB0 >= 0 ;
DA4 >= 0 ;          DB4 >= 0 ;

! All residual loads (except level 1) should take non-
negative values ;

EA2 >= 0 ;          EB2 >= 0 ;
EA3 >= 0 ;          EB3 >= 0 ;
EA4 >= 0 ;          EB4 >= 0 ;
EA5 >= 0 ;          EB5 >= 0 ;

! Net material flowrates (except first and last level) and
property load within each operator interval may take
negative values ;

@free(DA1);         @free (DB1) ;
@free(DA2);         @free (DB2) ;
@free(DA3);         @free (DB3) ;
@free(MA1);         @free (MB1) ;
@free(MA2);         @free (MB2) ;
@free(MA3);         @free (MB3) ;

END
```

The following is the solution report generated by LINGO:

Global optimal solution found.	
Objective value:	0.4117259
Total solver iterations:	0
Variable	Value
OC	0.411725
FW	4.571382
CL	1.628618
DA1	1.628618
DA0	1.628618
DA2	−8.371382
DA3	−4.571382
DA4	0.000000
WW	0.000000
EA1	0.000000
EA2	8.273379
EA3	0.8228488
EA4	0.000000
EA5	0.000000
DB1	4.571382
DB0	4.571382
DB2	8.371382
DB3	−1.628618
DB4	0.000000
EB1	0.000000
EB2	0.8228488
EB3	8.273379
EB4	0.000000
EB5	0.000000
MA1	8.273379
MA2	−7.450530
MA3	−0.8228488
MA4	0.000000
MB1	0.8228488
MB2	7.450530
MB3	−8.273379
MB4	0.000000

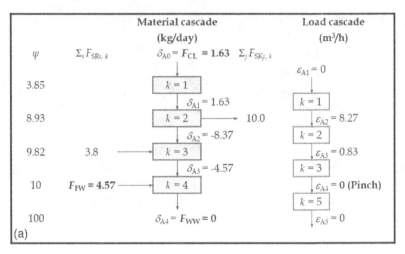

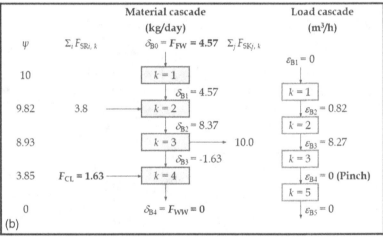

FIGURE 5.17
Results of ATM for Example 5.4: (a) superior cascade and (b) inferior cascade. (From Ng et al., 2009c. Reproduced with permission. Copyright 2009 Elsevier.)

The solution report of LINGO indicates that the OC of the RCN is determined as $0.412/day. Next, the revised cascade diagram takes the form as in Figure 5.17. As shown, the clay and freshwater flowrates are determined as 1.63 and 4.57 kg/day, respectively. Note that each cascade has its respective pinch operator and also the pinch-causing source.

Since the ATM is based on the MCA technique, insights to develop the material recovery scheme in the MCA can also be used for the ATM. For this case, the material recovery scheme for this RCN may be derived from either of the cascade diagrams in Figure 5.17.

As shown in Figure 5.17a, 1.63 kg/day of clay is cascaded from the first level to the sink (SK1, 10 kg/day) at level 2. Since the flowrate of clay is insufficient to fulfill that of the sink, sources of lower levels are used.

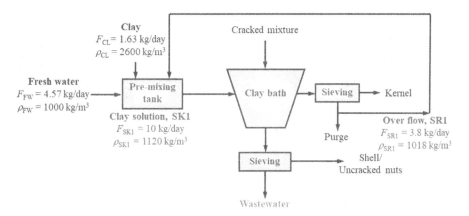

FIGURE 5.18
RCN for clay bath operation. (From Ng et al., 2009c. Reproduced with permission. Copyright 2009 Elsevier.)

Hence, SR1 at level 3 (3.8 kg/day) and freshwater at level 4 (4.57 kg/day) are sent to fulfill the requirement of SK1. The allocation of sources from a lower quality level to sink at a higher quality level is indicated by the negative net material flowrate between levels 2 and 4 (see Example 4.1 for discussion on other similar problems).[5] The RCN that achieves the established flowrate targets is shown in Figure 5.18.

PROBLEMS
Water Minimization Problems

5.1 Determine the minimum flowrates of freshwater and wastewater for the tire-to-fuel process in Problem 2.1. Note that the operating flowrates of the water sinks and sources may be added as process constraints in the ATM. Note also that the source flowrate of the seal pot has to be equal to its sink flowrate.

5.2 Revisit Example 3.6 with ATM. Determine the minimum freshwater and wastewater flowrate targets for this RCN, assuming that pure freshwater is used. Limiting data for the case are given in Table 3.5 (note: this is a threshold problem; see the answer in Figure 3.18).

5.3 Limiting water data for a water minimization study are given in Table 5.5 (Sorin and Bédard, 1999). Determine the minimum flowrate targets for its freshwater and wastewater requirement, assuming that pure freshwater is used. See results in Problem 8.12.

5.4 For the bulk chemical production case in Problem 3.5, solve the following tasks:

a. Determine the minimum freshwater and wastewater flowrates for the process, assuming that pure freshwater is used.

---

[5] Note that derivation of RCN structure from targeting insights only applies to cases where single process sink (with non-zero quality level) that exists in the HQR.

TABLE 5.5

Limiting Water Data for Problem 5.3

SK_j	F_{SK_j} (t/h)	C_{SK_j} (ppm)	SR_i	F_{SR_i} (t/h)	C_{SK_i} (ppm)
SK1	120	0	SR1	120	100
SK2	80	50	SR2	80	140
SK3	80	50	SR3	140	180
SK4	140	140	SR4	80	230
SK5	80	170	SR5	195	250
SK6	195	240			

Source: Sorin and Bédard (1999).

b. Synthesize an RCN based on the insights obtained from the ATM (refer to Section 3.2). Compare the network structure from that obtained using the nearest neighbor algorithm and superstructural approach (note to readers: solve Problems 8.6 and 9.4 to compare results).

5.5 Revisit the AN case in Example 5.1. Determine the minimum water cost ($/s) for the RCN, when a secondary impure freshwater source with 5 ppm ammonia content is available for use apart from the primary pure freshwater source. Unit costs for the pure (CT_{FW1}) and impure freshwater sources (CT_{FW2}) are given as \$0.001/kg and \$0.0005/kg, respectively (see answer in Example 3.4).

5.6 Revisit Problem 3.4 (Polley and Polley, 2000). Determine the minimum flowrates of freshwater and wastewater with ATM for the following scenarios (see answers in Example 8.1).

a. Determine the minimum flowrates of freshwater and wastewater, when a single pure freshwater feed (i.e., zero impurity content) is available.

b. Redo a when the freshwater feed has 10 ppm impurity content.

c. Determine the minimum OC when two freshwater feeds, i.e., pure (0 ppm, \$1/ton) and impure freshwater feeds (80 ppm, \$0.2/ton) are available.

d. Redo c when three freshwater feeds are available, i.e., single pure (0 ppm) and two impure freshwater sources (20 and 50 ppm).

5.7 A coal mine water system (Jia et al., 2015) is analyzed for its potential of water minimization. Data for the case study are given in Table 5.6. Determine the minimum flowrates for its freshwater and wastewater, assuming that pure freshwater is used (see answers in Problem 8.13).

5.8 Revisit the Kraft pulping process in Problem 2.2 (El-Halwagi, 1997). Use ATM to determine the following targets by incorporating

TABLE 5.6

Limiting Water Data for Coal Mine Water System in Problem 5.7

Water Sinks, SK$_j$			
j	Stream	Flowrate F_{SKj} (t/d)	COD Concentration C_{SKj} (mg/L)
1	Domestic Water	280.4	2
2	Excavation	291.6	10
3	Coal Preparation	116.9	10
4	Transportation	55.7	25

Water Sources, SR$_i$			
i	Stream	Flowrate F_{SRi} (t/d)	COD Concentration C_{SRi} (mg/L)
1	Domestic Water	280.4	15
2	Excavation	291.6	20
3	Coal Preparation	116.9	30
4	Transportation	55.7	35

Source: Jia et al., 2015.

the flowrate operating range of the chemical recovery unit (Equation 2.46):

a. Minimum freshwater flowrate (see answers in Problem 8.15).

b. Minimum *OC*, with unit costs of fresh water and wastewater given as $1.5/t and $10/t, respectively. Make comparison with solution in (a).

5.9 Revisit the classical water minimization problem in Problem 3.3. Determine the following targets for a direct water reuse/recycle scheme:

a. Minimum freshwater and wastewater flowrates, when a single pure freshwater feed is available (note: see answers in Example 8.1).

b. Total water cost ($/h), when two freshwater sources of 0 and 10 ppm are used. These freshwater sources have unit costs of $1.0/ton (0 ppm) and $0.8/ton (10 ppm), respectively.

c. Perform a sensitivity analysis to analyze the effect of unit cost of impure freshwater source (range between $0.8/t and $1.0/t) on the total water cost and the flowrates of individual freshwater sources.

5.10 Revisit the Kraft pulping process in Problem 2.8 (Parthasarathy and Krishnagopalan, 2001). Use ATM to determine the following targets by incorporating the flowrate operating range of the water sinks (Equations 2.71–2.80):

a. Minimum freshwater flowrate freshwater source available for service, which has a chloride content of 4.2 ppm (see answers in Problem 8.20).

b. Minimum *OC*, with unit costs of fresh water and wastewater given as $1/t and $5/t, respectively. Do a comparison with results in (a).

UTILITY GAS RECOVERY PROBLEMS

5.11 For the refinery hydrogen network in Examples 2.4 and 3.3 (Hallale and Liu, 2001), identify the minimum flowrates of the fresh hydrogen feed and purge stream for the RCN with ATM. The limiting data are found in Table 3.2.

5.12 Limiting data for a refinery hydrogen network (Deng et al., 2020) are given in Table 5.7 (see its process flow diagram in Figure 12.20). Determine its minimum flowrates of fresh hydrogen feed and purge stream, assuming that fresh hydrogen with 0.1% impurity is available for use (note: see answers in Problem 12.5).

5.13 For the refinery hydrogen network (Alves and Towler, 2002) in Problems 2.6 and 4.15, identify the following targets for the RCN. The limiting data are found in Table 4.31.

   a. Minimum flowrates of fresh hydrogen feed and purge stream (note: see answers in Problem 8.25).

   b. Total operating cost ($/s) of the RCN. The imported hydrogen has a unit cost of 2.0 US$/kmol, while the hydrogen stream from the SRU has a unit cost of 0.8 US$/kmol (Alves and Towler, 2002).

TABLE 5.7

Limiting Data for Hydrogen Network in Problem 5.12

j	Sinks, SK_j	F_{SK_j} (Nm³/h)	C_{SK_i} (mol%)	i	Sources, SR_i	F_{SK_j} (Nm³/h)	C_{SK_i} (mol%)
1	LHPH-I	86,200	9.50	1	LHPH-I	83,150	10.00
2	LHPH-II	81,670	9.00	2	LHPH-II	74,480	10.00
3	DHT	67,260	9.50	3	DHT	59,370	10.00
4	CNHT	55,660	8.00	4	CNHT	47,760	8.00
				5	LHPH-I	1,340	22.00
				6	LHPH-II	1,250	21.40
				7	DHT	908	31.00
				8	CNHT	530	21.00
				9	LHPH-I	650	60.50
				10	LHPH-II	650	42.60
				11	DHT	1,758	93.00
				12	CNHT	227	84.00

Source: Deng et al., 2020.

Note: LHPH-I and LHPH-II: lubricating oil hydrogenation processes I and II; DHT: diesel hydrogenation; CNHT: coking naphtha hydrogenation.

Property Integration Problems

5.14 For the metal degreasing case in Example 2.5, determine the minimum fresh solvent and solvent waste flowrates for the RCN, with limiting data in Table 2.4.

5.15 Determine the minimum flowrates of fresh resources (ultrapure water and municipal freshwater) and wastewater for both scenarios of the wafer fabrication process in Problem 4.23. Use the limiting data in Table 4.33.

5.16 Figure 5.19 is the flow diagram of a biorefinery (Martinez-Hernandez et al., 2018). Various units in this biorefinery consumes (as sink) and produce (source) bioethanol, Hence, a bioethanol network maybe synthesized to optimize its utilization. Determine the minimum fresh ethanol needed for the bioethanol network, given that the fresh ethanol has a purity of 0.996 (fraction). Limiting data of the bioethanol sinks and sources are given in Table 5.8.

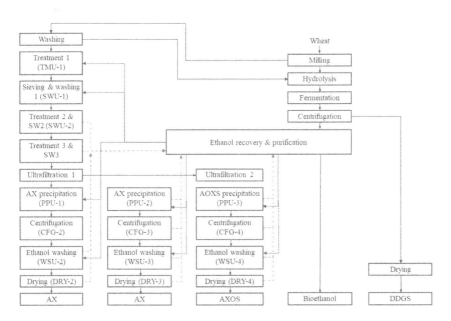

FIGURE 5.19
Process flow diagram for biorefinery in Problem 5.16 (Martinez-Hernandez et al., 2018.)

TABLE 5.8

Limiting Data for Bioethanol Network for Biorefinery in Problem 5.16

Bioethanol Sinks, SK_j			
j	Units	Flowrate F_{SK_j} (t/y)	Ethanol Purity (fraction)
1	PPU-1	11,888	0.9600
2	PPU-2	16,389	0.9600
3	PPU-3	26,063	0.9600
4	WSU-2	9503	0.9600
5	WSU-3	1055	0.9600
6	WSU-4	4216	0.9600
7	TMU-1	38,695	0.7000
8	SWU-1	18,380	0.7000

Bioethanol Sources, SR_i			
I	Units	Flowrate F_{SR_i} (t/y)	Ethanol Purity (Fraction)
1	WSU-3	2142	0.9345
2	RDY-3	203	0.9345
3	WSU-4	8558	0.9258
4	RDY-4	809	0.9258
5	PPU-2	16,065	0.9160
6	CFG-3	142	0.9137
7	CFG-4	568	0.8978
8	PPU-3	22,150	0.8978
9	WSU-2	19,285	0.8409
10	RDY-2	1823	0.8409
11	SWU-1	55,633	0.6822
12	SWU-2	80,304	0.0249

References

Alves, J. J. and Towler, G. P. 2002. Analysis of refinery hydrogen distribution systems, *Industrial and Engineering Chemistry Research*, 41, 5759–5769.

Deng, C., Zhu, M., Liu, J., and Xiao, F. 2020. Systematic retrofit method for refinery hydrogen network with light hydrocarbons recovery, *International Journal of Hydrogen Energy*, 45, 19391–19404.

El-Halwagi, M. M.1997. *Pollution Prevention Through Process Integration: Systematic Design Tools*. San Diego, CA: Academic Press.

Foo, D. C. Y. 2019. Optimize plant water use from your desk. *Chemical Engineering Progress*, 115, 45–51 (February 2019).

Hallale, N. and Liu, F. 2001. Refinery hydrogen management for clean fuels production. Advances in Environmental Research, 6, 81–98.

Jia, X., Li, Z., Wang, F., Foo, D. C. Y., and Tan, R. R. 2015. Integrating input-output models with pinch technology for enterprise sustainability analysis. *Clean Technologies and Environmental Policy*, 17(8): 2255–2265.

Martinez-Hernandez, E., Tibessart, A., and Campbell, G. M. 2018. Conceptual design of integrated production of arabinoxylan products using bioethanol pinch analysis, *Food and Bioproducts Processing*, 112, 1–8.

Ng, D. K. S., Foo, D. C. Y., and Tan, R. R. 2009a. Automated targeting technique for single-component resource conservation networks—Part 1: Direct reuse/recycle, *Industrial and Engineering Chemistry Research*, 48(16), 7637–7646.

Ng, D. K. S., Foo, D. C. Y., and Tan, R. R. 2009b. Automated targeting technique for single-component resource conservation networks—Part 2: Single pass and partitioning waste interception systems, *Industrial and Engineering Chemistry Research*, 48(16), 7647–7661.

Ng, D. K. S., Foo, D. C. Y., Tan, R. R., Pau, C. H., and Tan, Y. L. 2009c. Automated targeting for conventional and bilateral property-based resource conservation network, *Chemical Engineering Journal*, 149, 87–101.

Parthasarathy, G. and Krishnagopalan, G. 2001. Systematic reallocation of aqueous resources using mass integration in a typical pulp mill, *Advances in Environmental Research*, 5, 61–79.

Polley, G. T. and Polley, H. L. 2000. Design better water networks, *Chemical Engineering Progress*, 96(2), 47–52.

Sorin, M. and Bédard, S. 1999. The global pinch point in water reuse networks, *Process Safety and Environmental Protection*, 77, 305–308.

6

Automated Targeting Model for Material Regeneration and Pretreatment Networks

When the potential for flowrate reduction via material reuse/recycle is exhausted, a common mean to further reduce the flowrates of fresh resource and waste discharge for resource conservation network (RCN) is through material *regeneration*. The regenerated source possesses better quality and hence may be further reused/recycled to the process sinks. Doing this helps to reduce the fresh resource and waste flowrates of the RCN. In this chapter, the automated targeting model (ATM) introduced in Chapter 5 is extended for various RCNs with regeneration schemes. This includes the *material regeneration network* where process sources are purified for recovery; as well as pretreatment network where fresh resource is purified before it is used for process sinks. The ATM is demonstrated using MS Excel and LINGO through various examples.

6.1 Conceptual Understanding of Material Regeneration

In principle, there are two recovery schemes when material regeneration is involved, i.e., regeneration-reuse and regeneration-recycle. In the former, the material sources (i.e., effluent from a process unit) are sent for regeneration in the purification unit prior to their recovery to another process sink (inlet of another processing unit, see Figure 6.1a). The regenerated source does not re-enter the same process unit where it is produced, due to the concern of impurity build-up. On the other hand, when regeneration-recycle scheme is adopted, the regenerated material sources are permitted to re-enter the same unit where it is generated (Figure 6.1b). These two schemes are not differentiable in the current chapter.

Figure 6.1c shows a generic superstructure for a material regeneration network. As shown, the process sinks may be sent for regeneration with the purification unit, and may also be sent to the process sinks directly (i.e., without purification). In other words, not all process sources will be purified prior to their recovery to the process sinks. The main task for the synthesis of a material regeneration network is to identify the source(s) that will be sent for regeneration, apart from identifying the minimum fresh resource and waste discharge of the RCN. An illustration is shown using the material recovery pinch diagram (MRPD) in Figure 6.2. As shown, sources of different quality

DOI: 10.1201/9781003173977-7

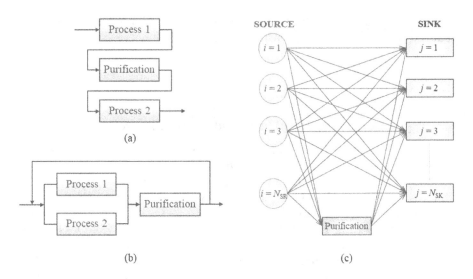

FIGURE 6.1
(a) Regeneration-reuse scheme; (b) regeneration-recycle scheme; (c) superstructure for material regeneration network.

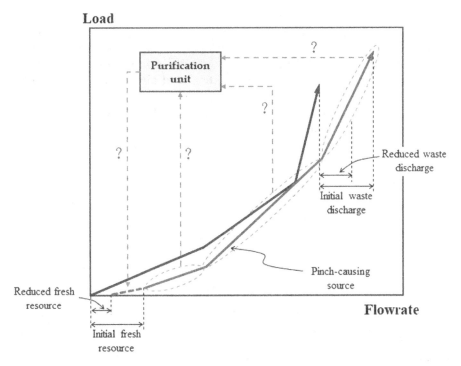

FIGURE 6.2
Selection of process sources for material regeneration network.

(indicated by its slope) may be selected for regeneration; doing so leads to reduced fresh resources and waste discharge.

6.2 Types of Purification Units and Their Characteristics

Figure 6.3 illustrates the two general types of purification units: *single pass* and *partitioning* units. In the former type (Figure 6.3a), it may be assumed that the feed and purified streams have the same flowrates, i.e., $F_{Rin} = F_{Rout}$. In other words, flowrate loss is negligible. The purified stream is of better quality compared to the feed stream, i.e., $q_{Rout} > q_{Rin}$.

In the partitioning type purification units, however, flowrate loss is significant. As shown in Figure 6.3b, the purification unit accepts a feed stream with quality q_{Rin} and produces two outlet streams: *purified* stream of higher quality q_{RP} (with flowrates F_{RP}) and *reject* stream of lower quality q_{RJ} (with flowrate F_{RJ}). Note that the quality of the streams is denoted as $q_{RP} > q_{Rin} > q_{RJ}$.

Both types of purification units may be classified as *fixed outlet quality* or *removal ratio*. Purification units of the former type upgrade the quality of the feed streams (with quality q_{Rin}) to a constant outlet quality, i.e., q_{Rout} or q_{RP} (e.g., 20 ppm, 5 wt%, etc.). Typical examples of this type of purification unit include sand filter (single pass) and membrane separation (partitioning type). On the other hand, purification units of the removal ratio type upgrade the quality of the given feed stream by removing a portion of its impurity/property load. Adsorption, stripping (single pass) and dissolved air flotation (partitioning units) are typical examples of this purification type.

In the following sections, the ATM for an RCN with different types of purification units is outlined.

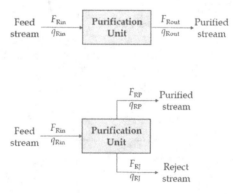

FIGURE 6.3
Types of purification units: (a) single pass and (b) partitioning units. (From Ng et al., 2009a. Reproduced with permission. Copyright 2009 American Chemical Society.)

6.2 ATM for RCN with Single Pass Purification Unit of Fixed Outlet Quality Type

The basic ATM framework for this RCN takes the revised form as that for the reuse/recycle network given in Chapter 5, as shown in Figure 6.4. The same procedure may be followed to formulate the ATM for the RCN (see detailed steps in Chapter 5). However, two minor modifications are needed:

1. The outlet quality of the purification unit (q_{Rout}) is added in the cascade diagram, if it is not present among the quality levels of the process sinks and sources.

2. For all quality levels that are lower than q_{Rout}, wherever a process source is found, a new sink is added to represent the potential flowrate to be sent for regeneration ($F_{RE,r}$, $r = 1, 2,..., RG$). Also, a new source is added at q_{Rout} level where the regenerated source (F_{REG}) is found. In other words, inlets to the purification unit are treated as water sinks, while the outlet stream of the purification unit is regarded as a source.

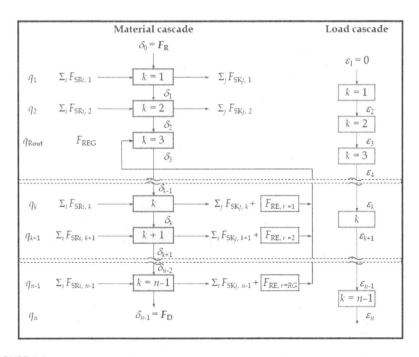

FIGURE 6.4
ATM framework for RCN with single pass purification unit(s) of fixed outlet quality. (From Ng et al., 2009a. Reproduced with permission. Copyright 2009 American Chemical Society.)

The formulation of the ATM also differs slightly from that of the direct reuse/recycle network. The following constraints describe the material and load cascades in the ATM framework:

$$\delta_k = \delta_{k-1} + F_{\text{BAL},k} \qquad k = 1, 2, \ldots, n-1 \tag{6.1}$$

$$F_{\text{BAL},k} = \sum_i F_{\text{SRi},k} - \sum_j F_{\text{SKj},k} + F_{\text{REG}} - F_{\text{RE},r} \quad k = 1, 2, \ldots, n-1 \tag{6.2}$$

$$F_{\text{REG}} = \sum_r F_{\text{RE},r} \tag{6.3}$$

$$F_{\text{RE},r} \leq F_{\text{SRi},k} \qquad q_k < q_{\text{Rout}}; \qquad \forall r \; \forall k \; \forall i \tag{6.4}$$

$$F_{\text{RE},r} \geq 0 \qquad \forall r \tag{6.5}$$

$$\varepsilon_k = \begin{cases} 0 & k = 1 \\ \varepsilon_{k-1} + \Delta m_k & k = 2, 3, \ldots, n \end{cases} \tag{6.6}$$

$$\Delta m_k = \delta_{k-1}\left(q_k - q_{k-1}\right) \tag{6.7}$$

$$\delta_{k-1} \geq 0 \qquad k = 1, n \tag{6.8}$$

$$\varepsilon_k \geq 0 \qquad k = 2, 3, \ldots, n \tag{6.9}$$

Equation 6.1 is an extended form of Equation 5.1 (ATM for reuse/recycle network—Chapter 5), which describe the *net material flowrate* of level k (δ_k) in the material cascade. Apart from the net material flowrate cascaded from an earlier quality level (δ_{k-1}), the equation also takes into account the *material flowrate balances* at level k ($F_{\text{BAL},k}$). The latter is calculated using Equation 6.2, which considers flowrate balances between total sink ($F_{\text{SKj},k}$) and total source flowrates ($F_{\text{SRi},k}$), as well as the regeneration flowrate (F_{REG}) and individual flowrate to be sent for regeneration ($F_{\text{RE},r}$). For quality levels other than q_{Rout}, the F_{REG} term is set to zero. Similarly, the potential regeneration flowrate terms ($F_{\text{RE},r}$) are set to zero for quality levels where no sources exist, as well as for levels of higher quality than the regeneration outlet quality, i.e., $q_k > q_{\text{Rout}}$. Equation 6.3 next describes the flowrate balance for the purification unit(s). Since the regeneration flowrate originates from the process source, the maximum flowrate to be regenerated is bound by the process source flowrate, as described by Equation 6.4. Equation 6.5 states that all flowrates sent for regeneration have to be non-negative values. Equations 6.6–6.9 are essentially the same as those used for the ATM for direct reuse/recycle network in Chapter 5 (Equations 5.3–5.6).

Since there are two variables to be optimized in this case, i.e., fresh resource and regeneration flowrates, we may set the optimization objective to minimize the flowrates of both fresh resource and regenerated source with the two-stage optimization approach. In stage 1, the flowrate of fresh resource is minimized using Equation 6.10, subject to the constraints in Equations 6.1–6.9. We then proceed to minimize the regenerated source flowrate in stage 2 using Equation 6.11. Apart from constraints in Equations 6.1–6.9, the fresh resource flowrate determined in stage 1 is also added as a new constraint, given as in Equation 6.12.

$$\text{Minimize } F_R = \delta_0 \tag{6.10}$$

$$\text{Minimize } F_{REG} = \sum_r F_{RE,r} \tag{6.11}$$

$$F_R = F_R^{max} \tag{6.12}$$

Apart from minimum flowrates, we can also minimize the cost of the RCN, or to be more specific the operating cost (OC; Equation 6.13); the latter is usually contributed by fresh resource, waste discharge, and regenerated source(s).

$$\text{Minimize } COST = OC \tag{6.13}$$

Since the ATM is an extended form of that given in Chapter 5, it is a linear program (LP) model that can be solved to obtain a global optimum solution.

Example 6.1 Synthesis of water regeneration network with minimum flowrates

The classical water minimization example of Wang and Smith (1994) is analyzed here, with limiting water data given in Table 6.1. A single pass purification unit with outlet concentration (C_{Rout}) of 5 ppm is used to regenerate water for further water recovery. Determine the minimum flowrates of freshwater (0 ppm) and regenerated sources for this *water regeneration network* using MS Excel. Identify also the structure of the RCN using the insights of the ATM.

TABLE 6.1

Limiting Water Data for Example 6.1

Sinks, SK_j	F_{SKj} (t/h)	C_{SKj} (ppm)	Sources, SR_i	F_{SRi} (t/h)	C_{SRi} (ppm)
SK1	20	0	SR1	20	100
SK2	100	50	SR2	100	100
SK3	40	50	SR3	40	800
SK4	10	400	SR4	10	800
$\sum_i F_{SKj}$	170		$\sum_i F_{SRi}$	170	

SOLUTION

The ATM structure for this example is shown in Figure 6.5. Although it is similar to those ATM structures for reuse/recycle network in Chapter 5, there are two additional features. First, a new concentration level of 5 ppm is added, corresponding to the outlet concentration (C_{Rout}) of the purification unit. Next, a column F is added where regeneration source and sinks are to be located. As shown in Figure 6.5, a new source is added at concentration level of 5 ppm for the flowrate of regenerated water (F_{REG}), while two new sinks are added at 100 ($F_{RE,1}$) and 800 ppm ($F_{RE,2}$), to allow the water sources to be sent for regeneration from these levels. As described earlier, inlets to purification unit are treated as water sinks, while its outlet as a source.

Since both freshwater and regenerated flowrates are to be minimized, the two-stage optimization approach is used. In this case, we will minimize the freshwater flowrate in stage 1, while the regenerated flowrate is minimized at stage 2. To determine the minimum freshwater flowrate (F_{FW}), the optimization objective in Equation 6.10 is revised as Equation 6.14, subject to the constraints in Equations 6.1–6.9 (see Figure 6.6a for the setting of these constraints).

$$\text{Minimize } F_{FW} = \delta_0 \tag{6.14}$$

By executing Solver, a solution is found with all constraints and optimality conditions met. The ATM result indicates that the minimum freshwater flowrate for the water network is determined as 20 t/h (result not shown for brevity). Since the objective was set to minimize freshwater, the regeneration flowrate obtained at stage 1 may not be necessary for the minimum value. Hence, the optimization objective is revised to

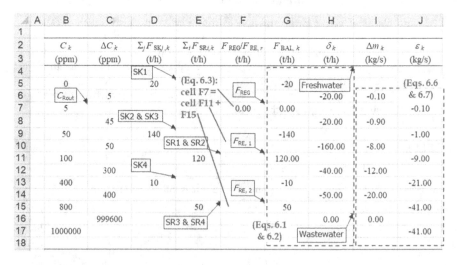

FIGURE 6.5
Cascade table for Example 6.1 in MS Excel.

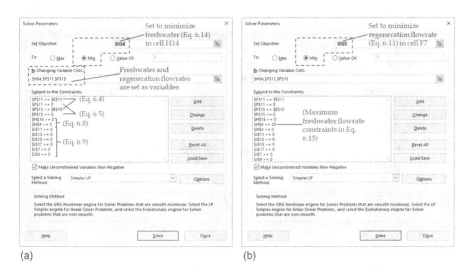

FIGURE 6.6
Solver setting for Example 6.1: (a) stage 1; (b) stage 2.

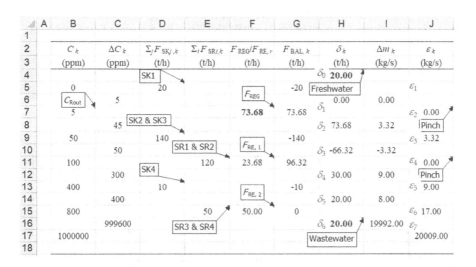

FIGURE 6.7
Results of ATM for stage 2 optimization—with minimum freshwater and regeneration flowrates (Example 6.1).

minimize the regenerated flowrate (F_{REG}) for stage 2 optimization using Equation 6.11 while the minimum freshwater flowrate obtained in stage 1 is included as a new constraint in Equation 6.15. (as revised Equation 6.12)

$$F_{FW} = 20 \tag{6.15}$$

The Solver setting for stage 2 optimization remains identical to that of stage 1, except that the objective is revised (given by Equation 6.11) and a new constraint for freshwater flowrate (given by Equation 6.15) is added, as shown in Figure 6.6b.

Executing Solver resulting a solution in Figure 6.7, with all constraints and optimality conditions met. As shown, the ATM result indicates that the minimum regenerated water flowrate is determined as 73.68 t/h, with flowrates of freshwater and wastewater remain identical as those in stage 1. Note that in some cases, the individual source flowrates that are sent for regeneration (cells F11 and F15) may be different for different computer and the version of MS Excel used, indicating that degenerate solutions exist for this problem.

An important characteristic for material regeneration network is worth mentioning—multiple pinches. As shown in Figure 6.7, two pinches are found where the residual load vanishes ($\varepsilon_2 = \varepsilon_4 = 0$), i.e., indicating that pinch concentrations correspond to 5 and 100 ppm.

For a simple case such as this problem, the structure for the RCN can be determined from the result of the ATM in Figure 6.7.[1] At 0 ppm level, there exists a sink (SK1) that requires 20 t/h of freshwater feed. Hence, the entire freshwater feed is consumed by this sink, which leaves zero net water flowrate from the first concentration level, i.e., $\delta_1 = 0$.

We then move to the next concentration level of 5 ppm, where 73.68 t/h of regenerated water source is available. Since no water sink is found at this level, the regenerated source is cascaded to the following level (50 ppm), i.e., $\delta_2 = 73.68$ t/h. Sinks SK2 and SK3 are at 50 ppm with a flowrate requirement of 120 t/h, for which it is insufficient to use the regenerated source alone. In order to fulfill the flowrate requirement, a flowrate of 66.32 t/h is drawn from the lower concentration level of 100 ppm (indicated by negative net water flowrate, i.e., $\delta_3 = -66.32$ t/h). The latter may be contributed by either SR1 or SR2, since both sources are found at 100 ppm. To minimize piping connections (which helps to reduce piping complexity), we shall utilize SR2 (100 t/h) which has sufficient flowrates for both sinks. The allocation of regenerated water and SR2 to both SK2 and SK3 are then determined based on the flowrate ratio of the sinks. Pairing SR2 to these sinks resulted a structure given in Figure 6.8.

We then move to the next concentration level, i.e., 100 ppm where sources SR1 and SR2 are found. Apart from supplying water to the sinks at 50 ppm, a flowrate of 23.68 t/h (F_{RE1}) is also sent for regeneration from these sources. We also observe that 30 t/h of these sources (δ_4) are cascaded down to the next concentration level at 400 ppm, from either of these sources ($\delta_4 = 30$ t/h). At 400 ppm level, the cascaded flowrate from 100 ppm ($\delta_4 = 30$ t/h) is used to fulfill the 10 t/h flowrate requirement of sink SK4 (either SR1 or SR2 may be used), while the leftover flowrate ($\delta_5 = 20$) will be cascaded to lower level, i.e., 800 ppm.

---

[1] There should only be sink that exists in each pinch region, with non-zero quality level, in order to apply this technique.

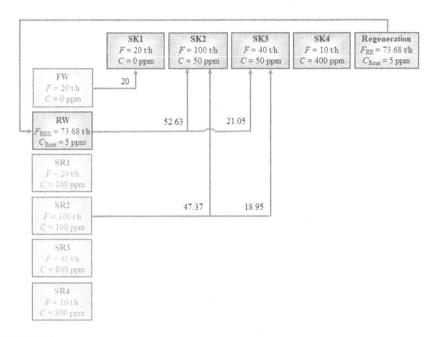

FIGURE 6.8
RCN for Example 6.1, after pairing SK1, SK2 and SK3 (all flowrate terms in t/h).

In the concentration level of 800 ppm, where sources SR3 and SR4 exist. The ATM indicates that both of these sources are sent for water regeneration (F_{RE2}). The cascaded flowrate from 800 ppm will leave as wastewater ($\delta_6 = F_{WW}$). A complete structure of the water regeneration network is shown in Figure 6.9. Note that there may exist different network structure when the design is performed using a formal design procedure (see Example 8.3),[2] indicating that a degenerate solution exists for the problem.

Example 6.2 Cost minimization with ATM

Revisit Example 6.1 (Wang and Smith, 1994). Solve the ATM using MS Excel to determine the minimum operating cost (OC) and the corresponding flowrates of the water regeneration network, for the following unit cost data.

1. Unit costs of freshwater, wastewater, and regenerated sources are given as 1 $/t, 1 $/t, and 0.8 $/t, respectively.
2. Unit cost of regeneration is raised to 2 $/t, while those of freshwater and wastewater are kept at 1 $/t, respectively.
3. Unit cost of regeneration for the cleaner source (100 ppm) is given as 1.7 $/t, while that of the dirtier source (800 ppm) is 2.3 $/t. Unit costs of freshwater and wastewater are kept at 1 $/t, respectively.

---

[2] See Example 8.3..

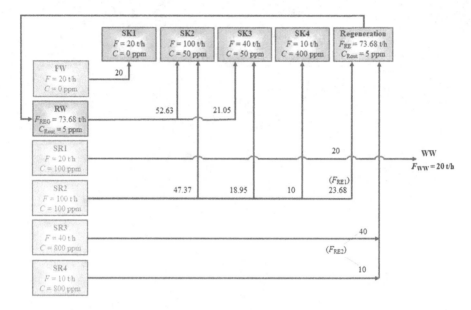

FIGURE 6.9
Final RCN for Example 6.1 (all flowrate terms in t/h).

SOLUTION

1. Unit costs of freshwater, wastewater and regenerated sources are given as 1 $/t, 1 $/t and 0.8 $/t, respectively.

   To determine the minimum cost solution, the optimization objective in Equation 6.13 takes the revised form as in Equation 6.16.

   $$minimize\ OC = F_{FW}CT_{FW} + F_{WW}CT_{WW} + F_{RW}CT_{RW} \qquad (6.16)$$

   where CT_{FW}, CT_{WW} and CT_{RW} are the unit costs of freshwater, wastewater, and regenerated water, respectively. The ATM structure for this case is similar to that in Example 6.1. The Solver setting is shown in Figure 6.10, where the objective is set to minimize OC (cell J23), given by Equation 6.16. The calculation of the OC is carried out below the cascade table, where the individual flowrates of freshwater, wastewater, and regeneration sources are linked (see Figure 6.11).

   Executing the Solver resulted in a solution in Figure 6.11, with all constraints and optimality conditions met. As shown, the ATM determines that the freshwater and wastewater each has a minimum flowrates of 20 t/h, while the regeneration flowrate is determined as 73.68 t/h. These flowrate targets are identical as those determined using the two-stage approach in Example 6.1 (see Figure 6.7). Similarly, two pinch concentrations are found at 5 and 100 ppm.

2. Unit cost of regeneration is raised to 2 $/t, while those of freshwater and wastewater are kept at 1 $/t, respectively.

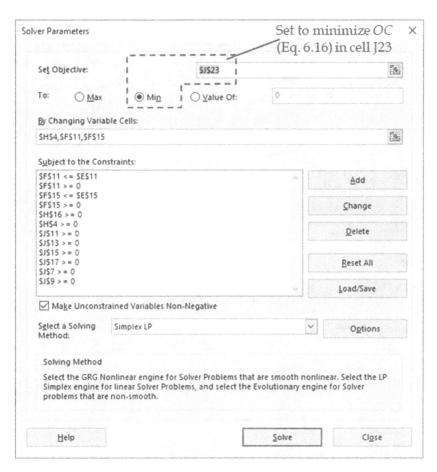

FIGURE 6.10
Solver setting for Example 6.2.

Solver setting for this case is identical to that of the previous case, and hence is not shown for brevity. Executing the Solver resulted in a solution in Figure 6.12, where ATM result indicates that the minimum freshwater and wastewater are both determined as 90 t/h, while that for regeneration water flowrate is reduced to 0 t/h. Note also that the ATM solution in Figure 6.12 shows a single pinch concentration at 100 ppm. This is a solution that is identical to the reuse/recycle network (readers may solve Problem 3.3 to verify this). In other words, the ATM advices users not to deploy the regeneration scheme as its operating cost is higher than that of the reuse/recycle network.

3. Unit cost of regeneration for the cleaner source (100 ppm) is given as 1.7 \$/t, while that of the dirtier source (800 ppm) is 2.3 \$/t. Unit costs of freshwater and wastewater are kept at 1 \$/t, respectively.

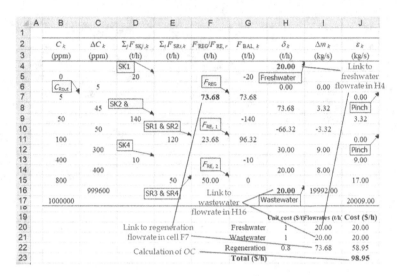

FIGURE 6.11
Results of ATM for minimum cost solution (case 1, Example 6.2).

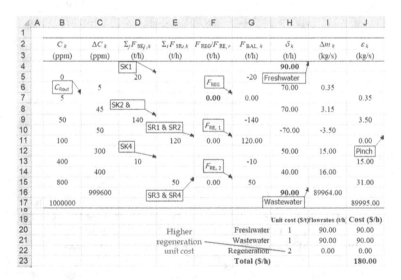

FIGURE 6.12
Results of ATM for minimum cost solution (case 2, Example 6.2).

Solving the ATM with new cost data yields the results in Figure 6.13, where the minimum freshwater and wastewater are both determined as 45.71 t/h, while the regenerated water flowrate is determined as 46.62 t/h. Note that the regenerated water is originated from the 100 ppm source, which is cheaper. In other words, the ATM advices users to deploy the

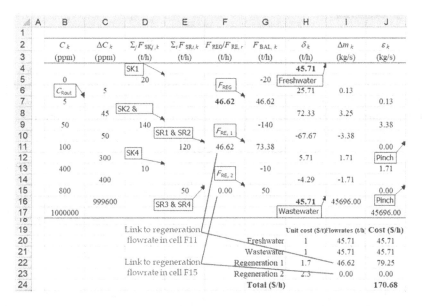

FIGURE 6.13
Results of ATM for minimum cost solution (case 3, Example 6.2).

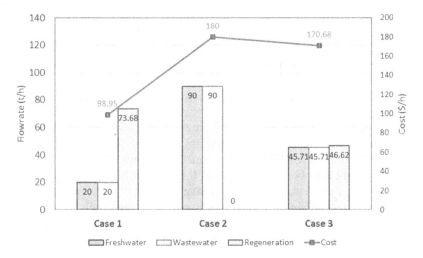

FIGURE 6.14
Overall comparison for all cases in Example 6.2.

regeneration unit for the cleaner water source in order to achieve a cost-effective solution. Besides, it is interesting to note that two pinch concentrations are found at 100 and 800 ppm, which are different from the earlier case where pinch concentrations at 5 and 100 ppm (see Figure 6.11).

A comparison for all cases in this example is shown in Figure 6.14.

6.3 Modeling of Mass Exchange Operation as Purification Unit

El-Halwagi (1997, 2006) defined mass exchange operation as a separation process that employs a mass-separating agent (MSA, termed as *lean stream*) to selectively remove certain components (e.g., impurities, pollutants, by-products, products) from a component-rich stream (e.g., waste, product, etc.). To perform the separation task, the MSA has to be immiscible (partially or completely) in the rich stream. Common examples of mass exchange operations include absorption, adsorption, stripping, ion exchange, extraction, leaching, etc. These units may be used as purification units in a material regeneration network. In this section, the modeling of a mass exchange operation is presented from the perspective of RCN synthesis. Since the mass exchange operations are mainly concentration-based processes, their quality indices are always given in concentration basis.

A schematic representation of a mass exchange operation that acts as an purification unit is shown in Figure 6.15, where the rich and lean streams pass through the mass exchange operation in countercurrent flow. In this case, the process source that acts as an impurity-rich stream (with feed quality C_{Rin}) is purified to higher quality (C_{Rout}) and exists at the top outlet stream of the purification unit. Its impurity load (Δm) is transferred to the MSA (lean stream, with flowrate F_{MSA}) and is represented by Equation 6.17. In turn, the impurity load raises the concentration of the MSA from $C_{MSA,in}$ to $C_{MSA,out}$. Note that flowrate losses for both rich and lean streams are negligible in most cases:

$$\Delta m = F_{Rout}\left(C_{Rin} - C_{Rout}\right) = F_{MSA}\left(C_{MSA,out} - C_{MSA,in}\right) \qquad (6.17)$$

Figure 6.16 shows a graphical representation of the mass exchange operation in a concentration versus load diagram similar to that in Figure 2.4. Note however that in this case, the process source becomes the rich stream, as its impurity content is to be removed to the MSA that acts as the lean stream.

For a feasible mass exchange operation, a minimum mass transfer driving force is needed to operate the mass exchange unit. As shown in Figure 6.16, the minimum driving force is found at both ends of the mass exchange operation. A few points are worth mentioning here. First, any process source to be

FIGURE 6.15
A mass exchange operation used as a purification unit in an RCN. (From El-Halwagi, 1997. Reproduced with permission. Copyright 1997 Elsevier.)

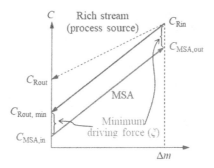

FIGURE 6.16
A mass exchange operation represented in a concentration versus load diagram. (From Wang and Smith 1994. Reproduced with permission. Copyright 1994 Elsevier.)

purified by the MSA has to lie in the region that is higher than the minimum driving force from the MSA line. In other words, a process source may be purified to any outlet concentration C_{Rout} that is higher than the minimum threshold value, i.e., $C_{Rout,min}$. Note that the latter is given as the summation of the MSA inlet concentration (to the mass exchange operation, $C_{MSA,in}$) with the minimum driving force (ξ), given as in Equation 6.5:

$$C_{Rout,min} = C_{MSA,in} + \xi \tag{6.18}$$

However, it is always desired to keep the regeneration flowrate of the process source to the minimum, as this leads to minimum operating cost of the purification unit, as well as reduced capital cost indirectly (which is often a function of regeneration flowrate). Since the source flowrate is given by the inverse slope of the plot in Figure 6.16, it should be purified to the minimum outlet concentration ($C_{Rout,min}$) in order to keep a minimum regenerated source flowrate.

Similarly, to keep the MSA flowrate (and hence its operating cost) to the minimum, the MSA line in Figure 6.16 should have the steepest slope. In other words, the MSA should be operated with minimum driving force difference from the process source, at the rich end of the mass exchange operation. This is conceptually similar to the case of *limiting water profile* (see Section 2.3 for detailed discussion).

In some cases, there are more than one MSAs that may be used for a particular regeneration task. In such cases, it will be good to evaluate the performance of these MSAs from thermodynamic perspective. In equilibrium condition, the concentration of process source SRi (C_{SRi}) may be related to that of MSA l $\left(C_{MSAl}^{*}\right)$ with a linear relationship (El-Halwagi, 1997, 2006), given as in Equation 6.19:

$$C_{SRi} = \alpha_l C_{MSAl}^{*} + \beta_l \tag{6.19}$$

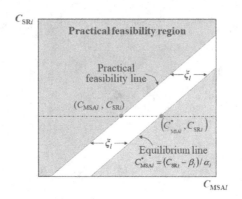

FIGURE 6.17
Equilibrium and practical feasibility lines. (From El-Halwagi, 1997. Reproduced with permission. Copyright 1997 Elsevier.)

where α_l and β_l are specific parameters for a given MSA l. As described earlier, in order for a feasible mass transfer operation to take place, the process source is to be operated with a minimum driving force difference from the MSA. This is depicted in Figure 6.17 (El-Halwagi, 1997).

As shown in Figure 6.17, the operating point with coordinate (C_{MSAl}, C_{SRi}) is separated from the equilibrium point $\left(C_{MSAl}^{*}, C_{SRi}^{*}\right)$ by the minimum driving force (ξ), given as in Equation 6.20. Note that the latter may have different values for different MSA l:

$$C_{MSAl}^{*} = C_{MSAl}^{*} + \xi_l \tag{6.20}$$

Combining Equations 6.19 and 6.20 yields the operating line equation for MSA l:

$$C_{SRi} = \alpha_l \left(C_{MSAl} + \xi_l \right) + \beta_l \tag{6.21}$$

When multiple MSA candidates are available for use, the selection of MSA will depend on factors such as overall cost, or the minimum flowrates of fresh resources and waste generation of the RCN; these may be set as optimization objective of the ATM.

Example 6.3 Water minimization for acrylonitrile (AN) production

The AN process in Examples 3.1, 4.1, and 5.1 is revisited. In this case, regeneration is considered in order to further reduce the water flowrates of the RCN. Limiting data for this case are given in Table 6.2 (reproduced from Table 4.2).

Resin adsorption may be used to purify sources SR2, SR3, and SR4 to an outlet concentration of 11.6 ppm so that they have better recovery potential (El-Halwagi, 1997, 2006). The specific parameters used for the

TABLE 6.2

Limiting Water Data for AN Production in Example 6.3

Water Sinks, SK_j			
j	Stream	Flowrate F_{SKj} (kg/s)	NH₃ Concentration C_{SKj} (ppm)
1	BFW	1.2	0
2	Scrubber inlet	5.8	10
Water Sources, SR_i			
i	Stream	Flowrate F_{SRi} (kg/s)	NH₃ Concentration C_{SRi} (ppm)
1	Distillation bottoms	0.8	0
2	Off-gas condensate	5	14
3	Aqueous layer	5.9	25
4	Ejector condensate	1.4	34

Source: El-Halwagi (1997).

operating line equation for this MSA (Equations 6.21) are 0.01 (α_l), 0 (β_l) and 5 ppm (ξ_l), respectively. The inlet ($C_{MSA,in}$) and outlet concentrations ($C_{MSA,out}$) of resin are given as 3 and 1100 ppm, respectively.

Solve the following tasks:

1. Determine the minimum *OC* solution. Identify the water source(s) to be purified, and the corresponding flowrates of freshwater, wastewater, and regeneration. Unit costs for freshwater (CT_{FW}) and wastewater (CT_{WW}) are given as \$0.001/kg each, while that for regenerated flowrate (CT_{RW}) is given based on its impurity removal, i.e., \$91 × 10⁻⁶/mg NH₃ removed (El-Halwagi, 1997).
2. Plot a concentration versus load diagram for the source purification in (a). Determine the resin flowrate needed to accomplish the purification task.

SOLUTION

1. Determine the minimum *OC* solution. Identify the water source(s) to be purified, and the corresponding flowrates of freshwater, wastewater, and regeneration.

   To determine the OC solution, the objective in Equation 6.22 is used.

$$\text{minimize } OC = F_{FW}CT_{FW} + F_{WW}CT_{WW} + \Delta m \, CT_{RW} \qquad (6.22)$$

   where Δm is the NH₃ impurity load removal, to be calculated using Equation 6.17. The setting in MS Excel and Solver is shown in Figure 6.18. Solving the ATM resulted in the solution in Figure 6.18. As shown, the minimum freshwater flowrate is determined as 1.2 kg/s, while the minimum regeneration flowrate is found as 5 kg/s, which is sourced from SR2 (off-gas condensate). In other words, only SR2 will be sent to the resin

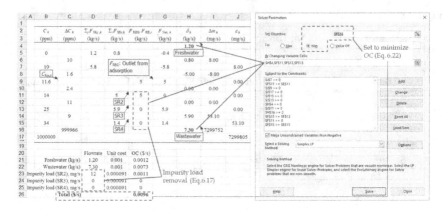

FIGURE 6.18
Setting in Excel spreadsheet and Solver in Example 6.3.

adsorption unit for purification, while SR3 and SR4 will not be purified. Figure 6.18 also shows that the corresponding wastewater flowrate is determined as 7.3 kg/s.

2. Plot a concentration versus load diagram for the source purification in (a). Determine the resin flowrate needed to accomplish the purification task.

Figure 6.19a illustrates purification of SR2 with the purification unit (resin adsorption). The removal of NH_3 load (Δm) and its resin flowrate (F_{MSA}) are both determined using Equation 6.17:

$$\Delta m = 5.0 \text{ kg/s} \left(14 - 11.6 \text{ ppm}\right) = 12 \text{ mg/s};$$

$$F_{MSA} = \frac{12 \text{ mg/s}}{(1100 - 0.08) \text{ ppm}} = 0.01 \text{ kg/s}$$

The concentration versus load diagram for this mass transfer operation is shown in Figure 6.19b.

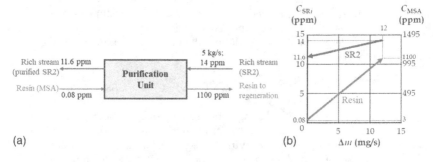

FIGURE 6.19
(a) The resin adsorption process, and its (b) concentration versus load diagram in Example 6.3.

6.4 ATM for RCN with Single Pass Purification Unit of Removal Ratio Type

Apart from the purification units that have a fixed outlet quality, another type of purification unit are those that are rated in terms of its *removal ratio index* (RR), which describes the fractional removal of impurity/property load from the feed stream of the purification unit. It may be mathematically described as follows:

$$RR = \frac{F_{Rin}\left(q^{max} - q_{Rin}\right) - F_{Rout}\left(q^{max} - q_{Rout}\right)}{F_{Rin}\left(q^{max} - q_{Rin}\right)} \tag{6.23}$$

where q^{max} is the highest possible value for a given type of purity index, e.g., fraction of 1, 100%, etc. A more commonly used version of Equation 6.23 is given for the quality index of concentration, as follows (Wang and Smith, 1994):

$$RR = \frac{F_{Rin}C_{Rin} - F_{Rout}C_{Rout}}{F_{Rin}C_{Rin}} \tag{6.24}$$

where C_{Rin} and C_{Rout} are the impurity concentrations of the feed and purified streams.

For a single pass purification unit without flowrate loss, i.e., $F_{Rin} = F_{Rout}$, Equation 6.24 may be simplified as follows:

$$RR = \frac{C_{Rin} - C_{Rout}}{C_{Rin}} \tag{6.25}$$

From Equation 6.25, we observe that the quality of the purified stream is dependent on the feed stream quality. For instance, for a purification unit of $RR = 0.9$, feed streams of 100 ppm will be purified to 10 ppm, while that of 200 ppm will end up at 20 ppm. In other words, new quality levels are to be included in the cascade diagram of the ATM for every source that is present, as all sources may be sent for regeneration (see Figure 6.20). Identifying the quality levels for the purification units helps to maintain the ATM as an LP model that ensures a global optimum solution.

The formulation for the ATM of this case is similar to that of the purification unit of fixed outlet quality, except that the constraints which involve

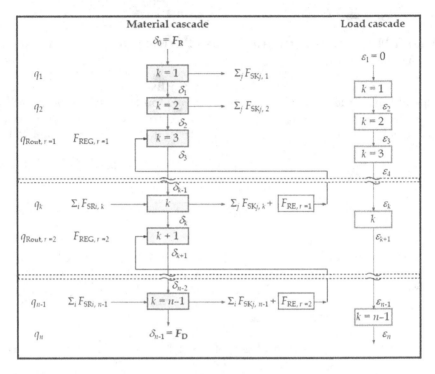

FIGURE 6.20
ATM framework for RCN with single pass purification unit(s) of RR type. (From Ng et al., 2009a. Reproduced with permission. Copyright 2009 American Chemical Society.)

flowrate terms of the purification units (i.e., Equations 6.2 and 6.3) are revised slightly as follows:

$$F_{\text{BAL},k} = \sum_i F_{\text{SRi},k} - \sum_j F_{\text{SKj},k} + F_{\text{REG},r} - F_{\text{RE},r} \quad k = 1, 2, \ldots, n-1 \qquad (6.26)$$

$$F_{\text{REG},r} = F_{\text{RE},r} \qquad (6.27)$$

where $F_{\text{RE},r}$ and $F_{\text{REG},r}$ are the inlet and outlet rates of the purification unit r.

Example 6.4 Water regeneration network (purification unit with RR = 0.9)

Example 6.1 is revisited (Wang and Smith, 1994). A single pass purification unit with $RR = 0.9$ is used to purify water source for further recovery. Determine the minimum operating cost (OC) for the water regeneration network, assuming that the unit costs for fresh and regenerated water are given as \$1/ton (CT_{FW}) and \$0.8/ton (CT_{RW}), respectively. Wastewater treatment cost is assumed to be insignificant here.

SOLUTION

From Table 6.1, it is observed that the water sources that may be sent for regeneration originate from 100 and 800 ppm. Equation 6.24 hence determines that the concentrations of the regenerated sources correspond to 10 and 80 ppm, respectively. Hence, two new concentration levels are added to the cascade diagram for the regenerated sources. The cascade diagram for this example is shown in Figure 6.21.

As wastewater treatment cost is ignored, the optimization objective in Equation 6.16 is revised to minimize its *OC* as follows:

$$\text{Minimize } OC = F_{FW} \, CT_{FW} + \left(F_{RE1} + F_{RE2} \right) CT_{RW} \qquad (6.28)$$

where $F_{RE1} + F_{RE2}$ are the two possible water sources that may be sent for regeneration.

The ATM result in Figure 6.22 indicates that the minimum *OC* is determined as \$85/h. The result also indicates that only 49.21 t/h of water source at 100 ppm is to be regenerated to 10 ppm, in order to reduce the freshwater and wastewater flowrates to 45.71 t/h. Note that two pinch concentrations are observed, i.e., 100 and 800 ppm, similar to those in Example 6.2 (case 3; Figure 6.13).

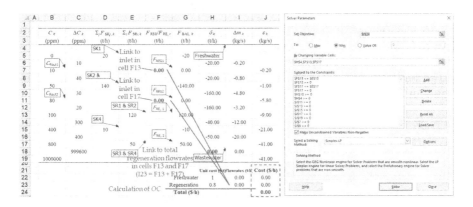

FIGURE 6.21
Setting in Excel spreadsheet and Solver for Example 6.4.

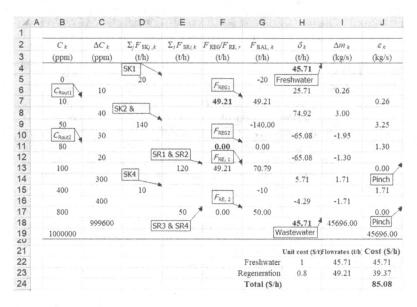

FIGURE 6.22
Result of ATM for Example 6.4.

6.5 Modeling for Partitioning Purification Unit(s) of Fixed Outlet Quality Type

Figure 6.23 shows a partitioning type purification unit reproduced from Figure 6.3b. In this section, a more detailed model is described, which takes into account the impurity/property load of the RCN.

The flowrate and material balances of the purification unit are given in Equations 6.29 and 6.30:

$$F_{Rin} = F_{RP} + F_{RJ} \tag{6.29}$$

$$F_{Rin}\, q_{Rin} = F_{RP}\, q_{RP} + F_{RJ}\, q_{RJ} \tag{6.30}$$

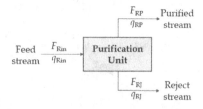

FIGURE 6.23
A partitioning purification unit.

where

F_{RP} and q_{RP} are the flowrate and quality index of the higher quality *purified* stream, respectively

F_{RJ} and q_{RJ} are flowrate and quality index of the lower quality *reject* streams, respectively

Since the purification unit possesses two outlet streams, an overall *product recovery factor (RC)* is used to define the recovery of the desired product from the inlet stream, as given in Equation 6.31:

$$RC = \frac{F_{RP}\ q_{RP}}{F_{Rin}\ q_{Rin}} \tag{6.31}$$

Combining Equations 6.28 through 6.30 leads to Equation 6.32:

$$q_{RJ} = \frac{q_{RP}\,(RC-1)}{RC - QI} \tag{6.32}$$

where QI is the *quality improvement* index for a given purification unit, given as follows:

$$QI = \frac{q_{RP}}{q_{Rin}} \tag{6.33}$$

Similar expressions may also be developed from the perspective of impurity concentration. In this case, Equation 6.30 takes the following form:

$$F_{Rin}\ C_{Rin} = F_{RP}\ C_{RP} + F_{RJ}\ C_{RJ} \tag{6.34}$$

where C_{RP} and C_{RJ} are the impurity concentrations of the purified and reject streams. Similarly, the product recovery factor in Equation 6.31 takes the revised form as follows:

$$RC = \frac{F_{RP}\left(C^{max} - C_{RP}\right)}{F_{Rin}\left(C^{max} - C_{Rin}\right)} \tag{6.35}$$

where C^{max} is the highest possible value for a given concentration index, e.g., 1,000,000 ppm, 100 wt%, etc. Combining Equations 6.29, 6.34, and 6.35 leads to Equation 6.36:

$$C_{RJ} = \frac{C_{RP}RC - C_{Rin}QI}{RC - QI} \tag{6.36}$$

Note that the quality improvement (QI) index in Equation 6.36 has been defined in Equation 6.33 from quality perspective. A revised form of this equation in impurity concentration is given as follows:

$$QI = \frac{\left(C^{max} - C_{RP}\right)}{\left(C^{max} - C_{Rin}\right)} \tag{6.37}$$

From Equations 6.32 and 6.37, it may be observed that the reject stream quality of a partitioning purification unit is dependent on the quality (or concentration) of its inlet and purified streams, as well as the product recovery factor (RC), but not on the stream flowrates. Hence, for a given purification unit with purified stream quality and product recovery factor, the quality of the reject stream can be determined, provided that inlet stream quality is also known.

6.6 Modeling for Partitioning Purification Unit(s) of Removal Ratio Type

Apart from the product recovery factor (RC), a partitioning purification unit may also be rated by removal ratio index (RR), as given by Equations 6.23 and 6.24. It will be useful to relate the purified stream quality (q_{RP}) to both RC and RR values for a given unit. Hence, combining Equations 6.23 and 6.31 leads to the following correlation:

$$q_{RP} = \frac{q_{Rin} q^{max} RC}{q^{max}(1 - RR) + q_{Rin}(RR + RC - 1)} \tag{6.38}$$

Similarly, combining Equations 6.24 and 6.32 leads to the revised version of Equation 6.38 as follows:

$$C_{RP} = \frac{C_{Rin} C^{max}(1 - RR)}{\left(C^{max} - C_{Rin}\right) RC + C_{Rin}(1 - RR)} \tag{6.39}$$

From Equations 6.38 and 6.39, we observe that the quality/concentration of the purified stream is dependent on the inlet stream quality, as well as the values of RR and RC of the purification unit. Once the purified stream quality is known, we can use Equations 6.38 or 6.39 to determine the reject stream quality/concentration.

Example 6.5 Mass balance calculation for a partitioning purification unit

Figure 6.24 shows a partitioning purification unit that regenerates an inlet stream of 100 kg/s (F_{Rin}) and 20% impurity concentration (C_{Rin}), to a purified stream with flowrate F_{RP} and impurity concentration C_{RP}, and a reject stream of flowrate F_{RJ} and impurity concentration C_{RJ}. Determine the missing flowrate and/or concentration parameters of the purified and reject streams for the following cases:

1. The purification unit is of the fixed outlet quality type, which produces an 80 kg/s (F_{RP}) purified stream of 10% impurity (C_{RP}).
2. The purification unit has an RR value of 0.9, and RC value of 0.8.

SOLUTION

Case 1—F_{RP} = 80 kg/s; C_{RP} = 10%

For the fixed outlet quality type purification unit, the product recovery factor and quality improvement index are determined using Equations 6.35 and 6.37:

$$RC = \frac{80(100-10)}{100(100-20)} = 0.9$$

$$QI = \frac{(100-10)}{(100-20)} = 1.125$$

In other words, the purification unit recovers 90% of the desired product from the inlet stream and improves the stream quality by 12.5%.

To determine the reject stream concentration (C_{RJ}), we can perform a mass balance calculation using Equation 6.34, since the other parameters are known values (F_{RJ} can be easily determined as 20 kg/s using Equation 6.29). Alternatively, we can also use Equation 6.36 to determine the value of C_{RJ}, without having to perform mass balance calculation of the streams:

$$C_{RJ} = \frac{10(0.9)-20(1.125)}{0.9-1.125} = 60\%$$

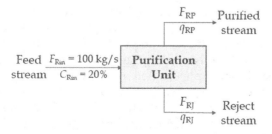

FIGURE 6.24
A partitioning purification unit for Example 6.5.

Case 2—$RR = 0.9$, $RC = 0.8$.

For this case, Equation 6.39 is used to determine the purified stream concentration (C_{RP}) of the purification unit:

$$C_{RP} = \frac{20(100)(1-0.9)}{(100-20)0.8+20(1-0.9)} = 3.03\%$$

Next, instead of using the mass balance equation, we can make use of Equations 6.36 and 6.37 to determine the reject stream concentration (C_{RJ}):

$$QI = \frac{(100-3.03)}{(100-20)} = 1.212$$

$$C_{RJ} = \frac{3.03(0.8)-20(1.212)}{0.8-1.212} = 52.94\%$$

Then, we can determine the flowrates of the purified (F_{RP}) and reject (F_{RJ}) streams by solving simultaneous equations with Equations 6.29 and 6.30:

$$100 \text{ kg/s} = F_{RP} + F_{RJ}$$

$$100 \text{ kg/s } (20\%) = F_{RP}(3.03\%) + F_{RJ}(52.94\%)$$

which leads to $F_{RP} = 66$ kg/s and $F_{RJ} = 34$ kg/s.

6.7 ATM for RCN with Partitioning Purification Unit(s)

To incorporate purification units into the ATM, the quality levels of the purified and reject streams are first determined using the model described in the previous sections. Similar to the earlier cases with a single pass purification unit, the quality levels of the purified and reject streams of the purification unit are added to the ATM framework, if they are not readily present among the quality levels of the process sinks and sources. An ATM framework for the incorporation of a partitioning purification unit is shown in Figure 6.25. Note that the ATM frameworks for RCNs with both types of partitioning purification units are quite similar. However, the quality level of the purified streams of all purification units is identical for the case of fixed outlet quality type units (see Example 6.6 for illustration), while the RCN with purification units of RR type will have a unique quality level for each of their purified and reject streams (depending on its inlet stream quality).

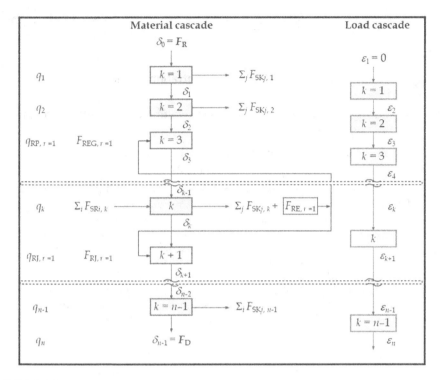

FIGURE 6.25
ATM framework for RCN with partitioning purification unit(s). (From Ng et al., 2009a. Reproduced with permission. Copyright 2009 American Chemical Society.)

Most constraints in the ATM formulation remain similar to the earlier cases, except that the constraints that involve purified and reject flowrates of the purification units (i.e., Equations 6.2 and 6.3) take the revised version as follows:

$$F_{BAL,k} = \sum_i F_{SRi,k} - \sum_j F_{SKj,k} + F_{REG,r} - F_{RE,r} + F_{RJ,r} \quad k = 1, 2, \ldots, n-1 \quad (6.40)$$

$$F_{RE,i} = F_{REG,r} + F_{RJ,r} \quad r = 1, 2, \ldots, RG \quad (6.41)$$

where $F_{RJ,r}$ is the reject stream flowrate of purification unit r.

Also, the product recovery factor in Equations 6.31 and 6.35 should be rearranged so as to avoid nonlinearity (division of variables $F_{REG,r}$ and $F_{RE,r}$) in the ATM. The revised forms of the equations are given as follows, in quality and concentration indices, respectively:

$$F_{RE,r} q_{RE,r} \ RC = F_{REG,r} \ q_{REG,r} \quad r = 1, 2, \ldots, RG \quad (6.42)$$

$$F_{RE,r}\left(C^{max} - C_{RE,r}\right)RC = F_{REG,r}\left(C^{max} - C_{REG,r}\right) \qquad r = 1, 2, \ldots, RG \quad (6.43)$$

where

$q_{RE,r}$ and $C_{RE,r}$ are the quality and concentration of the inlet stream of purification unit r

$q_{REG,r}$ and $C_{REG,r}$ are the quality and concentration of its purified stream

Example 6.6 Hydrogen network with partitioning purification unit of fixed outlet quality type

The hydrogen recovery problem in Examples 2.4 and 3.3 (Hallale and Liu, 2001) is revisited here, with the limiting data shown in Table 6.3 (reproduced from Table 2.2). The fresh hydrogen feed has an impurity concentration of 1%. A pressure swing adsorption (PSA) unit is used to purify the process sources for further recovery. The PSA unit can purify a hydrogen stream to 5% impurity (C_{RP}) and has a hydrogen recovery factor (RC) of 0.9 (Agrawal and Shenoy, 2006). Determine the minimum OC of the RCN with LINGO, given that the unit costs for fresh hydrogen (CT_{FH}) and the regenerated hydrogen sources (CT_{RH}) are \$2000/MMscf and \$1000/MMscf, respectively (cost of purged stream is ignored).

SOLUTION

Since two process sources exist in the problem, both sources may be sent for purification in the PSA units, for further recovery to the sinks. Hence, the concentrations of the purified (5%) and reject streams of the PSA units are to be added as new quality levels in the ATM. The quality improvement index and reject stream concentrations for both sources are calculated using Equations 6.36 and 6.37 as follows:

SR1: SR2:

$$QI = \frac{(100-5)}{100-9} = 1.044 \qquad QI = \frac{(100-5)}{(100-15)} = 1.118$$

$$C_{RJ} = \frac{5(0.9)-9(1.044)}{0.9-1.044} = 34\% \qquad C_{RJ} = \frac{5(0.9)-15(1.118)}{0.9-1.118} = 56.28\%$$

TABLE 6.3

Limiting Data for Example 6.6

Sinks, SK_j	F_{SK_j} (MMscfd)	C_{SK_j} (%)	Sources, SR_i	F_{SR_i} (MMscfd)	C_{SR_i} (%)
SK1	400	7.20	SR1	350	9.00
SK2	600	12.43	SR2	500	15.00

Source: Hallale and Liu (2001).

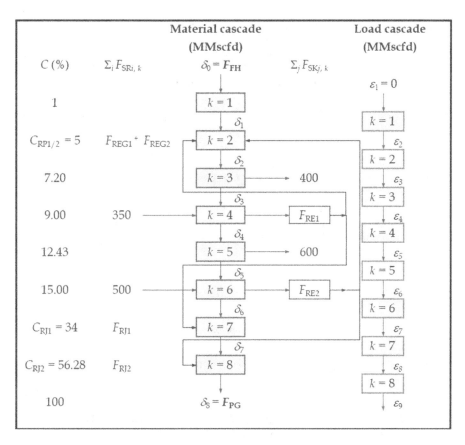

FIGURE 6.26
Cascade diagram for Example 6.6.

Hence, two new concentration levels (34% and 56.28%) are added to the cascade diagram, along with the concentration of the purified stream (5%), as shown in Figure 6.26.

The optimization objective is set to determine the minimum OC for the hydrogen network, using a revised form of Equation 6.16 as follows:

$$\text{Minimize } OC = F_{FH}\ CT_{FH} + (F_{RE1} + F_{RE2})\ CT_{RH} \tag{6.44}$$

where F_{FH} is the fresh hydrogen feed of the RCN. The optimization problem is subjected to the following constraints:

Hydrogen cascade across all quality levels, as given by Equations 6.1 and 6.2:

$$\delta_1 = \delta_0 + F_{BAL,1};\ F_{BAL,1} = 0$$

$$\delta_2 = \delta_1 + F_{BAL,2};\ F_{BAL,2} = F_{REG1} + F_{REG2}$$

$$\delta_3 = \delta_2 + F_{BAL,3}; \; F_{BAL,3} = -400$$

$$\delta_4 = \delta_3 + F_{BAL,4}; \; F_{BAL,4} = 350 - F_{RE1}$$

$$\delta_5 = \delta_4 + F_{BAL,5}; \; F_{BAL,5} = -600$$

$$\delta_6 = \delta_5 + F_{BAL,6}; \; F_{BAL,6} = 500 - F_{RE2}$$

$$\delta_7 = \delta_6 + F_{BAL,7}; \; F_{BAL,7} = F_{RJ1}$$

$$\delta_8 = \delta_7 + F_{BAL,8}; \; F_{BAL,8} = F_{RJ2}$$

The following equations are added to relate the net material flowrates that enter the first and the last concentration levels to the minimum fresh hydrogen (F_{FH}) and purge (F_{PG}) flowrates:

$$\delta_0 = F_{FH}$$

$$\delta_8 = F_{PG}$$

Equation 6.41 describes the flowrate balance of the PSA units:

$$F_{RE1} = F_{REG1} + F_{RJ1}$$

$$F_{RE2} = F_{REG2} + F_{RJ2}$$

The recovery of hydrogen ($RC = 0.9$) to the purified stream is described by Equation 6.43, which are nonlinear equations:

$$F_{RE1}(0.9)(100 - 9) = F_{REG1}(100 - 5)$$

$$F_{RE2}(0.9)(100 - 15) = F_{REG2}(100 - 5)$$

Source flowrate to be regenerated should not be more than what is available according to Equation 6.4:

$$F_{RE1} \leq 350$$

$$F_{RE2} \leq 500$$

Impurity load cascades across all quality levels are given by Equations 6.6 and 6.7:

$$\varepsilon_1 = 0$$

$$\varepsilon_2 = \varepsilon_1 + \Delta m_2; \; \Delta m_2 = \delta_1(5 - 1)/100$$

$$\varepsilon_3 = \varepsilon_2 + \Delta m_3; \ \Delta m_3 = \delta_2 \left(7.2 - 5\right)/100$$

$$\varepsilon_4 = \varepsilon_3 + \Delta m_4; \ \Delta m_4 = \delta_3 \left(9 - 7.2\right)/100$$

$$\varepsilon_5 = \varepsilon_4 + \Delta m_5; \ \Delta m_5 = \delta_4 \left(12.43 - 9\right)/100$$

$$\varepsilon_6 = \varepsilon_5 + \Delta m_6; \ \Delta m_6 = \delta_5 \left(15 - 12.43\right)/100$$

$$\varepsilon_7 = \varepsilon_6 + \Delta m_7; \ \Delta m_7 = \delta_6 \left(34 - 15\right)/100$$

$$\varepsilon_8 = \varepsilon_7 + \Delta m_8; \ \Delta m_8 = \delta_7 \left(56.28 - 34\right)/100$$

$$\varepsilon_9 = \varepsilon_8 + \Delta m_9; \ \Delta m_9 = \delta_8 \left(100 - 56.28\right)/100$$

Net material flowrates that enter the first and last levels should take non-negative values (Equation 6.8):

$$\delta_0 \geq 0$$

$$\delta_8 \geq 0$$

Finally, all residual loads (except at level 1) should take non-negative values (Equation 6.9):

$$\varepsilon_2 \geq 0 \qquad \varepsilon_3 \geq 0$$

$$\varepsilon_4 \geq 0 \qquad \varepsilon_5 \geq 0$$

$$\varepsilon_6 \geq \qquad \varepsilon_7 \geq 0$$

$$\varepsilon_8 \geq 0 \qquad \varepsilon_9 \geq 0$$

The LINGO formulation for this LP model is written as follows:

Model:
```
! Optimization objective: to determine the minimum operating
cost (OC) ;
min = FH*2000 + (RE1 + RE2)*1000 ;

! Hydrogen cascade ;

D1 = D0 + FBAL1 ; FBAL1 = 0 ;
D2 = D1 + FBAL2 ; FBAL2 = REG1 + REG2 ;
D3 = D2 + FBAL3 ; FBAL3 = -400 ;
D4 = D3 + FBAL4 ; FBAL4 = 350 - RE1 ;
```

```
D5 = D4 + FBAL5 ; FBAL5 = -600 ;
D6 = D5 + FBAL6 ; FBAL6 = 500 - RE2 ;
D7 = D6 + FBAL7 ; FBAL7 = RJ1 ;
D8 = D7 + FBAL8 ; FBAL8 = RJ2 ;

! The first and last net material flowrates corresponds to
minimum fresh hydrogen (FH) and purge (PG) flowrates ;

D0 = FH ;
D8 = PG ;

! Flowrate balance for purification unit(s) ;

RE1 = REG1 + RJ1 ;
RE2 = REG2 + RJ2 ;
RE1*0.9* (100 - 9)= REG1* (100 - 5) ;
RE2*0.9*(100 - 15)= REG2*(100 - 5) ;

! Flowrate constraints of regenerated hydrogen ;

RE1 <= 350 ;
RE2 <= 500 ;

! Impurity load cascade ;

E1 = 0 ;
E2 = E1 + DM2 ; DM2 = D1*(5 - 1)/100 ;
E3 = E2 + DM3 ; DM3 = D2*(7.2 - 5)/100 ;
E4 = E3 + DM4 ; DM4 = D3*(9 - 7.2)/100 ;
E5 = E4 + DM5 ; DM5 = D4*(12.43 - 9)/100 ;
E6 = E5 + DM6 ; DM6 = D5*(15 - 12.43)/100 ;
E7 = E6 + DM7 ; DM7 = D6* (34 - 15)/100 ;
E8 = E7 + DM8 ; DM8 = D7* (56.28 - 34)/100 ;
E9 = E8 + DM9 ; DM9 = D8* (100 - 56.28)/100 ;

! E1-E9 should take negative values ;

E1 >= 0 ; E2 >= 0 ; E3 >= 0 ; E4 >= 0 ; E5 >= 0 ;
E6 >= 0 ; E7 >= 0 ; E8 >= 0 ;  E9 >= 0 ;

! D1-D7 may take negative values ;

@free (D1) ; @free (D2) ; @free (D3) ;
@free (D4) ; @free (D5) ; @free (D6) ; @free (D7) ;
!FBAL1-FBAL8 may take negative values ;
@free (FBAL1) ; @free (FBAL2) ; @free (FBAL3) ;
@free (FBAL4) ; @free (FBAL5) ; @free (FBAL6) ;
@free (FBAL7) ; @free (FBAL8) ;

! DM1-DM8 may take negative values ;

@free (DM1) ; @free (DM2) ; @free (DM3) ; @free (DM4) ;
@free (DM5) ; @free (DM6) ; @free (DM7) ; @free (DM8) ;
@free (DM9) ;

END
```

In the above formulation, the last three sets of commands are meant to allow the variables of D1-D7, FBAL1-FBAL8, as well as DM1-DM9 to take negative values, as the default setting of LINGO assumes all variables to be non-negative values. The following is the solution report generated by LINGO:

Global optimal solution found.	
Objective value:	359554.7
Total solver iterations:	3
Variable	Value
FH	158.3467
RE1	0.000000
RE2	42.86133
D1	158.3467
D0	158.3467
FBAL1	0.000000
D2	192.8613
FBAL2	34.51465
REG1	0.000000
REG2	34.51465
D3	−207.1387
FBAL3	−400.0000
D4	142.8613
FBAL4	350.0000
D5	−457.1387
FBAL5	−600.0000
D6	0.000000
FBAL6	457.1387
D7	0.000000
FBAL7	0.000000
RJ1	0.000000
D8	8.346680
FBAL8	8.346680
RJ2	8.346680
PG	8.346680
E1	0.000000
E2	6.333867
E3	10.57682
E4	6.848320
E5	11.74846
E6	0.000000
E7	0.000000
E8	0.000000
E9	3.649168

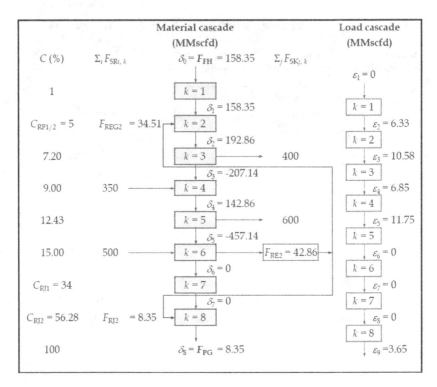

FIGURE 6.27
Result of ATM for Example 6.6.[3]

The optimization result indicates that the minimum OC is determined as 359,555 \$/day. The revised cascade diagram is shown in Figure 6.27. The results indicate that only the hydrogen source at the 15% level will be sent for purification in the PSA unit (F_{RE} = 42.86 MMscfd); the source at the 9% level will not be purified. Also, the minimum fresh hydrogen (F_{FH}) and purge flowrate (F_{PG}) corresponding to this scenario are identified as 158.35 and 8.35 MMscfd, respectively. The RCN for this example is found in Figure 1.6.

6.8 ATM for Pretreatment Networks

Even though in most RCN, fresh resources can fulfill the needs of its process sinks, there are cases where this assumption does not hold. For instance, water pretreatment units are always used in the process plant to produce dematerialized water (e.g., for boiler feed water, specialty chemical production) or

---

[3] Note that the targeting insights should not be used to drive the RCN structure here, as there are more than 1 sink in the pinch region (of different quality index).

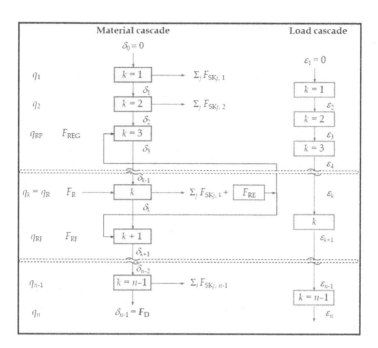

FIGURE 6.28
ATM framework for pretreatment network.

ultrapure water (e.g., semiconductor manufacturing). There are also hydrogen networks where its hydrogen sinks require higher quality of hydrogen gases than its fresh feed. In these cases, the purification units are used for purifying fresh resources before they are sent to the process sinks, which may be termed as *pretreatment network*. In most cases, partitioning purification unit is normally used in a pretreatment network. Hence, the ATM for a material regeneration network with partitioning purification unit (Section 6.7) may be extended for use here, with the revised framework given in Figure 6.28. It is observable that the latter resembles the framework in Figure 6.25. The main difference between the two frameworks is that the ATM framework for the pretreatment network in Figure 6.28 has only one purification unit, with its inlet located at the level where the fresh resource (with flowrate F_R) is found. Hence, a new sink is added to represent the potential flowrate to be sent for regeneration (F_{RE}) at this level. Also, the flowrate terms of the purified (F_{REG}) and reject (F_{RJ}) streams of the purification unit are added as sources in the ATM framework. Note also that the flowrate term of fresh resource is added at a lower level as it possesses a lower quality than some of the process sinks, and also the purified stream of the purification unit.

Most constraints in the ATM formulation remain similar to the case in Section 6.7.

Since fresh resources do not exist at the highest quality level, no net material flowrate should enter the first quality level. On the other hand, the net material flowrate that enters the last level should take non-negative value. Hence, Equation 6.8 is revised as follows:

$$\delta_{k-1} = \begin{cases} 0 & k = 1 \\ \geq 0 & k = n \end{cases} \tag{6.45}$$

Also, Equations 6.41 and 6.42 that determine the flowrate balance and allocation of the purification unit are revised as Equations 6.46 and 6.47, respectively, as only one purification unit is present in the RCN. Note that Equation 6.47 is a simplified form of Equation 6.24, which is applicable for a dilute system such as in a pretreatment network:

$$F_{RE} = F_{REG} + F_{RJ} \tag{6.46}$$

$$F_{RE}RC = F_{REG} \tag{6.47}$$

Equation 6.4 is also revised to impose a constraint where the flowrate for purification should not be greater than the fresh resource flowrate:

$$F_{RE} \leq F_R \tag{6.48}$$

Apart from the earlier-revised constraints, other constraints given by Equations 6.6, 6.7, and 6.9 remain identical as in the earlier cases.

Example 6.7 Water pretreatment network

A water pretreatment network problem (Tan et al., 2010) is analyzed, with limiting data for four water sinks given in Table 6.4. A freshwater feed with an impurity concentration of 25 ppm (C_R) is to be supplied to these sinks. A partitioning purification unit is used in the pretreatment network to purify the freshwater feed in order to fulfill sink requirements. The unit purifies the inlet stream to an impurity concentration of 5 ppm (C_{RP}) and has a recovery factor (RC) of 0.75. Determine the minimum freshwater flowrate needed for this pretreatment network.

TABLE 6.4

Limiting Data for Example 6.7

Sinks, SK$_j$	F_{SKj} (t/h)	C_{SKj} (ppm)
SK1	4	10
SR2	1	20
SR3	2	30
SR4	1	40

Source: Tan et al. (2010).

SOLUTION

With fresh concentration of 25 ppm and purified stream concentration (C_{RP}) of 5 ppm, the *QI* index for the purification unit is calculated using Equation 6.37 as 1.00 (= (1000000–5 ppm)/(1000000–25 ppm)). Equation 6.36 is next used to determine the reject stream concentration (C_{RJ}) of the purification unit (with *RC* = 0.75) as 85 ppm (= (5(0.75) – 25(1.0))/(0.75 – 1.00)). We can now construct the cascade diagram for the problem as given in Figure 6.29.

To determine the minimum freshwater flowrate (F_{FW}), the optimization objective in Equation 6.49 is used:

$$\text{Minimize } F_{FW} = \delta_0 \tag{6.49}$$

subject to the constraints that follow:

In the absence of process sources, and with a single purification unit, the water cascade across all quality levels is given by Equations 6.1 and 6.40:

$$\delta_1 = \delta_0 + F_{BAL,1}; \; F_{BAL,1} = F_{REG}$$

$$\delta_2 = \delta_1 + F_{BAL,2}; \; F_{BAL,2} = -4$$

$$\delta_3 = \delta_2 + F_{BAL,3}; \; F_{BAL,3} = -1$$

$$\delta_4 = \delta_3 + + F_{BAL,4}; \; F_{BAL,4} = F_{FW} - F_{RE}$$

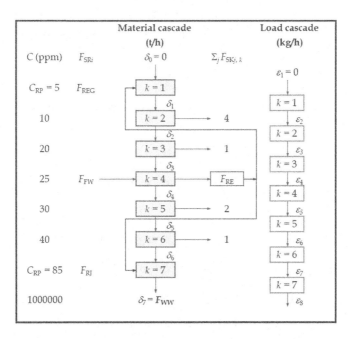

FIGURE 6.29
Cascade diagram for Example 6.7.

$$\delta_5 = \delta_4 + F_{BAL,5}; \; F_{BAL,5} = -2$$

$$\delta_6 = \delta_5 + F_{BAL,6}; \; F_{BAL,6} = -1$$

$$\delta_7 = \delta_6 + F_{BAL,7}; \; F_{BAL,7} = F_{RJ}$$

Equation 6.45 indicates that the net material flowrates that enter the first concentration level are set to zero, while those in the last level should have a value greater than zero. Also, it is useful to note that the latter corresponds to the minimum wastewater flowrate (F_{WW}) of the pretreatment network:

$$\delta_0 = 0$$

$$\delta_7 \geq 0$$

$$\delta_7 = F_{WW}$$

Equations 6.46 and 6.47 indicate the flowrate balance and allocation of the purification unit as follows (with $RC = 0.75$):

$$F_{RE} = F_{REG} + F_{RJ}$$

$$F_{RE}(0.75) = F_{REG}$$

The source flowrate to be purified should not be more than the freshwater flowrate (F_{FW}) available for the pretreatment network, according to Equation 6.48:

$$F_{RE} \leq F_{FW}$$

Impurity load cascades (in kg/h) across all quality levels are given by Equations 6.6 and 6.7 as follows:

$$\varepsilon_1 = 0$$

$$\varepsilon_2 = \varepsilon_1 + \Delta m_2; \; \Delta m_2 = \delta_1 (10 - 5)/1,000$$

$$\varepsilon_3 = \varepsilon_2 + \Delta m_3; \; \Delta m_3 = \delta_2 (20 - 10)/1,000$$

$$\varepsilon_4 = \varepsilon_3 + \Delta m_4; \; \Delta m_4 = \delta_3 (25 - 20)/1,000$$

$$\varepsilon_5 = \varepsilon_4 + \Delta m_5; \; \Delta m_5 = \delta_4 (30 - 20)/1,000$$

$$\varepsilon_6 = \varepsilon_5 + \Delta m_6; \; \Delta m_6 = \delta_5 (40 - 30)/1,000$$

$$\varepsilon_7 = \varepsilon_6 + \Delta m_7; \; \Delta m_7 = \delta_6 (85 - 40)/1,000$$

$$\varepsilon_8 = \varepsilon_7 + \Delta m_8; \; \Delta m_8 = \delta_7 (1,000,000 - 85)/1,000$$

Finally, all residual loads (except at level 1) should take non-negative values (Equation 6.9):

$$\varepsilon_2 \geq 0 \qquad \varepsilon_3 \geq 0$$

$$\varepsilon_4 \geq 0 \qquad \varepsilon_5 \geq 0$$

$$\varepsilon_6 \geq 0 \qquad \varepsilon_7 \geq 0$$

$$\varepsilon_8 \geq 0$$

The LINGO formulation for this LP model is written as follows:

```
Model
! Optimization objective: to minimize freshwater flowrate
(FW) ;
min = FW ;
! Water cascade ;
D1 = D0 + FBAL1 ; FBAL1 = REG ;
D2 = D1 + FBAL2 ; FBAL2 = - 4 ;
D3 = D2 + FBAL3 ; FBAL3 = - 1 ;
D4 = D3 + FBAL4 ; FBAL4 = FW - RE ;
D5 = D4 + FBAL5 ; FBAL5 = - 2 ;
D6 = D5 + FBAL6 ; FBAL6 = - 1;
D7 = D6 + FBAL7 ; FBAL7 = RJ ;

! No net flowrate enters the first level, net flowrate that
enters the last level should have non-negative value;

D0 = 0 ;
D7 >= 0;

! The last net material flowrates corresponds to the minimum
wastewater (WW) flowrate ;

D7 = WW ;

! Flowrate balance and allocation for purification unit ;

RE = REG + RJ ;
RE *0.75 = REG ;

! Flowrate constraints of regenerated source ;

RE <= FW ;

! Impurity load cascade ;

E1 = 0 ;
E2 = E1 + DM1 ; DM1= D1* (10 - 5)/1000 ;
E3 = E2 + DM2 ; DM2= D2*(20 - 10)/1000 ;
E4 = E3 + DM3 ; DM3= D3* (25 - 20)/1000 ;
E5 = E4 + DM4 ; DM4= D4*(30 - 25)/1000 ;
E6 = E5 + DM5 ; DM5= D5* (40 - 30)/1000 ;
E7 = E6 + DM6 ; DM6= D6*(85 - 40)/1000 ;
E8 = E7 + DM7 ; DM7= D7* (1000000 - 85)/1000 ;

! E2-E8 should take non-negative values ;

E2 >= 0 ; E3 >= 0 ; E4 >= 0 ; E5 >= 0 ;
E6 >= 0 ; E7 >= 0 ; E8 >= 0 ;
```

```
! D1-D6 may take negative values ;

@free (D1) ; @free (D2) ; @free (D3) ;
@free (D4) ; @free (D5) ; @free (D6) ;

! FBAL1-FBAL7 may take negative values ;

@free (FBAL1) ; @free (FBAL2) ; @free (FBAL3) ;
@free (FBAL4) ; @free (FBAL5) ; @free (FBAL6) ; @free (FBAL7) ;

! DM1-DM7 may take negative values ;

@free (DM1) ; @free (DM2) ; @free (DM3) ;
@free (DM4) ; @free (DM5) ; @free (DM6) ; @free (DM7) ;
```

END

Similar to earlier cases, the non-negativity constraints are excluded here, as LINGO assumes all variables take non-negative values by default. The following is the solution report generated by LINGO:

Global optimal solution found.	
Objective value:	8.666667
Total solver iterations:	0
Variable	Value
FW	8.666667
D1	3.250000
D0	0.000000
REG	3.250000
D2	−0.7500000
D3	−1.750000
D4	2.583333
RE	4.333333
D5	0.5833333
D6	−0.4166667
D7	0.6666667
RJ	1.083333
WW	0.6666667
E1	0.000000
E2	0.1625000E-01
E3	0.8750000E-02
E4	0.000000
E5	0.1291667E-01
E6	0.1875000E-01
E7	0.000000
E8	666.6100

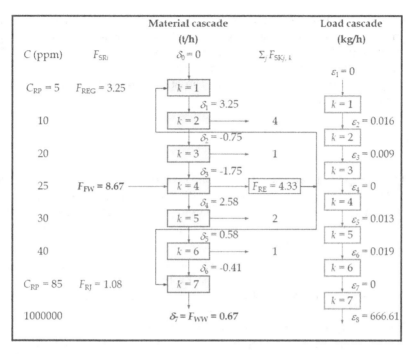

FIGURE 6.30
Result of ATM for Example 6.7.

The optimization result indicates that the minimum freshwater flow-rate needed for the pretreatment network is 8.67 t/h. The revised cascade diagram is shown in Figure 6.30. The results show that 4.33 t/h of the freshwater feed is sent for purification in the partitioning unit (F_{RE}). In other words, 4.34 t/h (= 8.67 − 4.33 t/h) of freshwater will bypass the purification unit. With the *RC* value of 0.75, the purification unit produces a purified stream with a flowrate of 3.25 t/h (F_{REG}) at 5% level, while a flowrate of 1.08 t/h is found in the reject stream (F_{RJ}). Also, the wastewater flowrate (F_{WW}) is determined as 0.67 t/h.

6.9 Additional Readings

For material regeneration networks that employ RR type purification units (single pass or partitioning units), there may be a possibility that more than one purification unit is to be used, as a result of the ATM. For cases where only one purification unit is desired (e.g., due to space consideration), additional constraints may be added to the ATM. In most cases, the ATM will become a nonlinear program (NLP). Readers can refer to the work of Ng et al. (2009a) for an extended model of this case.

Apart from ATM, regeneration targeting may also be performed using graphical (Agrawal and Shenoy, 2006; Bai et al., 2007) and algebraic approaches (Ng et al., 2007, 2008). One may also extend the targeting methods to cater for *total material network*, where waste treatment is considered along with material recovery. Detailed procedure for the above are also found in the first edition of this book (Foo, 2012).

PROBLEMS
Water Minimization Problems

6.1 The hypothetical case study from Problems 3.4 and 4.12 (Polley and Polley, 2000) is revisited, with limiting water data given in Table 3.11. Determine the minimum freshwater and regeneration flowrate targets with ATM when the following purification unit is used for regeneration-reuse/recycle scheme.

   a. A single pass purification unit with fixed outlet concentration (C_{Rout}) of 10 ppm (see answers in Problem 8.1).

   b. A single pass purification unit with C_{Rout} of 25 ppm. Compare the result with that obtained in (a).

6.2 Rework Example 6.2 (Wang and Smith, 1994), assuming that the unit cost for freshwater is $1/ton ($CT_{FW}$) and that for the regenerated water (CT_{RW}) are given in terms of impurity load removal, i.e., $0.8/kg impurity load removed. Impurity load removal for the purification unit r (Δm_r) is given as in Equation 6.50:

$$\Delta m_r = F_{RE,r}\left(C_{Rin,r} - C_{Rout,r}\right) \forall r \tag{6.50}$$

   Use the insights from the ATM to identify the structure of the RCN.

6.3 For the coal mine water system in Problem 5.7 (Jia et al., 2015), a single pass purification unit is used for water regeneration, with outlet concentration of 6.7 ppm (C_{Rout}). Data for the case study is given in Table 5.6. Determine the minimum freshwater and regeneration flowrates (see answers in Problem 8.13).

6.4 A water minimization study is carried out for an alumina plant (Deng and Feng, 2009), with limiting water data in Table 6.5. A single pass purification unit ($C_{Rout} = 20$ ppm) is used for water regeneration, so to reduce the overall freshwater and wastewater flowrates. Determine the possibility of achieving zero water discharge for the water network of this plant.

6.5 For the specialty chemical production process in Problem 4.6 (limiting data given in Table 4.22), perform the following tasks:

   a. Determine the freshwater and wastewater flowrate targets when a purification unit with $C_{Rout} = 60$ ppm is used (see answers in Problem 8.7).

TABLE 6.5

Limiting Water Data for Alumina Plant (Deng and Feng, 2009) in Problem 6.4

Sinks, SK_j	F_{SKj} (t/h)	C_{SKj} (ppm)	Sources, SR_i	F_{SRi} (t/h)	C_{SRi} (ppm)
SK1	40	20	SR1	30	0
SK2	30	60	SR2	25	100
SK3	200	20	SR3	150	80
SK4	160	100	SR4	80	200
SK5	60	20	SR5	40	60
SK6	150	20	SR6	120	80
SK7	70	80			

   b. Redo task a for another purification unit with C_{Rout} = 200 ppm (Deng et al., 2011; see answers in Problem 8.7).
   c. Do a comparison of the flowrate reduction for cases in (a) and (b) with that of direct reuse/recycle scheme (note: minimum flowrate targets for direct reuse/recycle scheme are found by solving Problem 4.6).

6.6 For the water minimization example from Sorin and Bédard (1999), with data given in Table 5.5. Determine the following:

   a. Minimum freshwater and wastewater flowrate targets when an purification unit with C_{Rout} = 10 ppm is used (see answers in Problem 8.12).
   b. Do a comparison of the flowrate reduction for cases in (a) with that of the direct reuse/recycle scheme (note: minimum flowrate targets for the direct reuse/recycle scheme are found by solving Problem 5.3).
   c. Identify reason why regeneration unit with $C_{Rout} \geq 100$ ppm should not be used for water regeneration for this case.

6.7 The paper milling process in Problem 4.5 is revisited here. A dissolved air flotation unit (DAF) is employed to remove the total suspended solid (TSS) content in the water sources, which is the most significant quality factor for water recovery in this process. The DAF unit takes the form of a partitioning purification unit, with a water recovery factor (RC) of 0.98 and TTS RR of 0.9 (Ng et al., 2009a). The unit costs for fresh and regenerated water are taken as $1/ton and $0.75/ton, respectively. Determine the minimum OC for this RCN, which is mainly contributed by fresh and regenerated water. Use the limiting water data in Table 4.21.

6.8 Rework the alkene production problem in Example 4.13 with regeneration-reuse/recycle scheme. Solve the following cases and observe the operating flowrates of the sinks (see answers in Problem 8.21)

   a. A partitioning purification unit with outlet concentration of 0.8% and recovery factor of 0.8 is used. Determine the minimum flowrates for fresh water, regenerated water and wastewater with two-stage optimization approach. To simplify the model, the reject stream from purification unit will not be recovered to the sinks (i.e., discarded as wastewater).

b. A single pass purification unit with RR of 0.8 is used. Determine the minimum cost solution when unit costs of fresh water and wastewater are both assumed as 1.0 $/t, while that of regeneration is taken as 1.2 $/t. Determine also what if the regeneration unit cost will rise to 1.5 or 2.0 $/t.

6.9 Rework the water minimization problem of the specialty chemical production case in Problem 4.6. For this case, the OC for using two purification units are to be compared. Unit A has an outlet concentrations of 60 ppm, while that of unit B is 200 ppm. The unit costs of fresh water and wastewater are given as 1 $/t, respectively, while the unit costs of purification units are given in the following scenarios:

a. Scenario 1—unit cost is a function of regeneration flowrate, i.e., 1.0 $/t for unit A, and 0.5 $/t for unit B.

b. Scenario 2—unit cost is a function of impurity removal (Δm), i.e., 1.0 $/kg Δm for unit A, and 0.5 $/kg Δm for unit B.

6.10 Rework the water minimization problem of the Kraft pulping process in Problems 2.8, 4.10 and 5.10 (Parthasarathy and Krishnagopalan, 2001). The freshwater resource has a chloride content of 4.2 ppm. Unit costs of fresh water (CT_{FW}) and wastewater (CT_{WW}) are given as $1/t and $5/t, respectively. In this case, two types of purification units may be used for water regeneration, both with outlet concentration (C_{Rout}) of 10 ppm. Use ATM to compare both types of purification units.

a. Purification unit A—with unit cost (CT_{REG}) of $1/t of regenerated flowrate (see solution in Problem 8.20).

b. Purification unit B—with unit cost (CT_{REG}) of $1/kg of chloride removal.

6.11 Rework the tricresyl phosphate manufacturing process (El-Halwagi 1997) in Problems 2.3 and 4.2. To further reduce fresh water and wastewater flowrates, regeneration-reuse/recycle scheme is considered. Two types of purification units may be used for water regeneration, i.e., oil treatment and light gas treatment. The former has an outlet concentration of 40 ppm (C_{Rout}), and unit treatment cost (CT_{REG}) of 116.6 $/kg cresol. On the other hand, the light gas treatment can purify water sources to better quality of 3 ppm (C_{Rout}); however, it has a higher unit treatment cost (CT_{REG}) of 434.7 $/kg cresol. Note that the impurity load in unit treatment cost (CT_{REG}) may be calculated from the product of regeneration flowrate and differences of its inlet and outlet concentrations. Determine which treatment unit should be employed in this case. Unit costs for freshwater (CT_{FW}) and wastewater (CT_{WW}) are given as $1.5/t and $30/t, respectively (see final answers in Problem 8.18).

6.12 For the brick manufacturing process in Problem 3.9 (Skouteris et al., 2018), it is desired to implement regeneration-reuse/recycle scheme by purifying the water sources. Assuming a purification unit of 10 ppm (C_{Rout}) is available, determine the minimum flowrates of freshwater and regeneration flowrates with the two-stage optimization approach (see final answers in Problem 8.19).

6.13 Rework the water minimization problem of the Kraft pulping process in Problems 4.1 and 5.8 (El-Halwagi, 1997). Solve the following cases.

a. Use Excel/LINGO to determine the minimum freshwater and wastewater flowrates for water regeneration network when an purification unit with C_{Rout} = 10 ppm is used (see answers in Problem 8.15).

b. Resolve (a) for another purification unit with C_{Rout} = 15 ppm (see answers in Problem 8.15).

c. Air stripping is used as a purification unit to regenerate water sources for further recovery, by removing its methanol content. The equilibrium relation for stripping of methanol using this mass exchange unit is given as follows:

$$C_{SRi} = 0.38 \ C_{MSA} \tag{6.51}$$

where C_{MSA} is the impurity (methanol) concentration in the MSA (stripping air). A minimum allowable concentration difference of 4.275 ppm should be incorporated to operate the mass exchange unit (El-Halwagi, 1997). It may be assumed that the ambient air is used as the MSA in this case, and hence has an inlet concentration of 0 ppm.

The operating (OC_{AS}) of the stripping unit are given as follows (El-Halwagi, 1997; Ng et al., 2010):

$$OC_{AS} (\$/h) = 1.5 \ F_{AS} \ RR_{AS} \tag{6.52}$$

Note that the operating cost of the stripping unit is a function of its regeneration flowrate (F_{AS}), as well as the RR of the methanol load from the water sources (RR_{AS}). Note further that only one stripping unit is to be used; hence, the RR of the stripping unit is determined by the following equation:

$$RR_{AS} = \frac{\Sigma_r \ F_{RE,r} \ C_{RE,r} - F_{REG} \ C_{REG}}{\Sigma_r \ F_{RE,r} \ C_{RE,r}} \tag{6.53}$$

Note that the unknown terms in Equation 6.53 lead to an NLP of the model.

Use LINGO to determine the minimum OC of the RCN, which consists of the freshwater and wastewater costs, as well as the operating costs of the purification unit. The unit costs of freshwater and wastewater are both taken as $1/ton. .

Utility Gas Recovery Problems

6.14 For the refinery hydrogen network in Problems 4.19 and 5.13 (Alves and Towler, 2002), a gas membrane unit with a hydrogen recovery factor (RC) of 0.95 is used for source regeneration. The membrane unit produces a purified stream of 2 mol % impurity (C_{RP}). The unit costs for fresh and regenerated hydrogen are given as 1.6×10^{-3}/mol and 1.0×10^{-3}/mol, respectively. The limiting data are given in Table 4.31. Determine the minimum OC for the RCN.

6.15 Consider the magnetic tape process in Problem 3.10. Three MSAs are available to purify nitrogen sources for further recovery. Data for the MSAs are given in Table 6.6 (El-Halwagi, 1997). The unit cost for fresh nitrogen gas (CT_{N2}) is given as $0.20/kg. Determine the minimum OC for the RCN, which is mainly contributed by the fresh and regenerated nitrogen sources.

6.16 For the refinery hydrogen network (Umana et al., 2014) in Problems 2.10 and 4.18, a purification unit installed for regeneration-reuse/recycle scheme, which will have an outlet impurity of 5% and recovery factor of 0.9 (may be approximated with its simplified form in Equation 6.46). Due to pressure constraints, the hydrogen sources are not allowed to mix. In other words, only one hydrogen source will be purified using the newly installed purification unit. Determine which source should be sent for purification. The unit costs for fresh hydrogen and the intercepted sources are given as $1.5/t and $0.8/t, respectively (purge stream to fuel system may be ignored; see answer in Problem 8.27).

6.17 For the hydrogen network of an aromatic plant (Shariati et al., 2013) in Problem 2.12, use ATM to determine the minimum flowrates of fresh hydrogen to be generated in the catalytic reformer, as well as

TABLE 6.6

Data of MSAs for Magnetic Tape Process (El-Halwagi, 1997) in Problem 6.15

MSA*l*	Inlet Concentration, $C_{MSA,in}$ (wt%)	Outlet Concentration, $C_{MSA,out}$ (wt%)	α_1	Minimum Driving Force, ξ_1 (wt%)	Unit Cost, $CT_{MSA l}$ ($/kg MSA)
1	0.014	0.040	0.4	0.001	0.002
2	0.020	0.080	1.5	0.001	0.001
3	0.001	0.010	0.1	0.001	0.002

TABLE 6.7

Limiting Data for Hydrogen Network in Problem 6.17

SK_j	Process	F_{SKj} (mol/s)	C_{SKj} (mol%)	SR_i	Process	F_{SRi} (mol/s)	C_{SRi} (mol%)
SK1	AF	40.0	8.55	SR1	AF	50.0	40.00
SK2	HT	406.2	7.67	SR2	HT	369.7	7.58
SK3	HG	209.1	13.37	SR3	HG	157.5	14.96
SK4	RG1*	28.0	8.55	SR4	TA	1626.3	33.00
SK5	RG2*	53.0	0.10	SR5	IS	4266.7	49.69
SK6	TA	1709.8	30.00	SR6	DP	1399.4	20.34
SK7	IS	4308.5	49.41				
SK8	DP	1421.0	20.00				

Source: Shariati et al. (2013).
Note: Feed streams to regeneration (RG) are to be kept as per existing network, while it outlet stream is excluded from recycling (i.e., not taken as source).

the hydrogen flowrate to be purified by a PSA unit. Note that the PSA unit will only purify hydrogen gas from the catalytic reformer but no other hydrogen-consuming processes. The fresh hydrogen generated by the catalytic reformer has an impurity of 8.55 mol%. The PSA unit has a recovery factor of 0.682, and its product and reject streams have impurity values of 0.10 and 22.00 mol%, respectively. As the hydrogen flowrates of catalytic reformer and PSA unit are to be minimized, they are not extracted as part of process sinks and sources. Limiting data of hydrogen sinks and sources are given in Table 6.7.

Property Integration Problems

6.18 The RCN problem of palm oil milling process in Example 5.4 is revisited. A filtration unit is installed to purify the bottom flow of the clay bath that is currently sent to wastewater treatment. The filtration unit is modeled as a partitioning purification unit of the fixed outlet quality type. It purifies the bottom flow to 1100 kg/m³, and produces a reject stream of 1600 kg/m³ (Ng et al., 2009b). However, to avoid accumulation of impurity load, 5% of wastewater flowrate is purged from the bottom stream, as shown in Figure 6.31. Besides, the reject stream of the filtration unit will not be recycled (i.e., sent as wastewater discharge). The revised limiting data are shown in Table 6.8 (reproduced from Table 13.4). Determine the minimum OC of the RCN, given that the unit costs for clay, freshwater, wastewater, and regenerated water are $0.25/kg, $0.001/kg, $0.0008/kg, and $0.0008/kg, respectively (Ng et al., 2010).

TABLE 6.8

Limiting Data for Palm Oil Milling Process (Ng et al., 2009b) in Problem 6.18

Sink, SK$_j$	F_{SKj} (kg/day)	Density, ρ (kg/m3)	$\psi_{SKi} \times 10^{-4}$ (m³/kg)
Clay solution (SK1)	10.0	1120	8.93
Source, SR$_i$	**F_{SRi} (kg/day)**	**Density, ρ (kg/m³)**	**$\psi_{SKj} \times 10^{-4}$ (m³/kg)**
Overflow (SR1)	3.8	1018	9.82
Bottom flow (SR2)	5.7	1200	8.33
Freshwater (FW)	To be determined	1000	10
Clay (CL)	To be determined	2600	3.85

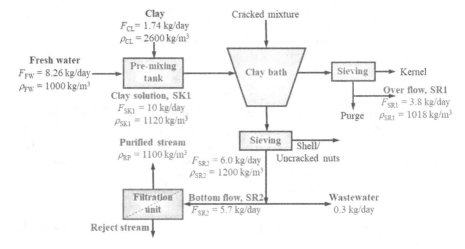

FIGURE 6.31

RCN for clay bath operation in palm oil milling process in Problem 6.18. (From Ng et al., 2009b.)

6.19 Revisit the wafer fabrication process in Problems 4.23 and 5.15. In order to enhance water recovery, a membrane treatment unit is installed. The membrane unit purifies water sources to a permeate quality (purified stream) of 15 MΩ m (hint: convert this to operator using Equation 2.85). The membrane unit is treated as a partitioning unit, with a water recovery factor of 0.95.

a. When water minimization is only considered for the wafer fabrication section (FAB), determine the minimum UPW and regeneration flowrates. Use the limiting data in Table 4.23a.

b. To consider water recovery in the entire plant, a minimum OC solution is to be identified. The unit costs for ultrapure (UPW), fresh, wastewater, and regenerated water are taken as $2/t, $1/t, $0.5/t, and $1.5/t, respectively (Ng et al., 2010). Use the limiting data in Table 4.23b.

TABLE 6.9

Limiting Data for Water Pretreatment Network
Problem in Problem 6.21

Sinks, SK_j	F_{SK_j} (t/h)	R_{SK_j} (MΩ m)	ψ_{SK_j} (MΩ m)$^{-1}$
SK1	20	5	0.2
SR2	10	3.33	0.3
SR3	10	0.05	20
SR4	2	0.017	60

Pretreatment Network Problems

6.20 Rework the water pretreatment network problem in Example 6.7, when
a partitioning purification unit with an *RR* of 0.9 is used. The purification unit produces a reject stream of 100 ppm. The purified stream
quality is unknown; however, it is expected to be of a higher quality
than all the sinks (note: this leads to the NLP model). Determine the
flowrate targets for freshwater, inlet, purified, and reject streams of the
purification unit, as well as the water recovery factor (*RC*) of the latter
(hint: since the purified stream concentration is an unknown, it is best
to include Equation 6.34 as a constraint in the ATM).

6.21 A pretreatment network is used to produce demineralized water for a
manufacturing process (Tan et al., 2010). Water quality is characterized
by electrical resistivity, *R* (MΩ m), which is an index that reflects the total
ionic content in the aqueous streams. The mixing rule of this property is
given by Equation 2.85. The freshwater supply has a resistivity value of
0.1 MΩ m. The pretreatment network uses multistage membrane filtration units as purification unit, which produces dematerialized water at a
fixed resistivity of 10 MΩ m and possesses a recovery factor (*RC*) of 0.8.
Limiting data for the water sinks are given in Table 6.9. Solve the following tasks:

a. Determine the resistivity operator of the reject stream (hint: make
use of Equation 2.85).

b. Determine the flowrate targets of the pretreatment network.

References

Agrawal, V. and Shenoy, U. V. 2006. Unified conceptual approach to targeting and
design of water and hydrogen networks, *AIChE Journal*, 52(3), 1071–1081.

Alves, J. J. and Towler, G. P. 2002. Analysis of refinery hydrogen distribution systems,
Industrial and Engineering Chemistry Research, 41, 5759–5769.

Bai, J., Feng, X., and Deng, C. 2007. Graphical based optimization of single-contaminant regeneration reuse water systems, *Chemical Engineering Research and Design*, 85(A8), 1178–1187.

Deng, C. and Feng, X. 2009. Optimal Water Network with Zero Wastewater Discharge in an Alumina Plant. In Proceedings of the 4th IASME/WSEAS international conference on Energy & environment; EE'09; World Scientific and Engineering Academy and Society (WSEAS): Cambridge, UK, 2009; pp 109–114.

Deng, C., Feng, X., Ng, D. K. S., and Foo, D. C. Y. (2011). Process-based graphical approach for simultaneous targeting and design of water network. *AIChE Journal*, 57(11): 3085–3104.

El-Halwagi, M. M.1997. *Pollution Prevention Through Process Integration: Systematic Design Tools.* San Diego, CA: Academic Press.

Foo, D. C. Y.2012. *Process Integration for Resource Conservation.* Boca Raton, Florida: CRC Press., US (Chapters 6 and 8).

Gabriel, F. B. and El-Halwagi, M. M. 2005. Simultaneous synthesis of waste interception and material reuse networks: Problem reformulation for global optimization, *Environmental Progress*, 24(2), 171–180.

Hallale, N. and Liu, F. 2001. Refinery hydrogen management for clean fuels production, *Advances in Environmental Research*, 6, 81–98.

Jia, X., Li, Z., Wang, F., Foo, D. C. Y., and Tan, R. R. 2015. Integrating input-output models with pinch technology for enterprise sustainability analysis. *Clean Technologies and Environmental Policy*, 17(8): 2255–2265.

Ng, D. K. S., Foo, D. C. Y., Tan, R. R., and Tan, Y. L. 2007. Ultimate flowrate targeting with regeneration placement, *Chemical Engineering Research and Design*, 85(A9), 1253–1267.

Ng, D. K. S., Foo, D. C. Y., Tan, R., and Tan, R. 2008. Extension of targeting procedure for 'Ultimate flowrate targeting with regeneration placement' by Ng et al., *Chemical Engineering Research and Design*, 85(A9), 1253–1267.; *Chemical Engineering Research and Design*, 86(10), 1182–1186.

Ng, D. K. S., Foo, D. C. Y., and Tan, R. R. 2009a. Automated targeting technique for single-component resource conservation networks—Part 2: Single pass and partitioning waste interception systems, *Industrial and Engineering Chemistry Research*, 48(16), 7647–7661.

Ng, D. K. S., Foo, D. C. Y., Tan, R. R., Pau, C. H., and Tan, Y. L. 2009b. Automated targeting for conventional and bilateral property-based resource conservation network, *Chemical Engineering Journal*, 149, 87–101.

Ng, D. K. S., Foo, D. C. Y., Tan, R. R., and El-Halwagi, M. M. 2010. Automated targeting technique for concentration- and property-based total resource conservation network, *Computers and Chemical Engineering*, 34(5), 825–845.

Parthasarathy, G. and Krishnagopalan, G. 2001. Systematic reallocation of aqueous resources using mass integration in a typical pulp mill, *Advances in Environmental Research*, 5, 61–79.

Shariati, M., Tahouni, N., and Panjeshahi, M. H. 2013. Investigation of different approaches for hydrogen management in petrochemical complexes, *International Journal of Hydrogen Energy*, 38, 3257–3267.

Skouteris, G., Ouki, S., Foo, D. C. Y., Saroj, D., Altini, M., Melidis, P., Cowley, B., Ells, G., Palmer, S., and O'Dell, S. 2018. Water footprint assessment and water pinch analysis techniques for sustainable water management in the brick-manufacturing industry, *Journal of Cleaner Production*, 172: 786–794.

Sorin, M. and Bédard, S. 1999. The global pinch point in water reuse networks. *Transactions of IChemE Part B*, **77**: 305–308.

Tan, R. R., Ng, D. K. S., Foo, D. C. Y., and Aviso, K. B. 2009. A superstructure model for the synthesis of single-contaminant water networks with partitioning regenerators, *Process Safety and Environmental Protection*, 87, 197–205.

Tan, R. R., Ng, D. K. S., and Foo, D. C. Y. 2010. Graphical approach to minimum flowrate targeting for partitioning water pretreatment unit, *Chemical Engineering Research and Design*, 88(4), 393–402.

Umana, B., Shoaib, A. Zhang, N., and Smith, R. 2014. Integrating hydroprocessors in refinery hydrogen network optimisation. *Applied Energy*, 133: 169–182.

Wang, Y. P. and Smith, R. 1994. Wastewater minimisation, *Chemical Engineering Science*, 49, 981–1006.

7

Process Changes for Resource Conservation Networks

In the previous chapters, we have seen how graphical, algebraic, and automated approaches can be used to determine minimum fresh resource, regeneration and waste discharge targets for a resource conservation network (RCN). These minimum targets can be further reduced if the operating conditions of the processes (e.g., flowrate, concentration, and temperature) may be altered. In this chapter, we shall see how process changes may lead to reduced flowrates of fresh resource and/or waste discharge for an RCN.

7.1 Plus–Minus Principle

The *plus–minus principle* is an established technique for heat exchanger network synthesis (Linnhoff et al., 1982[1994]; Smith, 1995, 2005). The same principle applies to RCN equally well. Feng et al. (2009) first extended this principle for water minimization cases where water is used as mass-separating agent in mass transfer operation (fixed load problem). Foo (2012) later generalized the principles for a wider range of RCN problems, including those for water minimization, hydrogen network, and property integration.

As discussed in earlier chapters, an RCN may be divided into higher (HQR) and lower quality (LQR) regions by the pinch point. This is shown from the perspective of *material recovery pinch diagram* (MRPD) and *material cascade analysis* (MCA) or *automated targeting model* (ATM)[1] in Figure 7.1. As shown, fresh resource is used in the HQR, as it is a region that experiences material deficit. On the other hand, as there is surplus material in the LQR, the excess material will be discarded as waste from the LQR. In principle, any attempt to maximize the use of process sources in the process sinks will lead to reduced fresh resource and/or waste discharge for the RCN.

To reduce fresh resource consumption for an RCN, the following process change strategies shall be implemented in the HQR:

1. Decrease/eliminate the flowrates of process sinks.
2. Reduce the quality requirement of process sinks.

---

[1] Examples in this chapter involve the use of these targeting methods. Readers are advised to have prior knowledge by reading Chapters 3–6.

DOI: 10.1201/9781003173977-8

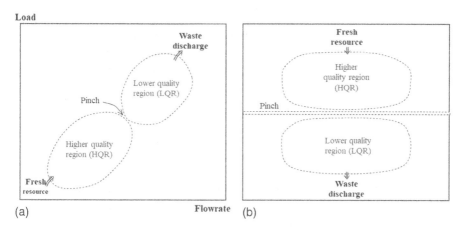

FIGURE 7.1
The higher quality region (HQR) and lower quality region (LQR) of an RCN, based on perspective of (a) MRPD; (b) MCA and ATM.

3. Improve the quality of process sources with/without increasing the flowrate of process sources.

    Note that reduced fresh resource consumption in the HQR may lead to reduced/increased waste discharge from the RCN, depending on the case-specific scenario (see examples that follow). On the other hand, if the intention was only to reduce waste discharge in the LQR (e.g., for cases where fresh resources have been reduced extensively), the following process change strategies can be implemented:

4. Increase the flowrates of process sinks in the LQR, with/without decreasing the quality levels of process sinks.

5. Decrease/eliminate the flowrates of process sources in the LQR, with/without increasing the quality levels of process sources.

6. Improve the quality of process sources in the HQR, with or without increasing the flowrate of process source(s).

Note that points (3) and (6) are essentially the same. In other words, improving the quality levels of process sources in the HQR (this includes the pinch-causing source) leads to a simultaneous reduction in fresh resources and waste discharge of the RCNs (see Figure 7.2). Hence, this point is taken as the highest priority for process change implementation. Note also that strategies 3, 4, and 6 are essentially the "plus principles"; while strategies 1, 2, and 5 correspond to "minus principles".

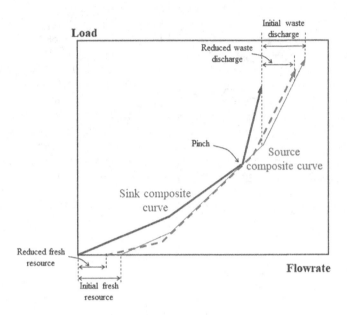

FIGURE 7.2
Decrease in the quality level of process source in the HQR leads to a simultaneous reduction in fresh resources and waste discharge.

Example 7.1 Solvent recovery for metal degreasing process

The metal degreasing process in Examples 2.5 and 3.7 is revisited here. The Reid vapor pressure (RVP) for Condensate I (SR1), i.e., RVP_{SR1}, is a function of the operating temperature (T, given in kelvin) of the thermal processing unit (Kazantzi and El-Halwagi, 2005), given as

$$RVP_{SR1} = 0.56e^{\left(\frac{T-100}{175}\right)} \tag{7.1}$$

The acceptable operating temperature range is between 430 and 520 K (Kazantzi and El-Halwagi, 2005). The thermal processing unit currently operates at 515 K, leading to an RVP of 6.0 atm. This operating condition can be modified to enable better solvent recovery between SR1 and the process sinks.

Determine the fresh solvent and solvent waste that can be further reduced for the RCN of this process. Identify the strategies associated with this process change.

SOLUTION

The MRPD for the direct solvent reuse/recycle case is shown in Figure 7.3 (reproduced from Figure 3.19). Both the fresh solvent (F_{FS}) and solvent waste (F_{WS}) for the RCN are targeted as 2.4 kg/s each. It is also observed that SR1 is the pinch-causing source that belongs to both the HQR and LQR. Hence, process modification for this source will lead to the

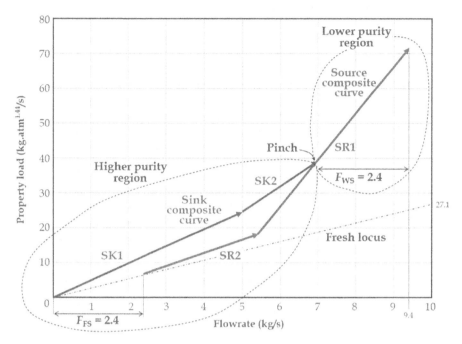

FIGURE 7.3
MRPD for direct solvent reuse/recycle case in Example 7.1. (From Kazantzi and El-Halwagi, 2005. Reproduced with permission. Copyright 2005 American Institute of Chemical Engineers [AIChE].)

simultaneous reduction in fresh solvent and solvent waste. This corresponds to points (3) and (6) of the process change strategies, i.e., improve the quality of process sources in the HQR, with or without increasing the flowrate of process source.

Based on Equation 7.1, the operating temperature range of the thermal processing unit of 430 and 520 K corresponds to the RVP values of 3.69 and 6.17 atm, respectively. Note that the property operator (ψ) of the RVP is given as (reproduced from Equation 2.20)

$$\psi = RVP_i^{1.44} \tag{7.2}$$

Following Equation 7.2, the RVP values of 3.69 and 6.17 atm are converted to the property operator (ψ) values of 6.56 and 13.74 atm$^{1.44}$, respectively. Obviously, for the case with fresh resource of superior operator value (see Section 2.5 for discussion of superior and inferior property operators of fresh resource), sources with a lower operator value are considered a better-quality feed. Figure 7.4 shows the MRPD with SR1 of $\psi = 6.56$ atm$^{1.44}$. This leads to the complete removal of both fresh solvent requirement and solvent waste, resembling the special case of *threshold problem*, as discussed in Chapters 3 and 4. The process that fulfills the recovery targets is shown in Figure 1.4.

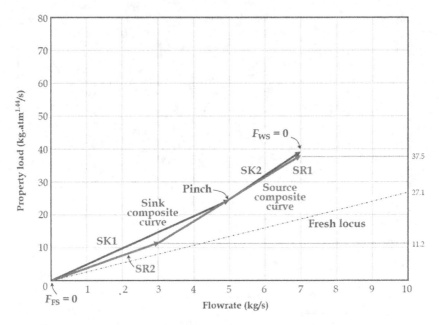

FIGURE 7.4
MRPD for Example 7.1 after process changes.

Example 7.2 Water minimization (Polley and Polley, 2000)

The classical water minimization example of Polley and Polley (2000) is analyzed here, with data shown in Table 7.1 (reproduced from Table 3.11). Evaluate if strategies 1 and 5 may be used to reduce the flowrates of freshwater and/or wastewater of the water network.

SOLUTION

Applying the algebraic targeting technique of MCA identifies that the freshwater and wastewater flowrates as 70 and 50 t/h, respectively, while the pinch concentration is reported as 150 ppm (see Figure 7.5). It may also be observed that process sinks SK1, SK2, and SK3, as well as process sources SR1 and SR2 are located in the HQR (with concentration lower

TABLE 7.1

Limiting Data for Example 7.2

j	Sinks, SK_j	F_{SKj} (t/h)	C_{SKj} (ppm)	i	Sources, SR_i	F_{SRi} (t/h)	C_{SRi} (ppm)
1	SK1	50	20	1	SR1	50	50
2	SK2	100	50	2	SR2	100	100
3	SK3	80	100	3	SR3	70	150
4	SK4	70	200	4	SR4	60	250

C_i (ppm)	ΔC_i (ppm)	$\Sigma_j F_{SKj}$ (t/h)	$\Sigma_j F_{SRj}$ (t/h)	$\Sigma_j F_{SRj}-\Sigma_j F_{SKj}$ (t/h)	$F_{C,i}$ (t/h)	Δm_i (kg/h)	Cum. Δm_i (kg/h) [Infeasible cascade]	$F_{PW,i}$ (t/h)	$F_{C,i}$ (t/h)	Δm_i (kg/h)	Cum. Δm_i (kg/h) [Feasible cascade]
0					0				70 (Freshwater)		
	20					0				1.4	
20		50 (SK1)		−50	0		0		70		1.4
	30					−1.5				0.6	
50		100 (SK2)	50 (SR1)	−50	−50		−1.5	−30	20		2.0
	50					−5				−1.5	
100		80 (SK3)	100 (SR2)	20	−100		−6.5	−65	−30		0.5
	50					−4				−0.5	
150			70 (SR3)	70	−80		−10.5	−70	−10		0.0 (Pinch)
	50					−0.5				3.0	
200		70 (SK4)		−70	−10		−11	−55	60		3.0
	50					−4				−0.5	
250			60 (SR4)	60	−80		−15	−60	−10		2.5
	999750					−19995				49987.5	
1000000					−20		−20010	−20.01	50 (Wastewater)		49990.0

FIGURE 7.5
MCA for direct water reuse/recycle scheme for Example 7.2.

than 150 ppm), while other sinks and sources are in the LQR. Note however that SR3 is exceptional as it is the pinch-causing source that belongs to both regions.[2]

We next evaluate strategy 1 (decrease/eliminate the flowrates of process sinks) for sinks SK1–SK3 of this example. Assuming that 30 t/h of water flowrate will be reduced from SK1, re-perform the targeting with MCA identifies that the flowrate of freshwater is reduced further to 44 t/h, but the wastewater flowrates are increased to 54 t/h (see Figure 7.6).

If the same amount of flowrate reduction (i.e., 30 t/h) is implemented on SK2, the MCA identifies that the flowrates of freshwater and wastewater are identified as 50 and 60 t/h (see Figure 7.7), i.e., higher than those when the same flowrate is reduced on SK1.

C_i (ppm)	ΔC_i (ppm)	$\Sigma_j F_{SKj}$ (t/h)	$\Sigma_j F_{SRj}$ (t/h)	$\Sigma_j F_{SRj}-\Sigma_j F_{SKj}$ (t/h)	$F_{C,i}$ (t/h)	Δm_i (kg/h)	Cum. Δm_i (kg/h) [Infeasible cascade]	$F_{PW,i}$ (t/h)	$F_{C,i}$ (t/h)	Δm_i (kg/h)	Cum. Δm_i (kg/h) [Feasible cascade]
0					0				44 (Freshwater)		
	20					0				0.9	
20		20 (SK1 flowrate is reduced by 30 t/h)		−20	0		0		44		0.9
	30					−0.6				0.7	
50		100	50	−50	−20		−0.6	−12	24		1.6
	50					−3.5				−1.3	
100		80	100	20	−70		−4.1	−41	−26		0.3
	50					−2.5				−0.3	
150			70	70	−50		−6.6	−44	−6		0.0 (Pinch)
	50					1				3.2	
200		70		−70	20		−5.6	−28	64		3.2
	50					−2.5				−0.3	
250			60	60	−50		−8.1	−32.4	−6		2.9
	999750					9997.5				53986.5	
1000000					10		9989.4	9.9894	54 (Wastewater)		53989.4

FIGURE 7.6
MCA for flowrate reduction in SK1 for Example 7.2.

---

[2] See Chapters 3 and 4 for detailed discussion.

C_t (ppm)	ΔC_t (ppm)	$\Sigma_j F_{SKj}$ (t/h)	$\Sigma_i F_{SRj}$ (t/h)	$\Sigma_i F_{SRj} - \Sigma_j F_{SKj}$ (t/h)	$F_{c,t}$ (t/h)	Δm_t (kg/h)	Cum. Δm_t (kg/h) [Infeasible cascade]	$F_{FW,t}$ (t/h)	$F_{c,t}$ (t/h)	Δm_t (kg/h)	Cum. Δm_t (kg/h) [Feasible cascade]
					0			50 Freshwater			
0									50	1.0	
	20	50 [SK2 flowrate is reduced by 30 t/h]		−50	0	0	0	0			1.0
20					−50	−1.5			0	0.0	
	30		50	−20			−1.5	−30			1.0
50		70			−70	−3.5			−20	−1.0	
	50		100	20			−5	−50			0.0
100		80			−50	−2.5			0	0.0	
	50		70	70			−7.5	−50			0.0 Pinch
150		70			20	1			70	3.5	
	50			−70			−6.5	−32.5			3.5
200		70			−50	−2.5			0	0.0	
	50		60	60			−9	−36			3.5
250		60			10	9997.5		60 Wastewater		59985.0	
	999750						9988.5	9.9885			59988.5
1000000											

FIGURE 7.7
MCA for flowrate reduction in SK2 for Example 7.2.

We further examine the same of flowrate reduction (i.e., 30 t/h) on SK3. Figure 7.8 shows that the MCA identifies that the flowrates of freshwater and wastewater are identified as 65 and 75 t/h, i.e., highest among the three cases of flowrate reduction. Hence, it may be concluded that flowrate reduction should be carried out for sinks of the highest quality, as it will lead to reduced freshwater flowrate, coupled with the lowest flowrate of wastewater.

Next, strategy 5 (decrease/eliminate the flowrates of process sources in the LQR) is tested for flowrate reduction with SR4. The same amount of flowrate reduction (i.e., 30 t/h) on SR4 resulted in wastewater flowrate being reduced to 20 t/h, while freshwater flowrate remained unchanged (i.e., 70 t/h; see Figure 7.9). A summary of various flowrate reduction options is shown in Figure 7.10.

C_t (ppm)	ΔC_t (ppm)	$\Sigma_j F_{SKj}$ (t/h)	$\Sigma_i F_{SRj}$ (t/h)	$\Sigma_i F_{SRj} - \Sigma_j F_{SKj}$ (t/h)	$F_{c,t}$ (t/h)	Δm_t (kg/h)	Cum. Δm_t (kg/h) [Infeasible cascade]	$F_{FW,t}$ (t/h)	$F_{c,t}$ (t/h)	Δm_t (kg/h)	Cum. Δm_t (kg/h) [Feasible cascade]
					0			65 Freshwater			
0									65	1.3	
	20	50		−50	0	0	0	0			1.3
20					−50	−1.5			15	0.5	
	30	100	50	−50			−1.5	−30			1.8
50					−100	−5			−35	−1.8	
	50	50	100	50			−6.5	−65			0.0 Pinch
100		50 [SK1 flowrate is reduced by 30 t/h]			−50	−2.5			15	0.8	
	50	70		50			−9	−60			0.8
150			70	70	20	1			85	4.3	
	50	70		−70			−8	−40			5.0
200					−50	−2.5			15	0.8	
	50		60	60			−10.5	−42			5.8
250		60			10	9997.5		75 Wastewater		74981.3	
	999750						9987	9.987			74987.0
1000000											

FIGURE 7.8
MCA for flowrate reduction in SK3 for Example 7.2.

C_t (ppm)	ΔC_t (ppm)	$\Sigma_j F_{SKj}$ (t/h)	$\Sigma_j F_{SRj}$ (t/h)	$\Sigma_j F_{SRj} - \Sigma_j F_{SKj}$ (t/h)	$F_{C,t}$ (t/h)	Δm_t (kg/h)	Cum. Δm_t (kg/h)	$F_{FW,t}$ (t/h)	$F_{C,t}$ (t/h)	Δm_t (kg/h)	Cum. Δm_t (kg/h)
							Infeasible cascade		70 Freshwater		Feasible cascade
0					0				70		
	20					0	0			1.4	
20		50		-50	0		0	0	70		1.4
	30					-1.5			20	0.6	
50		100	50	-50	-50		-1.5	-30			2.0
	50					-5			-30	-1.5	
100		80	100	20	-100		-6.5	-65			0.5
	50					-4			-10	-0.5	
150			70	70	-80		-10.5	-70			0.0 Pinch
	50					-0.5			60	3.0	
200			70	-70	-10		-11	-55			3.0
	50					-4			-10	-0.5	
250			30	30	-80		-15	-60			2.5
	999750		SR4 flowrate is reduced by 30 t/h			-49987.5			20	19995.0	
1000000					-50		-50002.5	-50.0025	Wastewater		19997.5

FIGURE 7.9
MCA for flowrate reduction in SR4 for Example 7.2.

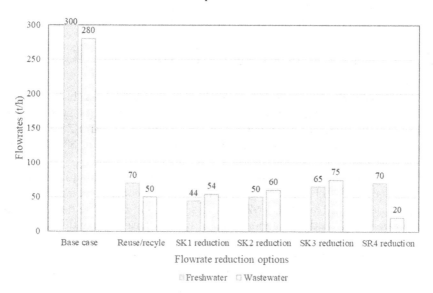

FIGURE 7.10
Summary for various options of flowrate reduction for Example 7.2.

7.2 Case Study – Water Minimization for AN Production

The acrylonitrile (AN) case study has been utilized in different chapters of this book for illustration. It is revisited here to demonstrate the impact of process changes.

In Examples 3.1 and 4.1, it has been demonstrated that with a direct reuse/recycle scheme, freshwater and wastewater flowrate of 2.1 and 8.2 kg/s can be achieved; this corresponds to 70.8% freshwater saving as compared to the

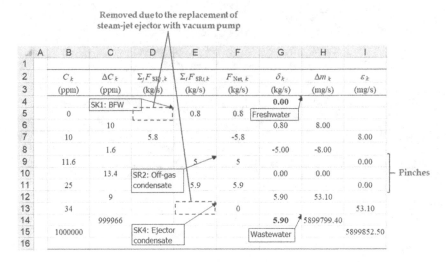

FIGURE 7.11
Flowrate targeting with ATM after process changes for AN case study.

base case (i.e., without water recovery). In Example 6.3, it was shown that the use of resin adsorption for purifying the off-gas condensate (SR2) can further reduce the flowrates of freshwater and wastewater to 1.2 and 7.3 kg/s. This corresponds to an additional 12.5% of freshwater saving, i.e., 83.3% relative to the base case.

In order to further reduce the freshwater and wastewater flowrates, one may consider a process change option, i.e., to replace the steam-jet ejector with a vacuum pump (El-Halwagi, 1997). Doing so will eliminate the need of boiler feed water (SK1) and the generation of steam-jet ejector condensate (SR4). Performing the flowrate targeting again with ATM yields the flowrates of freshwater and wastewater of 0 kg/s and 5.9 kg/s, respectively (see Figure 7.11). In other words, the requirement for freshwater is completely removed with this process change. However, one should note that carrying out such modification entails disturbance in current operation. It is likely that the involved process (i.e., distillation column in this case) will have to encounter process shutdown before such implementation can take place.

A comparison among the various water recovery schemes for the AN case study is shown in Figure 7.12. As shown, higher water-saving potential is achieved with direct reuse/recycle, which is then followed by a regeneration scheme, and finally process changes. The choice of deciding which option for implementation lies with other factors, such as cost consideration, space availability, etc. In general, one can expect higher acceptability with direct reuse/recycle scheme, as it is considered a *low-cost strategy* (El-Halwagi, 2006). Its implementation would usually involve minimal cost for piping and pumping (see Chapter 9 for capital cost calculations). This kind of *low-hanging fruit* (a common term used by the industrial practitioners) is the obvious solution

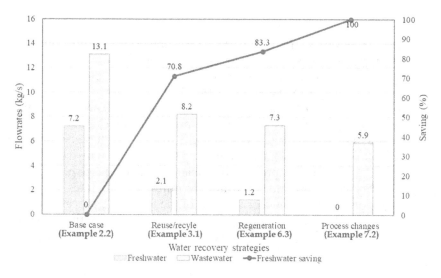

FIGURE 7.12
Summary of various water recovery schemes for AN case study.

that attracts attention for immediate implementation, and hence is taken as the highest priority in the hierarchy of resource conservation strategies.

If one were to have higher capital investment, the regeneration scheme would be the next immediate option, as it possesses higher resource conservation/waste reduction potential. Implementing such an option entails higher capital costs with the installation of the purification unit(s). Hence, one would expect lower acceptability with the regeneration option (relative to the direct reuse/recycle scheme). Finally, if one were to explore the option of process changes, higher cost, and lower acceptability are expected (see Figure 7.13).

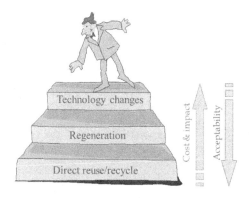

FIGURE 7.13
Hierarchy for resource conservation strategies (adapted from El-Halwagi, 2006).

PROBLEMS

7.1 Revisit the AN case in the Case Study Section. Evaluate the reduction of freshwater and wastewater flowrates if the replacement of the steam-jet ejector with a vacuum pump (El-Halwagi, 1997) is to be carried out for direct reuse/recycle case in Example 3.1.

7.2 For the wafer fabrication case study in Problems 2.13 and 4.23, demonstrate how plus/minus principles of process changes may be used to reduce fresh resources (UPW and municipal freshwater) and/or wastewater flowrates in this case. Suggest up to three alternative solutions.

7.3 Revisit the tire-to-fuel process in Problems 2.1 and 3.1 (see process flowsheet in Figure 7.14), and determine the minimum flowrates of freshwater (0 ppm) and wastewater for each of the following process changes. Determine the corresponding strategies associated with these changes. Limiting data for all changes are given in Table 7.2 (Noureldin and

TABLE 7.2

Limiting Data for Problem 7.3

		Base Case		Option 1		Option 2		Option 3	
j	Sinks, SK$_j$	F_{SKj}	C_{SKj}	F_{SKj}	C_{SKj}	F_{SKj}	C_{SKj}	F_{SKj}	C_{SKj}
1	Seal pot	0.15	200	0.15	200	0.15	200	0.15	200
2	Water jet makeup	0.25	50	0.25	50	0.20	50	0.20	50
i	Sources, SR$_i$	F_{SKi}	C_{SKi}	F_{SKi}	C_{SKi}	F_{SKi}	C_{SKi}	F_{SKi}	C_{SKi}
1	Decanter	0.27	500	0.27	500	0.24	500	0.20	500
2	Seal pot	0.15	200	0.15	500	0.15	500	0.15	500

Note: All flowrate terms in kg/s; concentration in ppm.

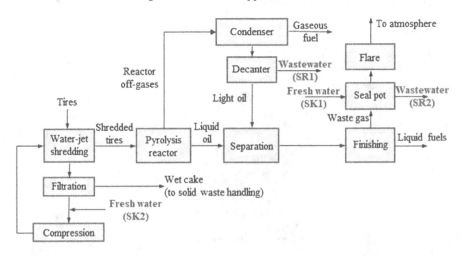

FIGURE 7.14
Process flowsheet for Problem 7.3.

El-Halwagi, 1999; Bandyopadhyay, 2006). Note that for the base case scenario, the minimum freshwater flowrate was targeted as 0.135 kg/s. Perform a comparison of the various scenarios.

a. Option 1—inlet concentration of the seal pot (SK1) is raised from 200 to 500 ppm.
b. Option 2 (to be implemented with Option 1)—reduction of makeup water to the compression station by increasing the compressor outlet pressure to 95 atm (current operating pressure is 70 atm; adjustable operating pressure range is 70–95 atm). The makeup water flowrate ($F_{make-up}$) is related to compressor outlet pressure (P_{comp}) by the following equation:

$$F_{make-up} = 0.47e^{-0.009\ P_{comp}} \tag{7.3}$$

Following this change, the makeup water is reduced to 0.20 kg/s (from 0.25 kg/s in Option 1), and decanter effluent is reduced to 0.24 kg/s (from 0.27 kg/s) in Option 1.

c. Option 3 (to be implemented with Option 2)—reduction of water generation in the reactor by increasing pyrolysis temperature to 710 K in the reactor (present temperature is set at 690 K, which leads to a water generation of 0.12 kg/s) (T_{pyro}, in kelvin), which leads to the change in water flowrate generation in the reactor (F_{gen}). The two variables are related as follows:

$$F_{gen} = 0.152 + \left(5.37 - 0.00784T_{pyro}\right)e^{27.4-0.04T_{pyro}} \tag{7.4}$$

Assuming that the same flowrate is separated in the decanter, the effluent of the latter will be reduced to 0.20 kg/s.

References

Bandyopadhyay, S. 2006. Source composite curve for waste reduction, *Chemical Engineering Journal*, 125, 99–110.

El-Halwagi, M. M. 1997. *Pollution Prevention through Process Integration: Systematic Design Tools*. San Diego, CA: Academic Press.

El-Halwagi, M. M. 2006. *Process Integration*. San Diego, CA: Elsevier Inc.

Feng, X., Liu, Y., Huang, L., and Deng, C. 2009. Graphical analysis of process changes for water minimization, *Industrial and Engineering Chemistry Research*, 48, 7145–7151.

Foo, D. C. Y. 2012. A generalised guideline for process changes for resource conservation networks. *Clean Technologies and Environmental Policy*, in press, DOI 10.1007/S10098–012–0475–4.

Kazantzi, V. and El-Halwagi, M. M. 2005. Targeting material reuse via property integration, *Chemical Engineering Progress*, 101(8), 28–37.

Linnhoff, B., Townsend, D. W., Boland, D., Hewitt, G. F., Thomas, B. E. A., Guy, A. R., and Marshall, R. H. 1982[1994]. *A User Guide on Process Integration for the Efficient Use of Energy*. Warwickshire, U.K: IChemE.

Noureldin, M. B. and El-Halwagi, M. M. 1999. Interval-based targeting for pollution prevention via mass integration, *Computers and Chemical Engineering*, 23, 1527–1543.

Polley, G. T. and Polley, H. L. 2000. Design better water networks. *Chemical Engineering Progress*, 96(2), 47–52.

Smith, R. 1995. *Chemical Process Design*. New York: McGraw-Hill.

Smith, R. 2005. *Chemical Process Design and Integration*. New York: John Wiley & Sons.

8

Network Design and Evolution Techniques

To synthesize a resource conservation network (RCN) using the pinch analysis technique, a two-step approach that consists of *targeting* and *design* is normally adopted. In the last few chapters, we have seen how minimum flowrate targets may be determined ahead of detailed design, for various direct reuse/recycle and material regeneration networks. For simple cases (e.g., Examples 3.2 and 4.1), we might determine the material recovery scheme based on insights from the targeting stage. However, for more complicated cases, a systematic design procedure is needed to synthesize an RCN that achieves the flowrate targets. In this chapter, a well-established technique called the *nearest neighbor algorithm* (NNA) is introduced for network synthesis. Besides, two evolution techniques are also introduced to simplify the preliminary RCN.

8.1 Procedure for Nearest Neighbor Algorithm

In order to synthesize an RCN that achieves the established flowrate targets, two criteria are to be met by all process sinks, i.e., *flowrate* (F_{SKj}) and *load* (m_{SKj}) requirements. The latter is given by the product of its flowrate and quality index (i.e., $m_{SKj} = F_{SKj}\, q_{SKj}$). To fulfill both of these requirements, one should follow the individual steps of the NNA, which are given as follows (Prakash and Shenoy, 2005a):

1. Arrange all material sinks and the sources in descending order of quality levels (i.e., ascending order of impurity concentration, for concentration-based problems). Note that the sources should include the external fresh resource (FR) and regenerated sources if any, with their respective flowrates obtained in the targeting stage. Start the design from the sink that requires the highest quality index (q_{SKj}).

2. Match the selected sink SK_j with source(s) SR_i of the same quality level, if any are found.

3. Mix two source candidates SR_i (with flowrate F_{SRi} and quality q_{SRi}) and SR_{i+1} (with flowrate F_{SRi+1} and quality q_{SRi+1}) to fulfill the flowrate and load requirements of sink SK_j. Note that the source candidates SR_i and SR_{i+1} are the nearest available "neighbors" to the sink SK_j,

with quality levels just lower and just higher than that of the sink, i.e., $q_{SRi} < q_{SKj} < q_{SRi+1}$. The respective flowrate between the source and the sink is calculated via the mass balance:[1]

$$F_{SRi,SKj} + F_{SRi,SKj} = F_{SKj} \tag{8.1}$$

$$F_{SRi,SKj}q_{SRi} + F_{SRi+1,SKj} q_{SRi+1} = F_{SKj}q_{SK_j} \tag{8.2}$$

where $F_{SRi,SKj}$ is the allocation flowrate sent from SR_i to SK_j. If SR_i has sufficient flowrate to be allocated to SK_j, i.e., $F_{SRi} \geq F_{SRi,SKj}$, go to Step 5 or else to Step 4.

A simple example is shown in Figure 8.1 for illustration. For Figure 8.1a, sink SK1 can tolerate a concentration of 50 ppm. So the two source candidates correspond to SR1 and SR2, with concentration just lower (0 ppm) and just higher (80 ppm) than that of SK1. On the other hand, SR2 and SR3 are identified as the source candidates for Figure 8.1b, since they fulfill the criteria of being just lower (20 ppm) and just higher (100 ppm) than the concentration of SK1.

4. If the source has insufficient flowrate to be used as the allocation flow-rate, i.e., $F_{SRi,SKj} > F_{SRi}$, then whatever is available of that source is used completely. A new pair of neighbor candidates is considered to satisfy the sink. For instance, sink SK1 in Figure 8.1c can tolerate a concentration of 50 ppm. Hence, the source candidates that serve as its neighbors are SR1 and SR2. It is then determined that 50 t/h of each of these sources is needed to fulfill the requirement of SK1. However, SR2 has a flowrate of 20 t/h, which is insufficient to fulfill the sink requirement. Hence, SR2 will fully be recovered to SK1. SR1 and SR3 next become the new pair of neighbor candidates for SK1.

5. Repeat Steps 2–4 for all other sinks. Once all sinks are fulfilled, the unutilized source(s) are discharged as waste.

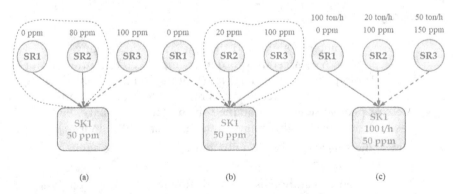

(a)   (b)   (c)

FIGURE 8.1
Identification of source candidates in NNA.

1 A spreadsheet for solving these equations are made available on book support website.
DOI: 10.1201/9781003173977-1

Note that for threshold problems, Equation 8.2 may be omitted in matching the last process sink in the RCN.[2]

8.2 Design for Direct Material Reuse/Recycle and the Matching Matrix

For a direct material reuse/recycle scheme, the NNA can be used to synthesize an RCN that achieves the minimum freshwater and wastewater targets. The latter may be identified with the graphical or algebraic approaches introduced in previous chapters.

A convenient representation for the RCN is the *matching matrix* (Prakash and Shenoy, 2005b), which takes the form in Figure 8.2. The sources are arranged as rows while the sinks as columns, both in descending order of quality index. Note that the external FR is located in the first row of the matching matrix, which acts as an external source along with other process sources of the RCN. In contrast, the waste discharge (WD) is placed in the last column of the matching matrix, which acts as an external sink. Variables in the cells indicate the allocation flowrates ($F_{SRi,SKj}$) of the sink-source matches. Note that the shaded area in the matching matrix indicates that matches between process sinks and sources across higher and lower quality regions are forbidden. Note also that the pinch-causing source (SR3 as in Figure 8.2) is an exception, as it belongs to both regions.

Example 8.1 Direct reuse/recycle network

A hypothetical example of Polley and Polley (2000) is illustrated, with limiting data given in Table 8.1). Flowrate targeting indicates that for direct water reuse/recycle schemes, the minimum freshwater (F_{FW}, 0 ppm)

F_{SRi}	q_{SRi}	F_{SKj} / q_{SKj} / SKj \ SRi	SK1 (F_{SK1}, q_{SK1})	SK2 (F_{SK2}, q_{SK2})	SK3 (F_{SK3}, q_{SK3})	SKj (F_{SKj}, q_{SKj})	WD (F_D)
F_R	q_R	FR	$F_{FW,SK1}$		$F_{FW,SK3}$		
F_{SR1}	q_{SR1}	SR1		$F_{SR1,SK2}$			
F_{SR2}	q_{SR2}	SR2			$F_{SR2,SK3}$		
F_{SR3}	q_{SR3}	SR3		$F_{SR3,SK2}$		$F_{SR3,SKj}$	
F_{SRi}	q_{SRi}	SRi				$F_{SRi,SKj}$	$F_{SRi,WW}$

FIGURE 8.2
The matching matrix. (From Prakash and Shenoy, 2005b. Reproduced with permission. Copyright 2005 Elsevier.)

---

2 See Sections 3.4 and 4.3 for types of threshold problems.

TABLE 8.1

Limiting Water Data for Example 8.1

Sinks, SK_j	F_{SKj} (t/h)	C_{SKj} (ppm)	Sources, SR_i	F_{SRi} (t/h)	C_{SRi} (ppm)
SK1	50	20	SR1	50	50
SK2	100	50	SR2	100	100
SK3	80	100	SR3	70	150
SK4	70	200	SR4	60	250
			FW	70	0

Source: Polley and Polley (2000).

and wastewater (F_{WW}) flowrate targets are determined as 70 and 50 t/h, respectively. Synthesize the RCN that achieves the flowrate targets and represent the RCN in a matching matrix.

SOLUTION

The NNA procedure is followed to synthesize the RCN, with the following detailed steps:

1. *Arrange all material sinks and the sources in descending order of quality levels. Start the design from the sink that requires the highest quality feed, i.e., q_{SKj} of highest quality.*

   As shown in Figure 8.3, the sinks are arranged horizontally, with the lowest concentration (highest quality) on the leftmost

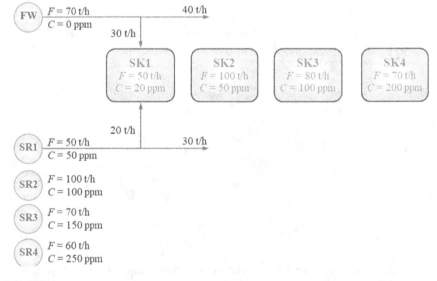

FIGURE 8.3
Results of allocating freshwater and SR1 to SK1. (From Prakash and Shenoy, 2005a. Reproduced with permission. Copyright 2005 Elsevier.)

to the highest concentration (lowest quality) on the rightmost side of the diagram. Similarly, all sources are arranged vertically, with the lowest concentration source being on the highest level, while the highest concentration is at the lowest level. We shall start the design with SK1, which requires the most stringent (highest quality) feed.

2. *Match the selected sink* SK_j *with source(s)* SR_i *of the same quality level, if any are available.*

3. *Mix two source candidates* SR_i *and* SR_{i+1} *to fulfill the flowrate and quality requirements of sink* SK_j. *Note that the source candidates* SR_i *and* SR_{i+1} *are the nearest available neighbors to the sink* SK_j, *with quality levels just lower and just higher than that of the sink. The respective flowrate between the source and the sink is calculated via Equations 8.1 and 8.2. If* SR_i *has sufficient flowrate to be allocated to* SK_j, *i.e.,* $F_{SRi} \geq F_{SRi,SKj}$, *go to Step 5 or else to Step 4.*

For the selected sink, i.e., SK1 (50 t/h, 20 ppm), there is no source with the same concentration. Hence we proceed to choose two sources, i.e., freshwater (FW) and SR1 that serve as its neighbor candidates, since they have concentrations just lower (FW, 0 ppm) and just higher (SR1, 50 ppm) than that of SK1. Equations 8.1 and 8.2 are then used to determine the allocation flowrates from freshwater ($F_{FW,SK1}$) and SR1 ($F_{SR1,SK1}$) to SK1:

$$F_{FW,SK1} + F_{SR1,SK1} = 50 \text{ t/h}$$

$$F_{FW,SK1}\left(0 \text{ ppm}\right) + F_{SR1,SK1}\left(50 \text{ ppm}\right) = 50 \text{ t/h}\left(20 \text{ ppm}\right)$$

Hence,

$$F_{FW,SK1} = 30 \text{ t/h}; \quad \text{and} \quad F_{SR1,SK1} = 20 \text{ t/h}$$

This means that the leftover flowrates of FW and SR1 are 40 t/h (=70 – 30 t/h) and 30 t/h (=50 – 20 t/h), respectively. Figure 8.3 shows the result of this step. Since both freshwater and SR1 have sufficient flowrate to be allocated to SK1, we then move to Step 5.

5. *Repeat Steps 2–4 for all other sinks.*

2. *Match the selected sink* SK_j *with source(s)* SR_i *of the same quality level, if any are available.*

We next look at the second sink, i.e., SK2 (100 t/h, 50 ppm). Since SR1 has the same concentration as SK2, its leftover flowrate (30 t/h) is sent to the latter. Hence, SK2 will only require a flowrate of 70 t/h (=100–30 t/h, Figure 8.4).

3. *Mix two source candidates* SR_i *and* SR_{i+1} *to fulfill the flowrate and quality requirements of sink* SKj. *Note that the source candidates* SR_i *and* SR_{i+1} *are the nearest available "neighbors" to the sink* SK_j, *with quality levels just lower and just higher than that of the sink. The respective flowrate between the source and the sink is calculated via Equations 8.1 and 8.2. If* SR_i *has sufficient flowrate to be allocated to* SKj, *i.e.,* $F_{SRi} > F_{SRi,SKj}$, *go to Step 5 or else to Step 4.*

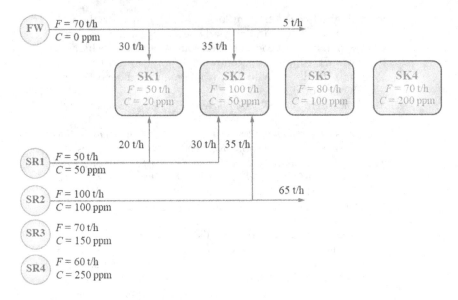

FIGURE 8.4
Results of allocating SR1, freshwater, and SR2 to SK2. (From Prakash and Shenoy, 2005a. Reproduced with permission. Copyright 2005 Elsevier.)

After SR1 is exhausted, the new neighbor candidates for SK2 are freshwater (with leftover flowrate of 40 t/h) and SR2 (100 t/h, 100 ppm). Equations 8.1 and 8.2 are used again to determine the allocation flowrates from freshwater ($F_{FW,SK2}$) and SR2 ($F_{SR2,SK2}$) to SK2:

$$F_{FW,SK2} + F_{SR2,SK2} = 70 \text{ t/h}$$

$$F_{FW,SK2} \left(0 \text{ ppm}\right) + F_{SR2,SK2} \left(100 \text{ ppm}\right) = 70 \text{ t/h} \left(50 \text{ ppm}\right)$$

Hence,

$$F_{FW,SK2} = F_{SR2,SK2} = 35 \text{ t/h}$$

Both freshwater and SR2 have sufficient flowrate to be allocated to SK2. However, the leftover flowrates now of freshwater and SR1 are 5 t/h (=40 − 35 t/h) and 65 t/h (=100 − 35 t/h), respectively. Figure 8.4 shows the result of this step. We then move to Step 5 again.

5. *Repeat Steps 2–4 for all other sinks.*

  2. *Match the selected sink SK_j with source(s) SR_i of the same quality level, if any are available.*

   We next proceed to sink SK3 (80 t/h, 100 ppm). The remaining flowrate from SR2 (65 t/h) is sent to SK3 since they have the same concentrations. This means that the flowrate requirement of SK2 is reduced to 15 t/h (= 80 − 65 t/h).

3. *Mix two source candidates* SR_i *and* SR_{i+1} *to fulfill the flowrate and quality requirements of sink* SK_j. *Note that the source candidates* SR_i *and* SR_{i+1} *are the nearest available neighbors to the sink* SK_j, *with quality levels just lower and just higher than that of the sink. The respective flowrate between the source and the sink is calculated via* Equations 8.1 and 8.2. *If* SR_i *has sufficient flowrate to be allocated to* SK_j, *i.e.,* $F_{SRi} \geq F_{SRi,SKj}$, *go to Step 5 or else to Step 4.*

Since SR2 has been fully utilized, the new neighbor candidates for SK3 are freshwater (with leftover flowrate of 5 t/h) and SR3 (70 t/h, 150 ppm). Equations 8.1 and 8.2 next determine the allocation flowrates from freshwater ($F_{FW,SK3}$) and SR3 ($F_{SR3,SK3}$) to SK3:

$$F_{FW,SK3} + F_{SR3,SK3} = 15 \text{ t/h}$$

$$F_{FW,SK3}(0 \text{ ppm}) + F_{SR3,SK3}(150 \text{ ppm}) = 15 \text{ t/h}(100 \text{ ppm})$$

Hence,

$$F_{FW,SK3} = 5 \text{ t/h}; \quad \text{and} \quad F_{SR3,SK3} = 10 \text{ t/h}$$

The leftover freshwater (5 t/h) is now sent to SK3 completely. On the other hand, the leftover flowrate of SR3 is determined as 60 t/h (=70 − 10 t/h). Figure 8.5 shows the result of this step. We then move to Step 5.

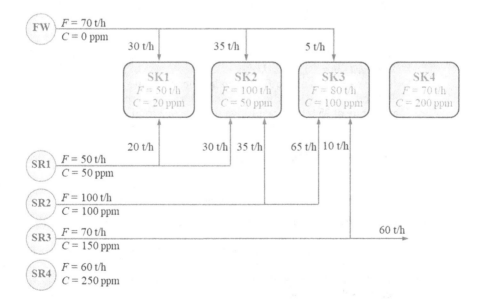

FIGURE 8.5
Results of allocating SR2, freshwater, and SR3 to SK3. (From Prakash and Shenoy, 2005a. Reproduced with permission. Copyright 2005 Elsevier.)

5. *Repeat Steps 2—4 for all other sinks.*
   2. *Match the selected sink SK_j with source(s) SR_i of the same quality level, if any are available.*
   3. *Mix two source candidates SR_i and SR_{i+1} to fulfill the flowrate and quality requirements of sink SK. Note that the source candidates SR_i and SR_{i+1} are the nearest available neighbors to the sink SK_j, with quality levels just lower and just higher than that of the sink. The respective flowrate between the source and the sink is calculated via Equations 8.1 and 8.2. If SR_i has sufficient flowrate to be allocated to SK, i.e., $F_{SRi} \geq F_{SRi,SKj}$, go to Step 5 or else to Step 4.*

   We next proceed to analyze the last sink, i.e., SK4 (70 t/h, 200 ppm). There is no source with the same concentration as this sink; hence, we proceed to identify the neighbor candidates for SK4. These correspond to SR3 (with leftover flowrate of 60 t/h, 150 ppm) and SR4 (60 t/h, 250 ppm). Equations 8.1 and 8.2 determine the allocation flowrates from SR3 ($F_{SR3,SK4}$) and SR4 ($F_{SR4,SK4}$) to SK4 as

$$F_{SR3,SK4} + F_{SR4,SK4} = 70 \text{ t/h}$$

$$F_{SR3,SK4}\left(150 \text{ ppm}\right) + F_{SR3,SK4}\left(250 \text{ ppm}\right) = 70 \text{ t/h}\left(200 \text{ ppm}\right)$$

Hence,

$$F_{SR3,SK4} = F_{SR4,SK4} = 35 \text{ t/h}$$

   Both SR3 and SR4 have sufficient flowrates for SK4. The flowrates of these sources are both reduced to 25 t/h (=60 − 35 t/h). We then move to Step 5.

5. *Repeat Steps 2–4 for all other sinks. Once all sinks are fulfilled, the unutilized source(s) are discharged as waste.*

   Since all sinks are fulfilled, the leftover flowrates in sources SR3 and SR4 are discarded as wastewater. This corresponds to a flowrate of 25 t/h for respective sources, or a total wastewater flowrate of 50 t/h that matches the identified target. Figure 8.6 shows the resulting RCN for the example, while its representation in the matching matrix is given in Figure 8.7.

Example 8.2 Property reuse/recycle network (Example 3.7 revisited)

The metal degreasing process in Examples 2.5 and 3.7 is revisited here. The main property of concern for a reuse/recycle scheme is the Reid vapor pressure (RVP), which is converted to the corresponding operator values using Equation 2.25. The limiting data are summarized in Table 8.2 (reproduced from Table 2.3). Flowrate targeting indicates that for direct reuse/recycle schemes, the minimum fresh solvent (F_{FS}) and waste solvent (F_{WS}) flowrate targets are determined as 2.4 kg/s, respectively (see Figure 3.18). Synthesize an RCN that achieves the flowrate targets, and represent the RCN in a conventional process flow diagram (PFD).

TABLE 8.2

Limiting Data for Example 8.2

Sink, SK$_j$	F_{SKj} (kg/s)	ψ_{SKj} (atm$^{1.44}$)	Source, SR$_i$	F_{SRi} (kg/s)	ψ_{SRi} (atm$^{1.44}$)
Degreaser (SK1)	5.0	4.87	Condensate I (SR1)	4.0	13.20
Absorber (SK2)	2.0	7.36	Condensate II (SR2)	3.0	3.74
			Fresh solvent	2.4	2.713

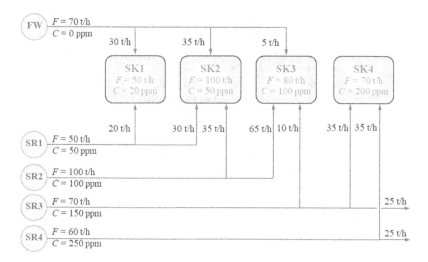

FIGURE 8.6
Final resulting RCN for example 8.1. (From Prakash and Shenoy, 2005a).

		F_{SKj} (t/h)	50	100	80	70	50
		C_{SKj} (ppm)	20	50	100	200	
F_{SRi} (t/h)	C_{SRi} (ppm)	SK$_j$ / SR$_i$	SK1	SK2	SK3	SK4	WW
70	0	FW	30	35	5		
50	50	SR1	20	30			
100	100	SR2		35	65		
70	150	SR3			10	35	25
60	250	SR4				35	25

FIGURE 8.7
Matching matrix for RCN in example 8.1. (From Prakash and Shenoy, 2005b. Reproduced with permission. Copyright 2005 Elsevier.)

SOLUTION

The NNA procedure is followed to synthesize the RCN for both cases, given as follows:

1. *Arrange all material sinks and the sources in descending order of quality levels, respectively. Start the design from the sink that requires the highest quality feed, i.e., q_{SKj} of highest quality.*

2. *Match the selected sink SK$_j$ with source(s) SR$_i$ of the same quality level, if any are available.*
3. *Mix two source candidates SR$_i$ and SR$_{i+1}$ to fulfill the flowrate and quality requirements of sink SK$_j$. Note that the source candidates SR$_i$ and SR$_{i+1}$ are the nearest available neighbors to the sink SK$_j$, with quality levels just lower and just higher than that of the sink. The respective flowrate between the source and the sink is calculated via Equations 7.1 and 7.2. If SR$_i$ has sufficient flowrate to be allocated to SK$_j$, i.e., $F_{SRi} \geq F_{SRi,SKj}$, go to Step 5 or else to Step 4.*
4. *If the source has insufficient flowrate to be used as the allocation flowrate, i.e., $F_{SRi,SKj} > F_{SRi}$, then whatever is available of that source is used completely. A new pair of neighbor candidates is considered to satisfy the sink.*

As shown in Figure 8.8, SK1 with the lowest property operator value (i.e., most stringent quality requirement) is placed on the left, while SK2 with a higher property operator value is placed on the right. Similarly, all sources are arranged vertically, with the highest quality external fresh solvent source (FS) being on the highest level while the lowest quality SR1 is at the lowest level.

Since no source and sink pairs have the same property operator values, we proceed to identify two sources, i.e., SR1 and SR2, with operators being just higher and lower than SK1, to serve as the neighbor candidates for the latter. Note that FS is not the neighbor candidate at this stage. Equations 8.1 and 7.2 are then used to determine the allocation flowrates from SR1 ($F_{SR1,SK1}$) and SR2 ($F_{SR2,SK1}$) to SK1:

$$F_{SR1,SK1} + F_{SR2,SK1} = 5 \text{ kg/s}$$

$$F_{SR1,SK1}\left(13.2 \text{ atm}^{1.44}\right) + F_{SR2,SK1}\left(3.74 \text{ atm}^{1.44}\right) = 5 \text{ kg/s}\left(4.87 \text{ atm}^{1.44}\right).$$

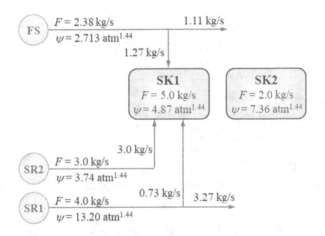

FIGURE 8.8
Results of allocating SR2, fresh solvent, and SR1 to SK1. (From Foo et al., 2006b.)

Hence,

$$F_{SR1,SK1} = 0.6 \text{ kg/s} \quad \text{and} \quad F_{SR2,SK1} = 4.4 \text{ kg/s.}$$

However, SR2 has a smaller flowrate of 3.0 kg/s. Hence, its entire flowrate is sent to SK1. This means that the leftover flowrate requirement of SK1 is now reduced to 2 kg/s (=5 − 3 kg/s, see Figure 8.8). We then identify the next pair of source candidates to fulfill the requirement of SK1. These correspond to FS and SR1. Equations 8.1 and 8.2 are used again to determine the allocation flowrates from FS ($F_{FS,SK1}$) and SR1 ($F_{SR1,SK1}$) to SK1:

$$F_{FS,SK1} + F_{SR1,SK1} = 2 \text{ kg/s}$$

$$F_{FS,SK1}\left(2.713 \text{ atm}^{1.44}\right) + F_{SR1,SK1}\left(13.2 \text{ atm}^{1.44}\right)$$
$$= 5 \text{ kg/s}\left(4.87 \text{ atm}^{1.44}\right) - 3 \text{ kg/s}\left(3.74 \text{ atm}^{1.44}\right).$$

Hence,

$$F_{FS,SK1} = 1.27 \text{ kg/s;} \quad \text{and} \quad F_{SR1,SK1} = 0.73 \text{ kg/s.}$$

Since both sources have sufficient flowrates, the requirement of SK1 is fulfilled. The leftover flowrates of FS and SR1 are determined as 1.11 kg/s (=2.38 − 1.27 kg/s) and 3.27 kg/s (=4 − 0.73 kg/s), respectively. Figure 8.8 shows the result of this step.

5. *Repeat Steps 2–4 for all other sinks.*
   a. *Match the selected sink SK_j with source(s) SR_i of the same quality level, if any are available.*
   b. *Mix two source candidates SR_i and SR_{i+1} to fulfill the flowrate and quality requirements of sink SK_j. Note that the source candidates SR_i and SR_{i+1} are the nearest available neighbors to the sink SK_j, with quality levels just lower and just higher than that of the sink. The respective flowrate between the source and the sink is calculated via Equations 8.1 and 8.2. If SR_i has sufficient flowrate to be allocated to SK_j, i.e., $F_{SRi} \geq F_{SRi,SKj}$, go to Step 5 or else to Step 4.*

6. Repeat Steps 2–4 for all other sinks. Once all sinks are fulfilled, the unutilized source(s) are discharged as waste.

   We next proceed to sink SK2. Similar to the case of SK1, there is no source with the same property operator value. Hence, we identify sources FS and SK1 as the neighbor candidates for this sink. Equations 81 and 8.2 determine the allocation flowrates from FS ($F_{FS,SK2}$) and SR1 ($F_{SR1,SK2}$) to SK2 as

$$F_{FS,SK2} + F_{SR1,SK2} = 2 \text{kg/s}$$

$$F_{FS,SK2}\left(2.713 \text{ atm}^{1.44}\right) + F_{SR1,SK2}\left(13.2 \text{ atm}^{1.44}\right) = 2 \text{ kg/s}\left(7.36 \text{ atm}^{1.44}\right).$$

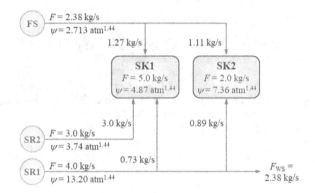

FIGURE 8.9
Final resulting RCN for Example 8.2. (From Foo et al., 2006b.)

Hence,

$$F_{FS,SK2} = 1.11 \text{ kg/s}; \quad \text{and} \quad F_{SR1,SK2} = 0.89 \text{ kg/s}.$$

Since both sources have sufficient flowrates, the requirement of SK2 is fulfilled. Note that the FS is used completely. The leftover flowrate of SR1, i.e., 2.38 kg/s (=3.27 – 0.89 kg/s), is discharged as waste solvent, matching the flowrate target identified earlier. Figure 8.9 shows the resulting RCN for the case, while Figure 8.10 shows the solvent recovery scheme in a conventional PFD.

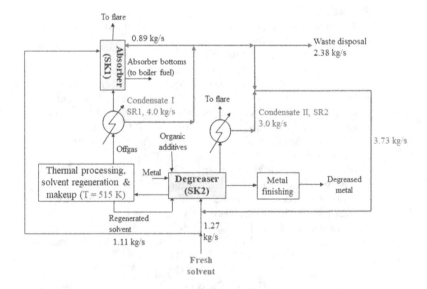

FIGURE 8.10
Direct solvent reuse/recycle scheme for metal degreasing problem (Example 8.2). (From Foo et al., 2006b.)

8.3 Design for Material Regeneration Network

The NNA may be used to synthesize an RCN with purification unit (in short *material regeneration network*) that achieves the minimum freshwater, wastewater, and regeneration flowrate targets. However, other insights gained from the targeting stage, particularly the identified sources to be sent for regeneration, should be incorporated to ease the network design stage.

Example 8.3 Water regeneration network (Example 6.1 revisited)

The classical water minimization case of Wang and Smith (1994) in Example 6.1 is revisited, with limiting water data given in Table 8.3. In Example 6.1, it was shown that the minimum flowrates of freshwater (F_{FW}), wastewater (F_{WW}), and regenerated water (F_{RW}) of the water network are targeted as 20, 20, and 73.68 t/h, respectively, when a purification unit with an outlet concentration (C_{Rout}) of 5 ppm is used for water regeneration (see Example 6.1). Synthesize a water regeneration network that achieves the flowrate targets, assuming pure freshwater (0 ppm) is used. Make good use of the insights gained from the targeting stage (Example 6.1).

SOLUTION

The results of automated targeting model from Example 6.1 are shown in Figure 8.11 (reproduced from Figure 6.7). As shown in column F, 73.68 t/h of source flowrate are purified to 5 ppm; this regeneration flowrate originates from sources of 100 ppm (23.68 t/h) and 800 ppm (50 t/h). Since the 800 ppm source is mainly contributed by SR3 and SR4 (see Table 8.3), it is obvious that both of these water sources are fully sent for purification. On the other hand, we may assume that the 100 ppm source (23.68 t/h) is fully originated from SR2. From these insights, the associated flowrates should first be allocated among the respective sinks, sources, and purification unit, as shown in Figure 8.12. Note that both SR3 and SR4 are now exhausted and cannot be paired with any water sink, while SR2 has a reduced flowrate of 76.32 t/h (= 100 − 23.68 t/h).

TABLE 8.3

Limiting Water Data for Example 8.3

Sinks, SK$_j$	F_{SKj} (t/h)	C_{SKj} (ppm)	Sources, SR$_i$	F_{SRi} (t/h)	C_{SRi} (ppm)
SK1	20	0	SR1	20	100
SK2	100	50	SR2	100	100
SK3	40	50	SR3	40	800
SK4	10	400	SR4	10	800
$\sum_i F_{SKj}$	170		$\sum_i F_{SRi}$	170	

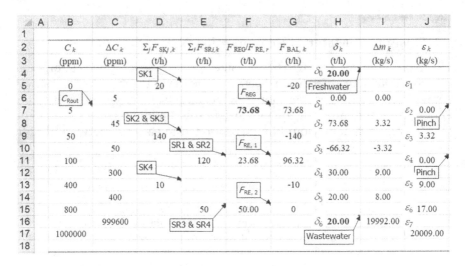

FIGURE 8.11
Insight gain from the targeting stage.

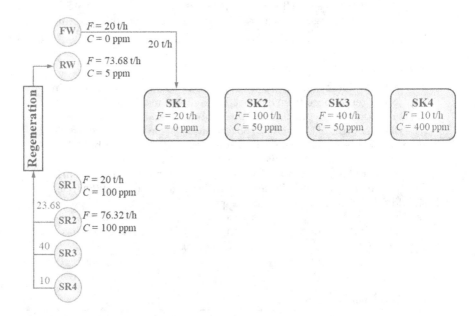

FIGURE 8.12
Results of allocating SR2, SR3, and SR4 to purification unit, and paring of FW to SK1.

Next, The NNA procedure is used to match the remaining sinks and sources in the RCN.

1. *Start the design from the sink that requires the highest quality feed, i.e., q_{SKj} of highest quality.*

2. *Match the selected sink SK$_j$ with source(s) SR$_i$ of the same quality level, if any are available.*

We first look at SK1 that requires 20 t/h of pure water (0 ppm). As pure freshwater (0 ppm) is available for use, it is sent to fulfill the requirement of SK1 (Figure 8.12). Freshwater is now exhausted. We move to SK2 that requires 100 t/h water of 50 ppm.

4. *Repeat Steps 2 – 4 for all other sinks.*

2. *Match the selected sink SKj with source(s) SRi of the same quality level, if any are available.*

3. *Mix two source candidates SR$_i$ and SR$_{i+1}$ to fulfill the flowrate and quality requirements of sink SK$_j$.*

As there are no water sources of the same quality level as SK2, we then identify the regenerated water RW (5 ppm) and SR1 (100 ppm) as the neighbor candidates for SK2. Equations 8.1 and 8.2 then determine the allocation flowrates as 52.63 and 47.37 t/h, respectively. As SR1 has a flowrate of 20 t/h, SR3 of the same quality (100 ppm) is also used, in which its flowrate of 27.37 t/h is fed to SK2 (see Figure 8.13). Doing this leads to the exhaustion of SR1, while the flowrates of RW and SR2 are reduced to 21.05 t/h (= 73.68 – 52.63 t/h) and 48.95 t/h (= 76.32 – 27.37 t/h), respectively. We then proceed to the next sink, i.e., SK3.

4. *Repeat Steps 2–4 for all other sinks.*

2. *Match the selected sink SK$_j$ with source(s) SR$_i$ of the same quality level, if any are available.*

3. *Mix two source candidates SR$_i$ and SR$_{i+1}$ to fulfill the flowrate and quality requirements of sink SK$_j$.*

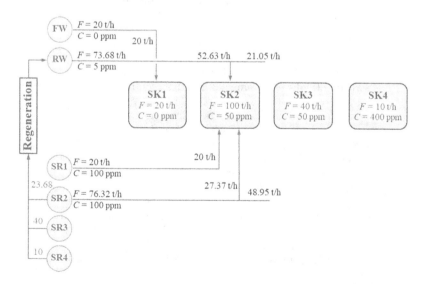

FIGURE 8.13
Results of allocating RW, SR1, and SR2 to SK2.

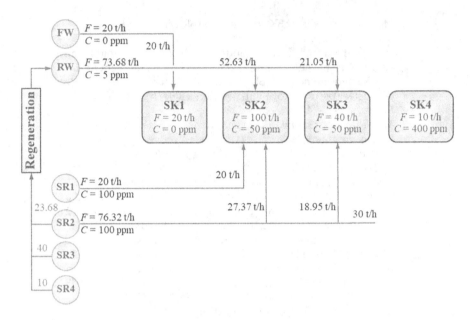

FIGURE 8.14
Results of allocating SR2 and RW to SK3.

The matching of sources for SK3 (100 t/h, 50 ppm) is similar to that of SK2, as RW (5 ppm) and SR2 (100 ppm) are taken as its neighbor candidates. With Equations 8.1 and 8.2, the allocated flowrates from these neighbors are determined as 21.05 (RW) and 18.95 t/h (SR2), respectively (see Figure 8.14). Doing this leads to the exhaustion of RW, while SR2 has a remaining flowrate of 30 t/h. We next move to SK4.

5. *Repeat Steps 2–4 for all other sinks. Once all sinks are fulfilled, the unutilized source(s) are discharged as waste.*

Since the leftover source is SR2, with flowrate of 30 t/h, it is sent to fulfill the requirement of SK4 that requires water of 10 t/h. As SR2 has much lower concentration (100 ppm) than SK4 (400 ppm), this pairing is feasible. The leftover flowrate of SR2 (20 t/h) will leave as wastewater discharge. The final RCN is shown in Figure 8.15, while its matrix representation is given in Figure 8.16.

Finally, note that for a material regeneration network, insights from the targeting stage serve as guidance in the network design stage. The NNA can equally be used even without insights from the targeting stage. For such a case, a *degenerate* solution exists for the problem (i.e., different network structure that achieves the same network targets).[3] However, it is a good practice to incorporate those insights, as they will certainly be useful when a more complicated RCN is to be synthesized.

---

[3] Readers may cross-check Figure 6.9 for a degenerate solution, and may also solve Problem 8.16b for another degenerate solution.

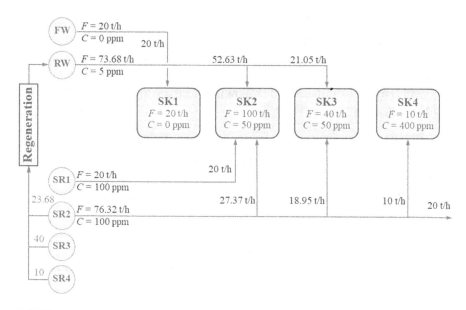

FIGURE 8.15
Final resulting RCN for Example 8.3.

F_{SRi} (t/h)	C_{SRi} (ppm)	F_{SKj} (t/h)	20	100	40	10	73.68	20
		C_{SKj} (ppm)	0	50	50	400	-	100
		SKj / SRi	SK1	SK2	SK3	SK4	RW	RW
20	0	FW	20					
73.68	5	RW		52.63	21.05			
20	100	SR1		20.00				
100	100	SR2		27.37	18.95	10	23.68	20
40	800	SR3					40	
10	800	SR4					10	

FIGURE 8.16
Matching matrix of the RCN in Example 8.3.

8.4 Network Evolution Techniques

In this section, two well-established network evolution techniques are introduced. These techniques are used to simplify a preliminary RCN that is synthesized using the NNA.

8.4.1 Source Shift Algorithm

In the first technique, i.e., the *source shift algorithm* (SSA), some matches between sinks and sources (e.g., established using the NNA) are removed, aiming to reduce their number of interconnections.

To evolve an RCN with SSA, the following steps are to be followed (Prakash and Shenoy, 2005b; Ng and Foo, 2006):

1. Identify a sink-source pair that fulfills both the following criteria:
   a. The chosen sink-source matches should have the same quality value, i.e., $q_{SKj} = q_{SRi}$.
   b. The source flowrate should be larger than or equal to that of the sink, i.e., $F_{SRi} \geq F_{SKj}$.

2. Feed the selected sink fully with the source.

3. Shift the excess flowrate experienced by the sink to the sink (where the source was matched to originally) that experiences a flowrate deficit.

A simple illustration is shown in Figure 8.17, where two process sinks are matched with an external FR as well as two process sources. As observed, a total of six interconnections were established between the sinks and sources (Figure 8.17a). Note that both sink-source pairs SK1–SR1 and SK2–SR1 fulfill both the aforementioned criteria (in Step 1). We shall first illustrate for the SK1–SR1 pair.

Following Step 2, SK1 has to be fully allocated by SR1 flowrate. Hence, SR1 that was earlier allocated to SK2 is shifted to SK1 (Figure 8.17b). However, this means that SK1 now has an excess flowrate of 50 t/h, while SK2 experiences a flowrate deficit of 50 t/h. To restore the flowrate deficit of the latter, 50t/h of excess source flowrate in SK1 is shifted to SK2, using the matches of

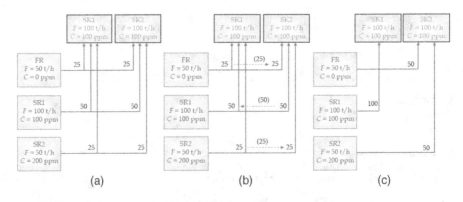

FIGURE 8.17
Application of SSA for SK1–SR1 pair. (From Ng and Foo, 2006. Reproduced with permission. Copyright 2006 American Chemical Society.)

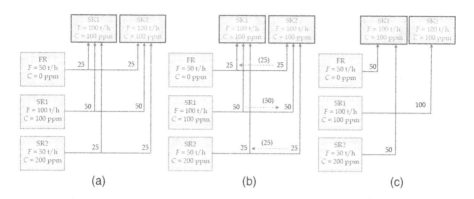

FIGURE 8.18
Application of SSA for SR2–SR1 pair. (From Ng and Foo, 2006. Reproduced with permission. Copyright 2006 American Chemical Society.)

SK2–FR and SK2–SR2, as shown in Figure 8.17b. A net result of these source shifts is the reduced number of interconnections. As shown in Figure 8.17c, the numbers of interconnections in the RCN are reduced by half, i.e., from six to three matches.

Similar steps may also be applied for the SK2–SR1 pair in Figure 8.18. Following Step 2, flowrate from SR1 is fully allocated to SK2. This causes a 50 t/h flowrate deficit in SK1, while excess flowrate is experienced in SK2. Hence, to restore the flowrate balances, 50 t/h flowrate is shifted from SK2 to SK1 using the SK1–FR and SK1–SR2 matches (Figure 8.18b). The net result is an RCN with three interconnections (Figure 8.18c). Note that the latter has a different configuration as observed in Figure 8.17c.

Apart from the earlier-described procedure, SSA may also be applied to interconnections that involve waste streams in the *lower quality region* of an RCN. To carry out the SSA for the lower quality region, the revised SSA steps are implemented:

1. Identify a set of sink-source pairs that fulfill both criteria:
    a. The sink-source pairs should consist of two sets of sink-source and source-waste matches.
    b. All matches should be served by the same source pairs.
2. Feed the sink fully with the source of higher quality.
3. Shift the excess flowrate experienced by the sink to the waste that was originally fed by the source.

A simple illustration is given in Figure 8.19, showing the lower quality region of an RCN as designed by NNA. A total of six interconnections are observed between three sources and two sinks, as well as the WD streams (Figure 8.19a). Following Step 1 of the revised procedure, the following

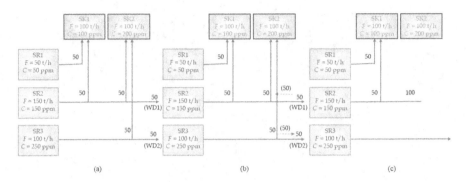

FIGURE 8.19
Application of SSA for lower quality region.

sink-source matches are identified, i.e., SK2–SR2, SK2–SR3, WD1–SR2, and WD2–SR3. SK2 is then fully fed with SR2 of higher quality (Step 2). The excess flowrate of 50 t/h is then discharged as waste using the WD2–SR3 match (Step 3; see Figure 8.19b). The net result of the shift is the reduced interconnection with four matches (Figure 8.19c). This may also be viewed as degenerate solutions for the problem.

Note that the revised SSA also works well for threshold problems with zero fresh resource.

8.4.2 Material Path Analysis

Material Path Analysis (MPA) aims to evolve an RCN by having additional FR usage (termed as *"penalty"*). In essence, it explores the tradeoff between the minimum external FR usage and network complexity. A material path is defined as *a continuous route that starts from an external fresh resource, linked with the sink—source connections, and ends at the waste discharge* (Ng and Foo, 2006). When a material path is observed, optimization may be performed to remove some interconnections by having FR penalty. In order to minimize the penalty, a useful heuristic is to remove the smallest sink-source match(es) along the material path (Ng and Foo, 2006).

A simple example is given in Figure 8.20. A material path is identified in Figure 8.20a that connects FR, SK1–SR1 match, and WD1. By sending an additional 50 t/h of FR to SK1, the original allocated flowrate between SR1 and SK1 is no longer necessary. Hence, the excess flowrate of SR1 is discharged as waste through WD1 (see Figure 8.20b). In other words, the removal of the SK1–SR1 match (i.e., the RCN with one reduced interconnection) is with an expense of 50 t/h FR penalty.

Note that SSA is normally applied before MPA, as the latter normally involves FR penalty, which is not the case for SSA. Note also that both SSA and MPA are easily implemented through the matching matrix (Prakash and Shenoy, 2005b; Ng and Foo, 2006).

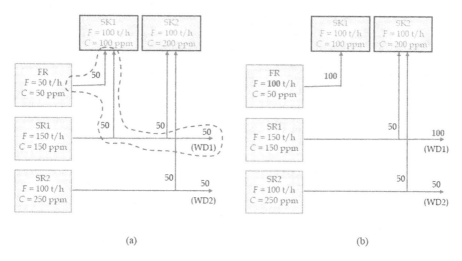

(a)   (b)

FIGURE 8.20
Example of MPA.

Example 8.4 Network evolution with SSA and MPA

Revisit the preliminary water network synthesized using NNA in Example 8.1. Reduce the number of interconnections in the RCN with the following techniques:

1. SSA—reduce interconnections in both higher and lower quality regions.
2. MPA—identify the water penalty incurred with the removal of interconnections.

SOLUTION

Case 1—Interconnection reduction using SSA

The matching matrix in Figure 8.7 is reproduced in Figure 8.21, which shows that the RCN has a total of 12 sink-source matches (indicated by cells with figures), including those that connect with the freshwater (FW) and wastewater (WW) streams. To evolve the RCN using the SSA, the three-step procedure is followed:

1. *Identify a sink-source pair that fulfills both the following criteria:*
   a. *The chosen sink-source matches should have the same quality value, i.e., $q_{SKj} = q_{SRi}$.*
   b. *The source flowrate should be larger than or equal to that of the smk, i.e., $F_{SRi} \geq F_{SKj}$.*

From Figure 8.21, it is observed that both SK2–SR1 and SK3–SR2 pairs fulfill the first criteria, i.e., concentrations of both sink and source are the same (indicated by the dashed boxes). However, only the SK3–SR2

		F_{SKj} (t/h)	50	100	80	70	50
		C_{SKj} (ppm)	20	50	100	200	
F_{SRi} (t/h)	C_{SRi} (ppm)	SKj \ SRi	SK1	SK2	SK3	SK4	WW
70	0	FW	30	35	5		
50	50	SR1	20	30			
100	100	SR2		35	65		
70	150	SR3			10	35	25
60	250	SR4				35	25

FIGURE 8.21
Identification of sink-source pairs for SSA.

pair fulfills the second criteria, i.e., $F_{SK2} \geq F_{SR3}$, and hence is selected to undergo Steps 2 and 3.

1. *Feed the selected sink fully with the source.*
2. *Shift the excess flowrate experienced by the sink to the sink (where the source was matched to originally) that experiences a flowrate deficit.*

Following Step 2, the flowrate requirement of SK3 should be fulfilled by SR2. However, at present, only 65 t/h of the SR2 flowrate is allocated to SK3, while the remaining (35 t/h) to SK2. To fully satisfy the SK3 flowrate with SR2, 15 t/h of flowrate is to be shifted from the current SK2–SK2 match. This causes a 15 t/h flowrate deficit for SK2. In contrast, SK3 experiences a flowrate surplus of 15 t/h due to the shift. To restore flowrate balance for both SK2 and SK3, 15 t/h of flowrate is to be shifted to SK2 from other existing sink-source matches associated with SK3. These correspond to the matches of SK3–FW and SK3–SR3, with allocated flowrates of 5 and 10 t/h, respectively (Figure 8.22).

The net result of the source shifts is shown in Figure 8.23 (newly allocated flowrates are highlighted in bold). Note that the SK2–FW match row has higher flowrate; while a new SK2–SR3 match emerges in the RCN. The total number of interconnections has now been reduced to 11 matches.

Besides, the revised SSA is also applied for the lower quality (higher concentration) region:

1. *Identify a set of sink-source pairs that fulfill both criteria:*
   a. *The sink-source pairs should consist of two sets of sink-source and source-waste matches.*
   b. *All matches should be served by the same source pairs.*
2. *Feed the sink fully with the source of higher quality.*
3. *Shift the excess flowrate experienced by the sink to the waste that was originally fed by the source.*

F_{SKj} (t/h)			50	100	80	70	50
		C_{SKj} (ppm)	20	50	100	200	
F_{SRi} (t/h)	C_{SRi} (ppm)	SKj \ SRi	SK1	SK2	SK3	SK4	WW
70	0	FW	30	35 ⟨5⟩ 5			
50	50	SR1	20	30			
100	100	SR2		35 ⟦15⟧ 65			
70	150	SR3		⟨10⟩ 10		35	25
60	250	SR4				35	25

FIGURE 8.22
Source shifts for Example 8.4. (From Ng and Foo, 2006. Reproduced with permission. Copyright 2006 American Chemical Society.)

F_{SKj} (t/h)			50	100	80	70	50
		C_{SKj} (ppm)	20	50	100	200	
F_{SRi} (t/h)	C_{SRi} (ppm)	SKj \ SRi	SK1	SK2	SK3	SK4	WW
70	0	FW	30	40			
50	50	SR1	20	30			
100	100	SR2		20	80		
70	150	SR3		10		35	25
60	250	SR4				35	25

FIGURE 8.23
Results of source shift between SK2 and SK3 matches. (From Ng and Foo, 2006. Reproduced with permission. Copyright 2006 American Chemical Society.)

Four matches that may undergo SSA in the higher concentration region are identified, i.e., SK4–SR3, SK4–SR4, WW–SR3, and WW–SR4. SK4 is then fully satisfied by SR3 (higher quality as compared with SR4), by shifting its flowrate that is originally allocated to WW (25 t/h). The excess flowrate in SK4 is then shifted to WW (Figure 8.24). Doing this leads to the removal of the WW–SR3 match. As shown in Figure 8.25, the resulting RCN has a total of 10 interconnections.

Case 2—Interconnection reduction using MPA
We next approach the problem with MPA. One of the *water paths* for this case is shown in Figure 8.26, which connects the matches of SK2–FW, SK2–SR3, SK4–SR3, SK4–SR4, and WW–SR4. It is then observed that this

F_{SRi} (t/h)	C_{SRi} (ppm)	F_{SKj} (t/h)	50	100	80	70	50
		C_{SKj} (ppm)	20	50	100	200	
		SKj \ SRi	SK1	SK2	SK3	SK4	WW
70	0	FW	30	40			
50	50	SR1	20	30			
100	100	SR2		20	80		
70	150	SR3		10		35 ◁25◻ 25	
60	250	SR4				35 ◻25▷ 25	

FIGURE 8.24
Source shift for lower quality region. (From Prakash and Shenoy, 2006.)

F_{SRi} (t/h)	C_{SRi} (ppm)	F_{SKj} (t/h)	50	100	80	70	50
		C_{SKj} (ppm)	20	50	100	200	
		SKj \ SRi	SK1	SK2	SK3	SK4	WW
70	0	FW	30	40			
50	50	SR1	20	30			
100	100	SR2		20	80		
70	150	SR3		10		60	
60	250	SR4				10	50

FIGURE 8.25
Results of SSA for lower quality region. (From Ng and Foo, 2006.)

F_{SRi} (t/h)	C_{SRi} (ppm)	F_{SKj} (t/h)	50	100	80	70	50
		C_{SKj} (ppm)	20	50	100	200	
		SKj \ SRi	SK1	SK2	SK3	SK4	WW
70	0	FW	30	40			
50	50	SR1	20	30			
100	100	SR2		20	80		
70	150	SR3		10		60	
60	250	SR4				10	50

FIGURE 8.26
Identification of material path. (From Ng and Foo, 2006.)

F_{SRi} (t/h)	C_{SRi} (ppm)	F_{SKj} (t/h) SKj SRi	50 SK1	100 SK2	80 SK3	70 SK4	50 WW
		C_{SKj} (ppm)	20	50	100	200	
70	0	FW	30	40			
50	50	SR1	20	30			
100	100	SR2		20	80		
70	150	SR3		10 ⟶ 10		60	
60	250	SR4				10 ⟶ 10	50

FIGURE 8.27
Shifting of source flowrate along material path. (From Ng and Foo, 2006.)

path consists of two matches (SK2–SR3 and SK4–SR4) with the smallest flowrates, i.e., 10 t/h. These matches may be removed from the path by adding 10 t/h of freshwater flowrates to the RCN.

As shown in Figure 8.27, 10 t/h of water is moved from the SK2–SR3 match along the water path to the SK4–SR3 match. Having fed with 10t/h of water, SK4 now experiences an excess of water. Hence, the SK4–SR4 match that supplies water to SK4 (10 t/h) is no longer needed. The excess water flowrate of SK4–SR4 match can then be moved along the path to the WW–SR4 match (Figure 8.27). This results in a larger wastewater flowrate of 60 t/h (from 50 t/h in Figure 8.26).

Next, we also notice that SK2 has a flowrate deficit of 10 t/h. The water deficit is supplemented by freshwater, which means that the freshwater flowrate is also increased by 10 t/h, i.e., to 80 t/h (from 70 t/h in Figure 8.26). In other words, two interconnections, i.e., SK2–SR3 and SK4–SR4, are removed from the network with the expense of 10 t/h of freshwater and wastewater penalties. The result of the MPA is shown in Figure 8.28, i.e., an RCN with eight interconnections.

Note that a new material path may also be defined even though some of its matches may not exist in the network yet. One such water path is shown for the problem in Figure 8.29. The path connects the matches of SK1–FW, SK1–SR1, SK2–SR1, SK2–SR2, and WW–SR2. For this path, the WW–SR2 match does not exist in the network yet, but may be easily added as the discharge flowrate for SR2.

As observed from Figure 8.29, the matches with smallest flowrate in this path correspond to SK1–SR1 and SK2–SR2. By adding an additional 20 t/h freshwater flowrate to SK1 (via SK1–FW match), the SK1–SR1 and SK2–SR2 matches are removed, coupled with the emergence of the new wastewater match, i.e., WW–SR2 of 20 t/h. Even though a new match is created for this case, it is offset by the removal of two other matches (SK1–SR1 and SK2–SR2), i.e., a net result of one match removal. Note that with the addition of 20 t/h freshwater, the total freshwater and

F_{SRi} (t/h)	C_{SRi} (ppm)	F_{SKj} (t/h)	50	100	80	70	60
		C_{SKj} (ppm)	20	50	100	200	
		SKj / SRi	SK1	SK2	SK3	SK4	WW
80	0	FW	30	50			
50	50	SR1	20	30			
100	100	SR2		20	80		
70	150	SR3				70	
60	250	SR4					60

FIGURE 8.28
Removal of SK2–SR3 and SK4–SR4 matches with 10 t/h freshwater penalty. (From Ng and Foo, 2006.)

F_{SRi} (t/h)	C_{SRi} (ppm)	F_{SKj} (t/h)	50	100	80	70	60
		C_{SKj} (ppm)	20	50	100	200	
		SKj / SRi	SK1	SK2	SK3	SK4	WW
80	0	FW	30	50			
50	50	SR1	20	30			
100	100	SR2		20	80		
70	150	SR3				70	
60	250	SR4					60

FIGURE 8.29
Path involving a new wastewater stream. (From Ng and Foo, 2006.)

wastewater flowrates are increased to 100 and 80 t/h, respectively (from 80 and 60 t/h, respectively, Figure 8.28). In other words, the removal of the sink-source match is at the expense of 20 t/h of water penalty. The result of this path analysis is a network with seven matches, as shown in Figure 8.30.

A simple comparison among all schemes is shown in Table 8.4. Network evolution with SSA does not lead to any flowrate penalty but can only reduce the number of interconnections to 10 matches. Hence, RCN synthesized by NNA and evolved by SSA are degenerate solutions with the same water flowrate targets. On the other hand, evolution via MPA reduces the number of interconnection further to seven matches but with additional water flowrate penalty. These options serve as alternatives in trading off water recovery and network complexity.

TABLE 8.4

Comparison of Various Network Evolution Schemes

Schemes	Number of Matches	Freshwater Flowrate (t/h)	Wastewater Flowrate (t/h)	Water Penalty (t/h)
NNA	12			
SSA	11	70	50	0
	10			
MPA	8	80	60	10
	7	100	80	30

F_{SRi} (t/h)	C_{SRi} (ppm)	F_{SKj} (t/h)	50	100	80	70	80
		C_{SKj} (ppm)	20	50	100	200	
		SKj / SRi	SK1	SK2	SK3	SK4	WW
100	0	FW	50	50			
50	50	SR1		30			
100	100	SR2			80		20
70	150	SR3				70	
60	250	SR4					60

FIGURE 8.30
Removal of SK1–SR1 and SK2–SR2 matches with additional 20 t/h freshwater penalty. (From Ng and Foo, 2006.)

8.5 Additional Readings

Apart from the NNA, there exist many other network design techniques for RCN synthesis, e.g., *water grid diagram* (Wang and Smith, 1994), water main method (Kuo and Smith, 1998; Feng and Seider, 2001), source-sink mapping diagram (El-Halwagi, 1997), water source diagram (Gomes et al., 2006), etc. A review that compares these techniques is found in the review paper by Foo (2009). Besides, the concepts of SSA and MPA are also extended into a mathematical optimization model (Das et al., 2009).

PROBLEMS
Water Minimization Problems

8.1 Revisit Example 8.1 (Polley and Polley, 2000), synthesize the water networks for the following cases:

a. Synthesize a direct reuse/recycle network, when single freshwater feed (10 ppm of impurity), and its minimum flowrate is determined as 75 t/h.

b. Synthesize a direct reuse/recycle network when two freshwater feeds are available for use, i.e., pure (0 ppm) and impure freshwater feeds (80 ppm), with targeted flowrates of 56.26 and 43.75 t/h, respectively.

c. Synthesize a direct reuse/recycle network when three freshwater feeds are available for the RCN, i.e., one pure (0 ppm) and two impure freshwater sources (20 and 50 ppm), with targeted flowrates of 0, 50, and 50 t/h, respectively.

d. Synthesize a water regeneration network when 20 t/h of freshwater feed (0 ppm), and 53.57 t/h of regenerated flowrate (10 ppm) are available for use. Note that the regenerated flowrate is originated from SR4 of 250 ppm (see Table 8.1).

e. Redo task (d) when 60 t/h of SR4 (250 ppm) is purified to 25 ppm (C_{Rout}). Note that freshwater feed (0 ppm) remains at 20 t/h.

8.2 For the acrylonitrile case study in Examples 3.1, 5.1, and 6.3, synthesize the RCN for the following cases. Show the RCN in a conventional PFD. For all cases, assume that pure freshwater feed is used.

a. A reuse/recycle network where recovery between process source(s) and sink(s) is given higher priority, before freshwater is used. The minimum freshwater and wastewater flowrates are targeted as 2.1 and 8.2 kg/s, respectively. Use the limiting data in Table 3.1.

b. Redo (a) by considering an additional process constraint, i.e., no water recovery between any process sources to boiler feed water sink (SK1). Note that the water flowrate targets remain the same as in (a) (see the resulting flowsheet in Figure 3.11).

c. An purification unit is added to the RCN in (c), with an outlet concentration of 11.6 ppm. Synthesize a water regeneration network that meets the freshwater and wastewater flowrate targets of 1.2 and 7.3 kg/s. (see Example 6.3 for details.

d. Reconstruct the RCN in (b) when the steam-jet ejector is replaced by a vacuum pump, which leads to a removal of SK1 and SR4. Targeting results show that freshwater is completely removed and wastewater flowrate is reduced to 5.9 kg/s. (see details in Section 7.2; the resulting flowsheet is given in Figure 1.2).

8.3 For the tire-to-fuel process in Problems 2.1, 3.1, and 7.3, synthesize direct water reuse/recycle networks for the following cases, assuming that pure freshwater is used.

a. Base case process as in Problem 2.1 (see limiting data in column 3 of Table 7.2). Targeting indicates that the minimum freshwater flowrate is 0.135 kg/s. Present the RCN in a conventional PFD.

b. Design for the "worst case" among the various process change scenarios in Problem 7.3.

8.4 Synthesize a water reuse/recycle network for the palm oil mill in Problem 4.8, assuming that pure freshwater is used. The minimum freshwater flowrate is determined as 1.63 t/h.

8.5 Synthesize a water reuse/recycle network for the steel plant in Problem 4.9, assuming that pure freshwater is used. The minimum freshwater flowrate is determined as 2234.21 t/h.

8.6 Synthesize a water reuse/recycle network for the bulk chemical production case in Problems 3.5 and 5.4. Use the limiting data in Table 3.12. Assume that pure freshwater is used. The minimum freshwater flowrate is determined as 11.1 kg/s. Evolve the preliminary network to remove interconnections of small flowrates (<1 t/d) but do not increase the flowrate targets. Present the RCN in a conventional PFD.

8.7 Synthesize the following water networks for the specialty chemical production process in Problems 4.6 and 6.5. Limiting data for water minimization are summarized in Table 4.22. Assume that pure freshwater is used.

a. A direct reuse/recycle network with minimum freshwater flowrate of 90.64 t/h. Can you synthesize another alternative reuse/recycle scheme to avoid the reuse of effluent from the cooling system to the reactor?

b. A water regeneration network with a purification unit (C_{Rout} = 60 ppm). The flowrate targeting procedure determines that the freshwater, regenerated water, and wastewater have a flowrate of 40.0, 55.4, and 0 t/h, respectively. Ignore the process constraint in (a), i.e., the reuse of effluent from cooling system to reactor is allowed.

c. A water regeneration network with a purification unit (C_{Rout} = 200 ppm). The flowrate targeting procedure determines that the freshwater, regenerated water, and wastewater have a flowrate of 42.25, 67.75, and 2.25 t/h, respectively. Ignore the process constraint in (a), i.e., the reuse of effluent from cooling system to reactor is allowed.

8.8 Synthesize the water reuse/recycle network for the organic chemical production in Examples 3.6 and 4.4 (limiting data are given in Table 4.12). The targeting step (see Figure 3.18 and Table 4.15) indicates that it is a zero-discharge network with a 60 t/h freshwater flowrate (0 ppm impurity content). (Hint: Equation 8.2 may be omitted when matching the last water sink). Is the network structure similar to that developed by insights from targeting insights (solve Problem 4.7 to compare answers)?

8.9 For the water minimization case in Examples 3.5 and 4.3 (threshold problem; limiting data are given in Table 4.10), complete the following tasks:

a. Synthesize the water network. The targeting step indicates that it is a zero FR network with a wastewater discharge of 20 t/h (Figure 3.16 and Table 4.11; hint: Equation 8.2 may be omitted when matching the last water sink). Is the network structure similar to that developed by insights from targeting insights (solve Problem 4.7 to compare answers)?

TABLE 8.5

Limiting Water Data for Problem 8.10

SK_j	F_{SK_j} (t/h)	C_{SK_j} (ppm)	SR_i	F_{SR_i} (t/h)	C_{SR_i} (ppm)
SK1	1200	120	SR1	500	100
SK2	800	105	SR2	2000	110
SK3	500	80	SR3	400	110
			SR4	300	60

Source: Jacob et al. (2002).

    b. Evolve the preliminary network with SSA to reduce one wastewater stream but do not increase its water flowrates (hint: treat this as a lower quality region).

8.10 Table 8.5 shows the limiting data for a water minimization case study (Jacob et al., 2002). Complete the following tasks:

    a. Synthesize the water reuse/recycle network for this case. The targeting stage indicates that this is a zero FR network, with a minimum wastewater flowrate of 700 t/h. (hint: Equation 8.2 may be omitted when matching the last water sink).

    b. Evolve the preliminary network with SSA to reduce network connection but do not increase its water flowrates (hint: treat this as a lower quality region).

    c. If we were to remove one additional interconnection, what would be the minimum freshwater flowrate penalty to be incurred? Which sink-source match should be removed to achieve this target?

8.11 Synthesize the water reuse/recycle network for the organic chemical production case in Problem 3.6 (limiting data are found in Figure 3.23). The targeting step indicates that this is a network with zero freshwater and zero wastewater flowrates. Verify this with the water network (hint: Equation 8.2 may be omitted when matching the last water sink). Present the RCN in a conventional PFD.

8.12 For the water minimization case in Problem 6.6 (Sorin and Bédard, 1999), perform the following tasks:

    a. Synthesize a water reuse/recycle network, assuming that pure freshwater feed is used. The minimum freshwater flowrate is reported as 200 t/h. Represent the RCN in matching matrix form.

    b. Synthesize a water regeneration network. The minimum freshwater and regenerated flowrates are reported as 120 (0 ppm) and 88.89 t/h (C_{Rout} = 10 ppm), respectively.

    c. Evolve the RCN obtained in (a) using MPA. How many interconnections are removed having 30 t/h of freshwater penalty?

    d. If the resulting RCN in (c) is to be evolved further, what is the minimum freshwater penalty that is to be incurred? How many interconnections are being removed in this case?

8.13 For the coal mine water system in Problems 5.7 and 6.3, synthesize the water network for the following case:

a. Direct water reuse/recycle network. The freshwater and wastewater flowrates are determined as 386.51 t/d.

b. Water regeneration network. With the use of purification unit (C_{Rout} = 6.7 ppm), the freshwater and wastewater flowrates are both determined as 196.7, while the regenerated flowrate is determined as 329.8 t/d, originated from the source SR1 (domestic water; 118 t/d) and SR2 (excavation; 211.8 t/d).

8.14 Obtain a different network configuration for Example 8.4 for the following case:

a. Identify another water path different from that given in Figure 8.26. Evolve the water network by adding 10 t/h of freshwater penalty.

b. Identify another water path different from that given in Figure 8.29. Evolve the water network by adding 20 t/h of freshwater penalty. Why does the number of interconnections remain the same?

8.15 For the Kraft paper process in Problems 2.2, 4.1, and 6.13 synthesize the water network for the following water schemes (assume that pure freshwater is used):

a. Direct water reuse/recycle—the minimum freshwater flowrate is targeted as 89.9 t/h.

b. Evolve network obtained in (a) with WPA (possible path: FW-SK3, SK3-W14, SK1-W13, W13-WW).

c. Water regeneration network with purification unit of C_{Rout} = 10 ppm—the ultimate flowrate targets are given as 0 t/h (freshwater), 101.96 t/h (wastewater), and 134.85 t/h (regenerated flowrate), respectively.

d. Water regeneration network with purification unit of C_{Rout} = 15 ppm—the ultimate flowrate targets are given as 2.73 t/h (freshwater), 104.69 t/h (wastewater), and 174.34 t/h (regenerated flowrate), respectively.

e. Water regeneration network with air stripping being used as purification unit with C_{Rout} = 1.625 ppm. The flowrate targets are given as 0 t/h (freshwater), 101.96 t/h (wastewater), and 95.05 t/h (regenerated water), respectively.

Note: For (b)–(d), sources to be sent for regeneration may be obtained from insights gained from the targeting stage (i.e., by solving Problem 6.13(c)).

8.16 For the classical water minimization case (Wang and Smith, 1994) in Example 8.3, synthesize the water network for the following water recovery schemes, assuming that pure freshwater is used. Limiting data are given in Table 8.3.

a. Direct water reuse/recycle—the minimum freshwater flowrate is targeted as 90 t/h.

b. Revisit the water regeneration network in Example 8.3, where purification unit of $C_{Rout} = 5$ ppm is used. Instead of purifying 23.68 t/h from SR2 alone, the flowrate is contributed by SR1 (20 t/h) and SR2 (3.68 t/h). Design the RCN that achieves the flowrate targets, which is the degenerate solution of that in Example 8.3.

8.17 For the water minimization case in Problem 4.4 (Savelski and Bagajewicz, 2001), perform the following tasks:

a. Synthesize a water reuse/recycle network, assuming that a pure freshwater feed is used. The minimum freshwater flowrate is reported as 166.266 t/h. Represent the RCN in matching matrix form.

b. Evolve the RCN obtained in (a) using SSA. How many interconnections are removed as a result of SSA?

8.18 For the tricresyl phosphate process (El-Halwagi, 1997) in Problems 4.2 and 6.11, perform the following tasks:

a. Synthesize a water reuse/recycle network, assuming that a pure freshwater feed is used. The minimum freshwater and wastewater flowrates are both reported as 3.02 kg/s.

b. Evolve the RCN obtained in (a) using SSA/MPA. How many interconnections are removed as a result of SSA?

c. Synthesize a water regeneration network, when light oil is used for purification. The minimum freshwater (0 ppm) and regeneration flowrates (40 ppm) are determined as 2.45 and 1.2 kg/s, respectively. Note that the regeneration flowrate is originated from Washer I (76.36 ppm; see Table 4.18).

8.19 For the brick manufacturing process in Problems 3.9 and 6.12, perform the following tasks:

a. Synthesize a water reuse/recycle network, assuming that a pure freshwater feed is used. The minimum flowrates for pure, impure freshwater (55 ppm) and wastewater flowrates are reported as 111.2, 569.4, and 611.2 t/d.

b. Synthesize a water regeneration network. The minimum freshwater (0 ppm) and regeneration flowrates (10 ppm) are determined as 111.2 and 418.7 t/d, respectively. Note that the regeneration flowrate is originated from treated process water (180 ppm).

8.20 For the Kraft pulping process in Problems 2.8, 4.11, and 5.10 (Parthasarathy and Krishnagopalan, 2001), perform the following tasks:

a. Synthesize a water reuse/recycle network, assuming that a freshwater feed with a chloride content of 4.2 ppm is used. The minimum flowrates for freshwater and wastewater flowrates are reported as 24,009 and 24,795 t/d.

b. Evolve the RCN obtained in (a) using SSA/MPA. How many interconnections are removed as a result of SSA?

c. Synthesize a water reuse/recycle network that achieves the minimum OC in Problem 5.10, assuming that the freshwater feed has a chloride content of 4.2 ppm. Please utilize the optimum operating flowrate of the water sinks identified in Problem 5.10.

d. Synthesize a water regeneration network. For this case, zero freshwater is needed and regeneration flowrates (10 ppm) is determined as 24,628 t/d. Note that the regeneration flowrate is originated from sources SR2, SR3, and SR4.

8.21 For the alkene production problem in Problems 4.14 and 6.8 (Foo, 2019), perform the following tasks:

a. Synthesize a water reuse/recycle network, assuming that a pure freshwater is used. The minimum flowrates for freshwater and wastewater flowrates are reported as 91.9 and 133.3 t/h.

b. Synthesize a water regeneration network for Problem 6.8 (task (a)). For this case, a partitioning purification unit with outlet concentration of 0.8% and recovery factor of 0.8 is used. The freshwater and regeneration flowrates are determined as 21.4 and 88.1 t/h. Note that the regeneration flowrate is originated from sources SR2 (Absorber 2).

c. Synthesize a water regeneration network for Problem 6.8 (task (b)). For this case, a single pass purification unit with removal ratio (RR) of 0.8 is used. The freshwater and regeneration flowrates are determined as 21.4 and 74.5 t/h, when the purification cost is 1.2 $/t. Note that the regeneration flowrate is originated from sources SR2 (Absorber 2).

Utility Gas Recovery Problems

8.22 For the magnetic tape manufacturing process in Problem 2.5, design the RCN and present the nitrogen recovery scheme in a conventional PFD. The minimum pure fresh nitrogen needed for the network is targeted as 5.6 kg/s. Use the limiting data in Table 3.12.

8.23 For Problem 3.8, design the RCN and present the oxygen recovery scheme in a conventional PFD. The targeting stage identifies that the minimum fresh oxygen (with 5% impurity) needed for the network is 30 kg/s (*hint:* Equation 8.2 may be omitted when matching the last two sinks).

8.24 Synthesize a reuse/recycle network for the refinery hydrogen network in Example 2.4. The limiting data are given in Table 2.2. For a fresh hydrogen feed with 1% impurity, the targeting stage identifies that its minimum flowrate is 182.86 MMscfd (see Figure 3.12). Design also its regeneration network when a purification unit (outlet concentration of 5%) is used, where the flowrate targets are given in Figure 6.27 (Example 6.6). Present the RCNs in conventional PFDs (see Figure 1.6 for result).

8.25 Synthesize a reuse/recycle network for the refinery hydrogen network in Problems 2.6 and 4.10. The fresh hydrogen feed has an impurity content of 5%. The minimum flowrate for the hydrogen feed is identified as 268.82 mol/s.

8.26 Synthesize a reuse/recycle network for the hydrogen recovery in Problem 4.16. For the fresh hydrogen feed of 0.1% impurity content, the minimum flowrate for hydrogen feed is identified as 125.2 mol/s.

8.27 For the hydrogen recovery in Problems 2.10, 4.18, and 6.16, solve the following tasks:

a. Synthesize a reuse/recycle network, where 6.36 t/h of fresh hydrogen feed is available with 7.44% of impurity content.

b. Synthesize a water regeneration network when the hydrogen source from CCR is purified to an outlet impurity of 5%, and the unit has a recovery factor of 0.9 (may be approximated with its simplified form in Equation 6.46). The flowrates for fresh hydrogen feed and purification flowrates are identified as 5.07 and 3.17 t/h.

Property Integration Problems

8.28 Revisit the metal degreasing process in Example 8.2. Synthesize the RCN for the case where the thermal processing unit is operated at a lower temperature, i.e., 430 K, which in turn leads to lower RVP value as well as lower operator value (see discussion in Example 7.1). Represent the RCN in a conventional PFD (see answer in Figure 1.4).

8.29 Synthesize an RCN for fiber recovery in the paper-making process in Examples 2.6 and 4.5. The limiting data are given in Table 4.16. The targeting stage identifies that the minimum fresh fiber needed for the RCN is 14.98 t/h (see Example 4.5). Present the RCN in a conventional PFD. Hint: be careful with the inferior property operator value of the fresh fiber feed.

8.30 For the microelectronics manufacturing facility in Problem 3.13, synthesize the RCN for the recovery of its ultrapure water. The ultrapure water (FR) has a resistivity value of 18,000 kΩ/cm. The targeting stage indicates that the minimum flowrate for the ultrapure water is 972 gal/min. Present the RCN in a conventional PFD (refer to Figure 3.26).

8.31 For the wafer fabrication process in Problem 4.23, synthesize the reuse/recycle network for the following section of the plant:

a. Fabrication section (FAB) only—flowrate targeting indicates that the minimum ultrapure water (UPW, 18 MΩ m) requirement is 1516.47 t/h (Problem 4.23a). Present the RCN in a conventional PFD (refer to Figure 2.25).

b. Entire plant—this includes cleaning, cooling water makeup, and scrubber operations (see Figure 4.6). Note that apart from ultrapure water, reject streams from ultrafiltration (UF) and reverse osmosis (RO) units may also be used. The targeting stage indicates that the minimum flowrates for these sources are 1516.47, 928.45, and 649.92 t/h (Problem 4.23b). Develop two alternative designs for this case, with and without the recovery of UF and RO reject streams, respectively.

References

Das, A. K., Shenoy, U. V., and Bandyopadhyay, S. 2009. Evolution of resource allocation networks, *Industrial and Engineering Chemistry Research*, 48, 7152–7167.

El-Halwagi, M. M.1997. *Pollution Prevention Through Process Integration: Systematic Design Tools*. San Diego, CA: Academic Press.

Feng, X. and Seider, W. D. 2001. New structure and design method for water networks, *Industrial and Engineering Chemistry Research*, 40, 6140–6146.

Foo, D. C. Y. 2009. A state-of-the-art review of pinch analysis techniques for water network synthesis, *Industrial and Engineering Chemistry Research*, 48(11), 5125–5159.

Foo, D. C. Y. 2019. Optimize plant water use from your desk. *Chemical Engineering Progress*, 115, 45–51 (February 2019).

Gomes, J. F. S., Queiroz, E. M., and Pessoa, F. L. P. 2006. Design procedure for water/wastewater minimization: Single contaminant, *Journal of Cleaner Production*, 15, 474–485.

Jacob, J., Kaipe, H., Couderc, F., and Paris, J. 2002. Water network analysis in pulp and paper processes by pinch and linear programming techniques, *Chemical Engineering Communications*, 189(2), 184–206.

Kuo, W. C. J. and Smith, R. 1998. Designing for the interactions between water-use and effluent treatment, *Chemical Engineering Research and Design*, 76, 287–301.

Ng, D. K. S. and Foo, D. C. Y. 2006. Evolution of water network with improved source shift algorithm and water path analysis, *Industrial and Engineering Chemistry Research*, 45(24), 8095–8104.

Parthasarathy, G. and Krishnagopalan, G. 2001. Systematic reallocation of aqueous resources using mass integration in a typical pulp mill. *Advances in Environmental Research*, 5, 61–79.

Polley, G. T. and Polley, H. L. 2000. Design better water networks, *Chemical Engineering Progress*, 96(2), 47–52.

Prakash, R. and Shenoy, U. V. 2005a. Targeting and design of water networks for fixed flowrate and fixed contaminant load operations, *Chemical Engineering Science*, 60(1), 255–268.

Prakash, R. and Shenoy, U. V. 2005b. Design and evolution of water networks by source shifts, *Chemical Engineering Science*, 60(7), 2089–2093.

Savelski, M. J. and Bagajewicz, M. J. 2001. Algorithmic procedure to design water utilization systems featuring a single contaminant in process plants, *Chemical Engineering Science*, 56, 1897–1911.

Sorin, M. and Bédard, S. 1999. The global pinch point in water reuse networks, *Process Safety and Environmental Protection*, 77, 305–308.

Wang, Y. P. and Smith, R. 1994. Wastewater minimisation. *Chemical Engineering Science*, 49, 981–1006.

9

Synthesis of Resource Conservation Networks: A Superstructural Approach

In this chapter, the most commonly used mathematical optimization technique, i.e., *superstructural approach*, is used to synthesize resource conservation networks (RCNs). These include direct reuse/recycle and material regeneration network. The model may also be extended to include process constraints such as forbidden/compulsory matches, reduced piping connection, and cost consideration. A simple extension is also shown for the synthesis of inter-plant resource conservation networks (IPRCNs).

9.1 Superstructural Model for Direct Reuse/Recycle Network

To synthesize an RCN, a superstructural representation is first constructed. For a direct reuse/recycle network, a common representation is shown in Figure 9.1 (reproduced from Figure 1.17). As shown, all sources are matched

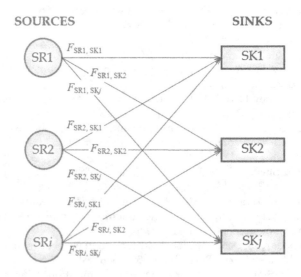

FIGURE 9.1
Superstructural model for direct reuse/recycle.

DOI: 10.1201/9781003173977-10

F_{SRi}	q_{SRi}	F_{SKj} / q_{SKj} / SKj ∖ SRi	F_{SK1} / q_{SK1} / SK1	F_{SK2} / q_{SK2} / SK2	F_{SK3} / q_{SK3} / SK3	F_{SKj} / q_{SKj} / SKj	F_D / q_D / WD
F_{R1}	q_{R1}	FR	$F_{R1, SK1}$	$F_{R1, SK2}$	$F_{R1, SK3}$	$F_{R1, SKj}$	-
F_{Rf}	q_{Rf}	FRf	$F_{Rf, SK1}$	$F_{Rf, SK2}$	$F_{Rf, SK3}$	$F_{Rf, SKj}$	-
F_{SR1}	q_{SR1}	SR1	$F_{SR1, SK1}$	$F_{SR1, SK2}$	$F_{SR1, SK3}$	$F_{SR1, SKj}$	$F_{SR1, D}$
F_{SR2}	q_{SR2}	SR2	$F_{SR2, SK1}$	$F_{SR2, SK2}$	$F_{SR2, SK3}$	$F_{SR2, SKj}$	$F_{SR2, D}$
F_{SR3}	q_{SR3}	SR3	$F_{SR3, SK1}$	$F_{SR3, SK2}$	$F_{SR3, SK3}$	$F_{SR3, SKj}$	$F_{SR3, D}$
F_{SRj}	q_{SRj}	SRi	$F_{SRj, SK1}$	$F_{SRj, SK2}$	$F_{SRj, SK3}$	$F_{SRj, SKj}$	$F_{SRj, D}$

FIGURE 9.2
Superstructural model for direct reuse/recycle in matching matrix form.

to the sinks via connections. Note that a fresh resource is also considered as a source, while waste discharge is considered as a sink. Each connection is represented by a flowrate variable ($F_{SRi,SKj}$), where its value is to be determined as the result of the optimization model.

Another representation for ease of model formulation takes the form of a *matching matrix* as introduced in Chapter 7, with the revised form shown in Figure 9.2. In the matching matrix, all sources including fresh resource (FR) are arranged as rows, while the sinks including waste discharge (WD) as columns, both in a descending order of quality index. Figures in the cells indicate the allocation flowrates ($F_{SRi,SKj}$) between sources and sinks in the RCN. The unutilized source flowrate will be sent for waste discharge ($F_{SRi,D}$). Note however that no match should be built between fresh resource and waste discharge (i.e., $F_{FR,WD} = 0$). As mentioned earlier, the values of these allocation flowrates are the results of the optimization model.

Next, the optimization model is formulated. Different optimization objective may be set. For a direct reuse/recycle network with single fresh resource, the objective can be set to determine the minimum flowrate of fresh resource (F_R), given by Equation 9.1:

$$\text{Minimize } F_R = \sum_j F_{R,SKj} \tag{9.1}$$

where $F_{R,SKj}$ is the flowrate of fresh *resource* that is allocated to sink j, which is the sum of the individual terms in the fourth row of the matching matrix in Figure 9.2.

For RCN with single or multiple fresh resources, the optimization objective can be set to minimize the operating cost (*OC*) given by

$$\text{Minimize } OC = \sum_r \sum_j F_{Rf,SKj} CT_{Rf} + F_D CT_D \qquad (9.2)$$

where $F_{Rf,SKj}$ is the flowrate of fresh resource f allocated to sink j, with unit cost of CT_{Rf}; F_D and CT_D are flowrate and unit cost of waste discharge, respectively.

When capital cost are considered, the *total annualized cost* (*TAC*) of the RCN can be calculated as follows:

$$\text{Minimize } TAC = OC(AOT) + CC(AF) \qquad (9.3)$$

where *AOT* is the annual operating time, *AF* is the annualizing factor for the capital cost (*CC*) correlation.

The previous optimization objective(s) is subject to the following constraints:

$$\sum_i F_{SRi,SKj} + \sum_r F_{Rf,SKj} = F_{SKj} \quad \forall j \qquad (9.4)$$

$$\sum_i F_{SRi,SKj} q_{SRi,c} + \sum_r F_{Rf,SRj} q_{Rf,c} \geq F_{SKj} q_{SKj,c} \quad \forall j \; \forall c \qquad (9.5)$$

$$\sum_j F_{SRi,SKj} + F_{SRi,D} = F_{SRi} \quad \forall i \qquad (9.6)$$

$$F_{SRi,SKj}, F_{Rr,SKj}, F_{SRi,D} \geq 0 \; \forall i, j \qquad (9.7)$$

The constraint in Equation 9.4 describes that flowrate requirement of sink j is fulfilled by various process source i ($F_{SRi,SKj}$) as well as the fresh resource f ($F_{Rf,SKj}$). Equation 9.5 describes that the process sources and fresh resource feed must fulfill the minimum quality requirement of the process sinks. Note that Equation 9.5 is applicable for multiple impurity problems which is commonly encountered. The flowrate balance in Equation 9.6 indicates that the flowrate of each process source may be allocated to sinks, while the unutilized source will be sent for waste discharge ($F_{SRi,D}$). Equation 9.7 indicates that all flowrate variables of the superstructural model should take non-negative values. Since the flowrates and quality of the process sink and source as well as the fresh resource feed are parameters, Equations 9.4 through 9.7 are linear. In other words, the superstructural model is a linear program (LP) which can be solved for global optimal solution, if it exists.

TABLE 9.1

Limiting Water Data for AN Production Case Study

Water Sinks, SK_j			
j	Stream	Flowrate F_{SKj} (kg/s)	NH$_3$ Concentration C_{SKj} (ppm)
1	Boiler feed water	1.2	0
2	Scrubber inlet	5.8	10
Water Sources, SR$_i$			
i	Stream	Flowrate F_{SRi} (kg/s)	NH$_3$ Concentration C_{SRi} (ppm)
1	Distillation bottoms	0.8	0
2	Off-gas condensate	5	14
3	Aqueous layer	5.9	25
4	Ejector condensate	1.4	34

Source: El-Halwagi (1997).

Example 9.1 Direct water reuse/recycle network (single contaminant)

The Acrylonitrile (AN) production case study in Examples 3.1 and 4.1 (El-Halwagi, 1997) is revisited. Use MS Excel to synthesize a direct water reuse/recycle network with minimum freshwater consumption. The limiting water data for the case study are shown in Table 9.1 (reproduced from Table 4.2).

SOLUTION

A superstructure for the water network is constructed in the matching matrix form using MS Excel spreadsheet in Figure 9.3. All sources

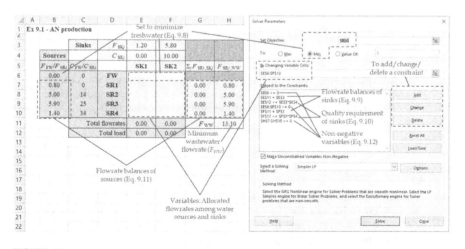

FIGURE 9.3

Matching matrix for direct reuse/recycle scheme in Example 9.1 (all flowrate terms in kg/s; concentration in ppm).

including freshwater (FW) are arranged as rows 6–10, while sinks including wastewater discharge (WW) are arranged as columns E–H of the spreadsheet.

To formulate an optimization model to minimize the total freshwater flowrate (F_{FW}) of the water network, the objective in Equation 9.1 takes the revised form of Equation 9.8:

$$\text{Minimize } F_{FW} = \sum_j F_{FW,SKj} \tag{9.8}$$

where $F_{FW,SKj}$ is the fresh water flowrate allocated to sink j. One may easily derive this flowrate balance from row 6 of the matching matrix in Figure 9.3.

Similarly, the other constraints in Equations 9.4–9.7 take their revised form as in Equations 9.9–9.12.

Equation 9.9 describes the flowrate balances for process sinks. Similarly, one can derive these flowrate balances by observing columns 4 and 5 of Figure 9.3:

$$\sum_i F_{SRi,SKj} + F_{FW,SKj} = F_{SKj} \ \ \forall j \tag{9.9}$$

The quality requirement of the sinks is described by Equation 9.10. Note however that the inequality sign takes a reverse form as that in Equation 9.5, as the quality index is described by concentration of impurity c. In other words, impurity sent from the process sources (and those from freshwater source[s], if any) must not be more than the maximum impurity load that can be tolerated in the process sink:

$$\sum_i F_{SRi,SKj} C_{SRi,c} + F_{FW,SRj} C_{FW,c} \leq F_{SKj} C_{SKj,c} \ \ \forall j \ \forall c \tag{9.10}$$

Equation 9.11 describes the flowrate balance of the process sources. As shown, each process source may be allocated to any process sinks, while the unutilized source will be sent for wastewater discharge ($F_{SRi,WW}$):

$$\sum_j F_{SRi,SKj} + F_{SRi,WW} = F_{SRi} \ \ \forall i \tag{9.11}$$

Equation 9.12 indicates that all flowrate variables take non-negative values. These include the flowrate terms for the matches between source i ($F_{SRi,SKj}$) and freshwater to sink j ($F_{FW,SKj}$), as well as the wastewater stream from source i ($F_{SRi,WW}$):

$$F_{SRi,SKj}, F_{FW,SKj}, F_{SRi,WW} \geq 0 \ \ \forall i, j \tag{9.12}$$

It is also convenient to determine the total flowrate of the wastewater streams (F_{WW}), given by the last column of the matching matrix in Figure 9.3:

$$F_{WW} = \sum_i F_{SRi,WW} \qquad (9.13)$$

Since the model is an LP, it may be solved for global solution with MS Excel with Solver add-in[1]. The configuration of the model in MS Excel and its solution is shown in Figure 9.4, where the optimized variables are shown. As shown, the minimum freshwater flowrate of the water network is determined as 2.06 kg/s (cell B6), while the wastewater flowrate is determined as 8.16 kg/s (cell H11). This matches the answer in Examples 3.1 and 4.1 (see Figure 3.4 or Table 4.5).

Finally, note that the concentration of the wastewater discharge is not determined as part of the optimization model. This is to avoid the bilinear term (product of unknown flowrate and concentration of wastewater stream) that would be present in the optimization model, which in turn leads to nonlinear program (NLP). Once the optimization model is solved, we can now determine the concentration of the wastewater discharge via mass balance. For this case, the concentration of the wastewater stream (C_{WW}) is calculated as follows:

$$C_{WW} = \frac{0.86(14) + 5.9(25) + 1.4(34)}{0.86 + 5.9 + 1.4} = 25.4 \text{ ppm}$$

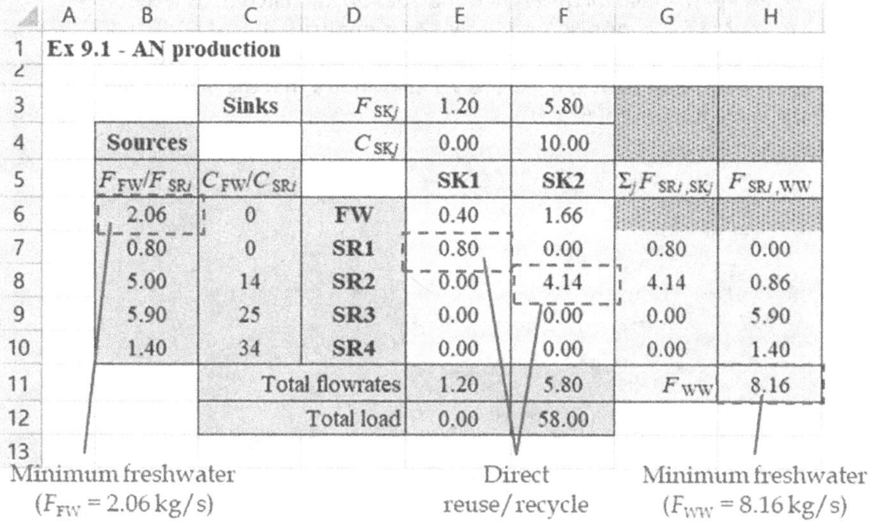

	A	B	C	D	E	F	G	H
1	**Ex 9.1 - AN production**							
2								
3			**Sinks**	F_{SKj}	1.20	5.80		
4		**Sources**		C_{SKj}	0.00	10.00		
5		F_{FW}/F_{SRj}	C_{FW}/C_{SRj}		**SK1**	**SK2**	$\Sigma_j F_{SRj,SKj}$	$F_{SRj,WW}$
6		2.06	0	FW	0.40	1.66		
7		0.80	0	SR1	0.80	0.00	0.80	0.00
8		5.00	14	SR2	0.00	4.14	4.14	0.86
9		5.90	25	SR3	0.00	0.00	0.00	5.90
10		1.40	34	SR4	0.00	0.00	0.00	1.40
11				Total flowrates	1.20	5.80	F_{WW}	8.16
12				Total load	0.00	58.00		
13								

Minimum freshwater (F_{FW} = 2.06 kg/s) — Direct reuse/recycle — Minimum freshwater (F_{WW} = 8.16 kg/s)

FIGURE 9.4
Network structure for RCN of AN case in Example 9.1 (all flowrate terms in kg/s; concentration in ppm; impurity load in mg/s).

[1] Solver add-in may be activated with the following path: File/Options/Add-ins/Excel Add-ins.

TABLE 9.2

Limiting Water Data for Multiple-Impurities Example

j	Sinks, SK_j	F_{SKj} (t/h)	$C_{SKj,A}$ (ppm)	$C_{SKj,B}$ (ppm)
1	SK1	40	0	25
2	SK2	35	80	30
i	Sources, SR_i	F_{SRi} (t/h)	$C_{SRi,A}$ (ppm)	$C_{SRi,B}$ (ppm)
1	SR1	40	100	75
2	SR2	35	240	90

Source: Wang and Smith (1994).

Example 9.2 Direct water reuse/recycle network (multiple contaminants)

A multiple-impurities example from Wang and Smith (1994) is analyzed for its water-saving potential. Synthesize a direct water reuse/recycle network with minimum freshwater consumption. The limiting water data for the case study are shown in Table 9.2. Note that the data have been converted into a *fixed flowrate problem* from the original *fixed load problem* (Wang and Smith, 1994).

SOLUTION

The superstructure model for this example is similar to that in Example 9.1, except that load of different impurities is calculated in different rows, i.e., row 11 for impurity A and row 12 for impurity B.

Solving the model resulted in the solution in Figure 9.5, where freshwater is identified as 61 t/h. Figure 9.5 also shows that the wastewater has the same flowrate of 61 t/h.

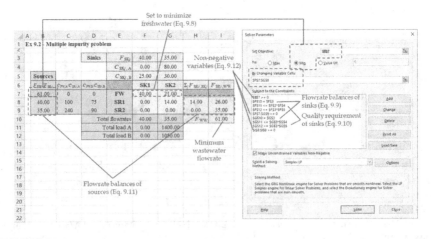

FIGURE 9.5

Solution for multiple-impurities problem in Example 9.2 (all flowrate terms in t/h; concentration in ppm; impurity load in g/s).

9.2 Incorporation of Process Constraints

The advantage of the superstructural approach is its flexibility to incorporate various process constraints, such as forbidden or compulsory matches between process sources and sinks, etc. The reasons for imposing forbidden matches may include impurity buildup (due to local recycling), geographical factors (process sink and source that are too far apart), contamination issues (e.g., for certain processes, the source contains other contaminants that may poison the catalyst in the reactor), etc. In contrast, one may impose compulsory matches to connect the process source that consists of a given species (e.g., solvent, catalyst, etc.) to a sink where the species is needed (e.g., solvent extraction unit, reactor, etc.).

Example 9.3 Incorporation of process constraints

Example 9.1 is revisited here. Figure 9.6 shows that the superstructural model in Example 9.1 determines that water source SR1 (distillation bottoms) is matched to sink SK1 (boiler feed water). In most cases, this is not a favorable solution, as boiler feed water would normally prefer freshwater feed, rather than recycled water. Besides, it is decided that SR1 is to be fully recovered to SK2 (scrubber inlet). Incorporate these process constraints into the superstructural model to synthesize an alternative network structure for the case study.

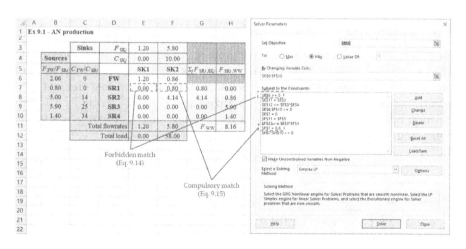

FIGURE 9.6
Network structure for RCN in Example 9.3 (all flowrate terms in kg/s; concentration in ppm; impurity load in mg/s).

SOLUTION

The superstructural model for this case remains similar to that in Example 9.1. Two additional constraints are added to impose the forbidden (Equation 9.14) and compulsory (Equation 9.15) match criteria:

$$F_{SR,SK1} = 0 \tag{9.14}$$

$$F_{SR1,SK2} = 0.8 \tag{9.15}$$

The setup for MS Excel and Solver, along with the optimization results are shown in Figure 9.6.

It is observed that the minimum freshwater and wastewater flowrates are determined as 2.06 and 8.15 kg/s, respectively, which is identical for the case in Example 9.1. This means that *degenerate* solutions exist for this problem, in which more than one network structure achieves the same objective, i.e., minimum fresh water flowrate for this case. The optimized variables are also shown in Figure 9.6. Similar to Example 9.1, the concentration for wastewater stream (C_{WW} = 25.4 ppm) is determined after the optimization model is solved.

Comparing the network structures in Figures 9.5 and 9.6, it is observed that only flowrates that involve matches among freshwater feed (FW), SR1, SK1, and SK2 have been changed, while flowrates of other matches remain identical.

9.3 Capital and Total Cost Estimations

Another advantage of the superstructural approach is its ability to estimate the capital cost (CC) of various RCN elements, e.g., piping connection and storage tank (for batch processes). The estimation of capital cost element then enables us to calculate the total annualized cost (TAC) of the RCN. CC correlation often includes variation and fixed items, such as those given in the following equation:

$$CC = aF^b + c \tag{9.16}$$

where

a and b are parameters for the variation terms, and are always expressed as a function of flowrate (F)

c is the parameter for the fixed term.

In order to incorporate capital (e.g., piping) cost correlation, the "big *M*" formulation is often used. This allows the activation of the fixed cost component by introducing a binary variable *(BIN)* in the optimization model. Hence, Equation 9.17 is revised as follows:

$$CC = aF^b + c \ BIN \qquad (9.17)$$

For instance, in a direct material reuse/recycle problem where the piping capital cost is to be minimized, binary term $B_{SRi,SKj}$ [0, 1] may be used to define the presence of a material recovery stream between source SR_i and sink SK_j, with flowrate $F_{SRi,SKj}$. The binary term is activated with Equation 9.18:

$$\frac{F_{SRi,SKj}}{M} \le B_{SRi,SKj} \qquad (9.18)$$

where M is a parameter with arbitrary large number (and hence the name "big M" formulation). Any value that is bigger than the potential flowrate term in Equation 9.18 may be used. When the optimization model determines that the flowrate term takes a value of zero, the fixed cost term in Equation 9.17 is forced to be zero, in order to minimize capital cost. In contrast, when the optimization model determines that the $F_{SRi,SKj}$ term has a nonzero value, the binary term will be forced as unity, thus activating the fixed cost component in Equation 9.17.

To determine the *TAC*, the capital cost is to be annualized so that it can be added to the operating cost *(OC)* following Equations 9.3. A commonly used expression for the annualized factor *(AF)* is given by Equations 9.19 (Smith, 2005):

$$AF = \frac{IR(1+IR)^{YR}}{(1+IR)^{YR} - 1} \qquad (9.19)$$

where

IR is the annual fractional interest rate

YR is the number of years considered for the analysis

Note, however, that introducing the binary term (Equations 9.17 and 9.18) leads the LP model (direct reuse/recycle scheme) into a mixed-integer linear problem (MILP). Instead of using MS Excel, commercial optimization software (e.g., LINGO, GAMS, etc.) should be used to solve the MILP model.

Example 9.4 Minimum cost solutions for direct reuse/recycle scheme

Example 9.3 is revisited. Determine the minimum *OC*, *CC* and the *TAC* for a direct water reuse/recycle network. Additional information for relevant cost calculation is given as follows.

1. For *OC* calculation, both freshwater (CT_{FW}) and wastewater (CT_{WW}) are assumed to have a unit cost of \$0.001/kg (\$1/ton).

TABLE 9.3

Distance (m) Between Water Sources/
Freshwater Feed with Water Sinks of
AN Case in Example 9.4

	SK1	SK2
FW	20	30
SR1	15	40
SR2	30	50
SR3	40	20
SR4	30	10

2. CC of the RCN is mainly contributed by the piping cost (*PC*) of reuse/recycle streams as well as the connections between freshwater feed and the sinks (wastewater streams are omitted). *PC* (in USD) for a stream of flowrate *F* is given as the revised form of Equation 9.16 (Kim and Smith, 2004):

$$PC = (2F + 250)\,DIST \tag{9.20}$$

where *DIST* is the individual distance between water sinks (SK1, SK2) and water sources (SR1–SR4), and freshwater feed (FW), respectively, given as in Table 9.3.

3. For calculation of the *TAC*, the AF in Equation 9.19 is to be used. The piping cost is annualized to a period of 5 years (*YR*), with an interest rate of 5% (*IR*). Assume an *AOT* of 8000 h.

SOLUTION

To determine the operating cost (*OC*), Equation 9.2 is used:

$$OC(\$/s) = F_{FW}CT_{FW} + F_{WW}CT_{WW} = (F_{FW} + F_{WW})\,(\$0.001/kg)$$

To determine the *PC*, Equation 9.20 is used in conjunction with Equation 9.18, to examine the presence/absence of a stream connection (see binary variables in Figure 9.7). These are given by the next 10 lines of constraint sets. Piping costs for matches between various sources (SR1-SR4) with sink SK1 are given in lines 1–4, while those with SK2 are given in lines 5–8, respectively. The constraint sets in lines 9 and 10 next determine the piping costs for matches between freshwater feed with both sinks.

$$PC_{SR1,SK1} = (2F_{SR1,SK1} + 250B_{SR1,SK1})15; \quad F_{SR1,SK1} \le 1000B_{SR1,SK1}$$

$$PC_{SR2,SK1} = (2F_{SR2,SK1} + 250B_{SR2,SK1})30; \quad F_{SR2,SK1} \le 1000B_{SR2,SK1}$$

$$PC_{SR3,SK1} = (2F_{SR3,SK1} + 250B_{SR3,SK1})40; \quad F_{SR3,SK1} \le 1000B_{SR3,SK1}$$

		F_{SKj}	1.2	5.8	F_{WW}
		C_{SKj}	0	10	
F_{SRi}	C_{SRi}	SKj ⟍ SRi	SK1	SK2	WW
F_{FW}	0	FW	$B_{FW, SK1}$	$B_{FW, SK2}$	
0.8	0	SR1	$B_{SR1, SK1}$	$B_{SR1, SK2}$	
5	14	SR2	$B_{SR2, SK1}$	$B_{SR2, SK2}$	
5.9	25	SR3	$B_{SR3, SK1}$	$B_{SR3, SK2}$	
1.4	34	SR4	$B_{SR4, SK1}$	$B_{SR4, SK2}$	

FIGURE 9.7
Binary variables of AN case in Example 9.4.

$$PC_{SR4,SK1} = \left(2F_{SR4,SK1} + 250B_{SR4,SK1}\right)30; \ F_{SR4,SK1} \le 1000B_{SR4,SK1}$$

$$PC_{SR1,SK2} = \left(2F_{SR1,SK2} + 250B_{SR1,SK2}\right)40; \ F_{SR1,SK2} \le 1000B_{SR1,SK2}$$

$$PC_{SR2,SK2} = \left(2F_{SR2,SK2} + 250B_{SR2,SK2}\right)50; \ F_{SR2,SK2} \le 1000B_{SR2,SK2}$$

$$PC_{SR3,SK2} = \left(2F_{SR3,SK2} + 250B_{SR3,SK2}\right)20; \ F_{SR3,SK2} \le 1000B_{SR3,SK2}$$

$$PC_{SR4,SK2} = \left(2F_{SR4,SK2} + 250B_{SR4,SK2}\right)10; \ F_{SR4,SK2} \le 1000B_{SR4,SK2}$$

$$PC_{FW,SK1} = \left(2F_{FW,SK1} + 250B_{FW,SK1}\right)20; \ F_{FW,SK1} \le 1000B_{FW,SK1}$$

$$PC_{FW,SK2} = \left(2F_{FW,SK2} + 250B_{FW,SK2}\right)30; \ F_{FW,SK2} \le 1000B_{FW,SK2}$$

Note that the arbitrary term M in Equation 9.18 takes the value of 1000 in this case. Note also that the binary terms in Equation 9.18 for all reuse/ recycle streams as well as those between freshwater feed and the sinks are represented in Figure 9.6.

CC of the RCN is dominated by the PC, given by the summation of those individual matches, i.e.,

$$CC = PC = PC_{SR1,SK1} + PC_{SR2,SK1} + PC_{SR3,SK1} + PC_{SR4,SK1} + PC_{SR1,SK2} + PC_{SR2,SK2}$$

$$+ PC_{SR3,SK2} + PC_{SR4,SK2} + PC_{FW,SK1} + PC_{FW,SK2}$$

Next, AF in Equation 9.19 is calculated as

$$AF = \frac{0.05(1+0.05)^5}{(1+0.05)^5 - 1} = 0.231$$

To determine the minimum *TAC*, Equation 9.3 is used:

$$\text{Minimize } TAC(\$/y) = OC(3600 \text{ s/h})(8000 \text{ h}) + CC(0.231)$$

The flowrate constraints are given as follows.
Equation 9.9 is used to describe the flowrate balances for process sinks:

$$F_{SR1,SK1} + F_{SR2,SK1} + F_{SR3,SK1} + F_{SR4,SK1} + F_{FW,SK1} = 1.2$$

$$F_{SR1,SK2} + F_{SR2,SK2} + F_{SR3,SK2} + F_{SR4,SK2} + F_{FW,SK2} = 5.8$$

Two additional constraints are added to impose forbidden (Equation 9.14) and compulsory (Equation 9.15) match criteria:

$$F_{SR,SK1} = 0$$

$$F_{SR1,SK2} = 0.8$$

The quality requirement of the sinks is described by Equation 9.10:

$$F_{SR1,SK1}(0) + F_{SR2,SK1}(14) + F_{SR3,SK1}(25) + F_{SR4,SK1}(34) + F_{FW,SK1}(0) \leq 1.2(0)$$

$$F_{SR1,SK2}(0) + F_{SR2,SK2}(14) + F_{SR3,SK2}(25) + F_{SR4,SK2}(34) + F_{FW,SK2}(0) \leq 5.8(0)$$

Equation 9.11 describes the flowrate balance of the process sources.

$$F_{SR1,SK1} + F_{SR1,SK2} + F_{SR1,WW} = 0.8$$

$$F_{SR2,SK1} + F_{SR2,SK2} + F_{SR2,WW} = 5$$

$$F_{SR3,SK1} + F_{SR3,SK2} + F_{SR3,WW} = 5.9$$

$$F_{SR4,SK1} + F_{SR4,SK2} + F_{SR4,WW} = 1.4$$

Equation 9.12 indicates that all flowrate variables take non-negative values.

$$F_{SR1,SK1}, F_{SR2,SK1}, F_{SR3,SK1}, F_{SR4,SK1} \geq 0$$

$$F_{SR1,SK2}, F_{SR2,SK2}, F_{SR3,SK2}, F_{SR4,SK2} \geq 0$$

$$F_{FW,SK1}, F_{FW,SK2} \geq 0$$

$$F_{SR1,WW}, F_{SR2,WW}, F_{SR3,WW}, F_{SR4,WW} \geq 0$$

It is convenient to determine the total wastewater flowrate (F_{WW}) with Equation 9.13:

$$F_{WW} = F_{SR1,WW} + F_{SR2,WW} + F_{SR3,WW} + F_{SR4,WW}$$

LINGO formulation for this MILP problem is given as[2]:

Model:
```
!Optimization objective: To minimize total annual cost (TAC);

min = TAC;
TAC = OC*3600*8000 + PC*0.231;
OC = (FW + WW)*0.001; !Unit cost of FW and WW = $1/ton;
PC = PFW1 + PFW2 + P11 + P21 + P31 + P41 + P12 + P22 + P32 +
P42;

!Total freshwater flowrate (FW);

FW = FW1 + FW2;

!Flowrate balance for process sinks;

F11 + F21 + F31 + F41 + FW1 = 1.2;
F12 + F22 + F32 + F42 + FW2 = 5.8;

!Process constraints;

F11 = 0; !Forbidden match between SR1 and SK1;
F12 = 0.8; !Compulsory match between SR1 and SK2;

!Impurity constraints for process sinks;

F11*0 + F21*14 + F31*25 + F41*34 + FW1*0 <= 1.2*0;
F12*0 + F22*14 + F32*25 + F42*34 + FW2*0 <= 5.8*10;

!Flowrate balance for process sources;

F11 + F12 + F1W = 0.8;
F21 + F22 + F2W= 5;
F31 + F32 + F3W =5.9;
F41 + F42 + F4W=1.4;

!Total wastewater flowrate (WW);

WW = F1W + F2W + F3W + F4W;

!All variables should take non-negative values;

F11 >= 0; F21 >= 0; F31 >= 0; F41 >= 0;
F12 >= 0; F22 >= 0; F32 >= 0; F42 >= 0;
FW1 >= 0; FW2 >= 0;
F1W >= 0; F2W >= 0; F3W >= 0; F4W >= 0;

!Capital cost estimation;

P11 = (2*F11 + 250*B11)*15;
P21 = (2*F21 + 250*B21)*30;
P31 = (2*F31 + 250*B31)*40;
P41 = (2*F41 + 250*B41)*30;
P12 = (2*F12 + 250*B12)*40;
P22 = (2*F22 + 250*B22)*50;
P32 = (2*F32 + 250*B32)*20;
P42 = (2*F42 + 250*B42)*10;
PFW1 = (2*FW1 + 250*BFW1)*20;
PFW2 = (2*FW2 + 250*BFW2)*30;
```

[2] The command of @ BIN (variable) is used to define binary variables in LINGO

```
!Binary variables that represent present/absent of a stream;

F11 <= M*B11; F21 <= M*B21; F31 <= M*B31; F41 <= M*B41;
F12 <= M*B12; F22 <= M*B22; F32 <= M*B32; F42 <= M*B42;
FW1 <= M*BFW1; FW2 <= M*BFW2;
M = 1000; !Arbitrary value;

!Defining binary variables;

@BIN (B11); @BIN (B21); @BIN (B31); @BIN (B41);
@BIN (B12); @BIN (B22); @BIN (B32); @BIN (B42);
@BIN (BFW1); @BIN (BFW2);
```

END

The following is the solution report generated by LINGO:

Global optimal solution found.	
Objective value:	302389.9
Extended solver steps:	0
Total solver iterations:	1

Variable	Value
TAC	302389.9
OC	0.1021429E-01
PC	35577.71
FW	2.057143
WW	8.157143
PFW1	5048.000
PFW2	7551.429
P11	0.000000
P21	0.000000
P31	0.000000
P41	0.000000
P12	10064.00
P22	12914.29
P32	0.000000
P42	0.000000
F11	0.000000
B11	0.000000
F21	0.000000
B21	0.000000
F31	0.000000
B31	0.000000

(Continued)

Variable	Value
F41	0.000000
B41	0.000000
F12	0.8000000
B12	1.000000
F22	4.142857
B22	1.000000
F32	0.000000
B32	0.000000
F42	0.000000
B42	0.000000
FW1	1.200000
BFW1	1.000000
FW2	0.8571429
BFW2	1.000000
M	1000.000
F1W	0.000000
F2W	0.8571429
F3W	5.900000
F4W	1.400000

The various cost elements for the water network are summarized in Table 9.3. Note that all process variables (i.e., reuse/recycle, freshwater, and wastewater flowrates) remain the same as those in Example 9.2. Hence, the water network takes the same structure as in Figure 9.6.

9.4 Reducing Network Complexity

In some cases, it is desired to reduce network complexity by having less piping connections. Hence, the maximum number of connections may be imposed in the optimization problem. For instance, the maximum number of reuse/recycle connections between source SR_i and sink SK_j may be given by the following constraint:

$$\sum_j B_{SR_i,SK_j} \leq B^{Max} \quad \forall i \tag{9.21}$$

where B^{Max} is the maximum number of connections.

Another realistic consideration in reducing network complexity is to impose minimum flowrate constraints for the piping connections. For instance, the reuse/recycle connections may be forced to take a flowrate that is bigger than the minimum value (F^{Min}), given by Equation 9.22:

$$F^{Min} B_{SRi,SKj} \leq F_{SRi,SKj} \quad \forall i, j \tag{9.22}$$

In Equation 9.22, the binary variable will be forced to take a zero value, if the reuse/recycle flowrate is smaller than the minimum flowrate value.

Example 9.5 Reduced network complexity

Example 9.3 is revisited. For this case, the forbidden match between any process sources with boiler feed water (SK1) is to be maintained, while the compulsory between distillation bottoms (SR1) and scrubber inlet (SK2) is to be relaxed. Determine the various cost elements such as those in Example 9.3, with maximum reuse/recycle connection limited to one; and the minimum flowrates in the reuse/recycle connections are to be bigger than 1 kg/s.

SOLUTION

In order to impose a maximum number of reuse/recycle connections in the RCN, Equation 9.21 takes the following form:

$$B_{SR1,SK1} + B_{SR2,SK1} + B_{SR3,SK1} + B_{SR4,SK1} + B_{SR1,SK2} + B_{SR2,SK2} + B_{SR3,SK2} + B_{SR4,SK2} \leq 1$$

Besides, all reuse/recycle connections in the RCN should have a minimum flowrate that is larger than 1 kg/s, which may follow Equation 9.22 as follows:

$$F^{Min} B_{SR1,SK1} \leq F_{SR1,SK1}; F^{Min} B_{SR2,SK1} \leq F_{SR2,SK1}$$

$$F^{Min} B_{SR3,SK1} \leq F_{SR3,SK1}; F^{Min} B_{SR4,SK1} \leq F_{SR4,SK1}$$

$$F^{Min} B_{SR1,SK2} \leq F_{SR1,SK2}; F^{Min} B_{SR2,SK2} \leq F_{SR2,SK2}$$

$$F^{Min} B_{SR3,SK2} \leq F_{SR3,SK2}; F^{Min} B_{SR4,SK2} \leq F_{SR4,SK2}$$

where F^{Min} takes a value of 1 kg/s.

Note that the rest of the constraints remain identical with those in Example 9.3 (except the compulsory match between SR1 and SK2 is now relaxed). LINGO formulation for this MILP problem is written as follows:

```
Model:
!Optimization objective: To minimize total annual cost (TAC);

min = TAC;
TAC = OC*3600*8000 + PC*0.231;
OC = (FW + WW)*0.001; !Unit cost of FW and WW = $1/t;
PC = PFW1 + PFW2 + P11 + P21 + P31 + P41 + P12 + P22 + P32 +
P42;
```

```
!Total freshwater flowrate (FW);

FW = FW1 + FW2;

!Flowrate balance for process sinks;

F11 + F21 + F31 + F41 + FW1 = 1.2;
F12 + F22 + F32 + F42 + FW2 = 5.8;

!Process constraints;

F11 = 0; !Forbidden match between SR1 and SK1;

!Impurity constraints for process sinks;

F11*0 + F21*14 + F31*25 + F41*34 + FW1*0 <= 1.2*0;
F12*0 + F22*14 + F32*25 + F42*34 + FW2*0 <= 5.8*10;

!Flowrate balance for process sources;

F11 + F12 + F1W = 0.8;
F21 + F22 + F2W= 5;
F31 + F32 + F3W =5.9;
F41 + F42 + F4W=1.4;

!Total wastewater flowrate (WW);

WW = F1W + F2W + F3W + F4W;

!All variables should take non-negative values;

F11 >= 0; F21 >= 0; F31 >= 0; F41 >= 0;
F12 >= 0; F22 >= 0; F32 >= 0; F42 >= 0;
FW1 >= 0; FW2 >= 0;
F1W >= 0; F2W >= 0; F3W >= 0; F4W >= 0;

!Capital cost estimation;

P11 = (2*F11 + 250*B11) *15;
P21 = (2*F21 + 250*B21)*30;
P31 = (2*F31 + 250*B31)*40;
P41 = (2*F41 + 250*B41)*30;
P12 = (2*F12 + 250*B12)*40;
P22 = (2*F22 + 250*B22)*50;
P32 = (2*F32 + 250*B32)*20;
P42 = (2*F42 + 250*B42)*10;
PFW1 = (2*FW1 + 250*BFW1)*20;
PFW2 = (2*FW2 + 250*BFW2)*30;

!Binary variables that represent present/absent of a
streams;

F11 <= M*B11; F21 <= M*B21; F31 <= M*B31; F41 <= M*B41;
F12 <= M*B12; F22 <= M*B22; F32 <= M*B32; F42 <= M*B42;
FW1 <= M*BFW1; FW2 <= M*BFW2;
M = 1000; !Arbitrary value;

!Minimum flowrate constraint in piping connections;

FMin*B11 <= F11 ; FMin*B21 <= F21 ; FMin*B31 <= F31 ;
FMin*B41 <= F41 ;
```

```
FMin*B12 <= F12 ; FMin*B22 <= F22 ; FMin*B32 <= F32 ;
FMin*B42 <= F42 ;
FMin= 1; !Minimum flowrate constraint;

!Maximum piping connections;

B11 + B21 + B31 + B41 + B12 + B22 + B32 + B42 <= 1;

!Defining binary variables;

@BIN (B11); @BIN (B21); @BIN (B31); @BIN (B41);
@BIN (B12); @BIN (B22); @BIN (B32); @BIN (B42);
@BIN (BFW1); @BIN (BFW2);
```

END

The following is the solution report generated by LINGO:

Global optimal solution found.	
Objective value:	346156.2
Extended solver steps:	0
Total solver iterations:	4

Variable	Value
TAC	346156.2
OC	0.1181429E-01
PC	25561.71
FW	2.857143
WW	8.957143
PFW1	5048.000
PFW2	7599.429
P11	0.000000
P21	0.000000
P31	0.000000
P41	0.000000
P12	0.000000
P22	12914.29
P32	0.000000
P42	0.000000
FW1	1.200000
FW2	1.657143
F11	0.000000
F21	0.000000
F31	0.000000
F41	0.000000

(Continued)

Variable	Value
F12	0.000000
F22	4.142857
F32	0.000000
F42	0.000000
F1W	0.8000000
F2W	0.8571429
F3W	5.900000
F4W	1.400000
B11	0.000000
B21	0.000000
B31	0.000000
B41	0.000000
B12	0.000000
B22	1.000000
B32	0.000000
B42	0.000000
BFW1	1.000000
BFW2	1.000000
M	1000.000
FMIN	1.000000

The various cost elements for the RCN are also summarized in Table 9.4 (last column). As compared with Example 9.3 (column 2 of Table 9.3), there is a significant increase in all cost elements for this case. The OC and TAC were increased by 20.0% and 14.5%, respectively, even though the PC has a reduction of 28.2%.

The network structure of the RCN is shown in Figure 9.8. As compared with cases in Examples 9.1 through 9.4, reuse/recycle streams have been reduced from two to one connection. This leads to an increase in both freshwater and wastewater flowrates, i.e., a rise of 38.8% (from 2.06 to 2.86 kg/s) and 9.8% (from 8.16 to 8.96 kg/s), respectively. Note that the reuse/recycle connection has a flowrate that is more than 1 kg/s.

TABLE 9.4

Cost Elements of AN Case Water Network in Examples 9.4 and 9.5

Cost Elements	Example 9.4	Example 9.5
Operating cost (OC)	$0.010/s	$0.012/s
Capital/piping cost (PC)	$35,577	$25,562
Total annual cost (TAC)	$302,389/year	$346,156/year

		F_{SKj}	1.2	5.8	8.96
		C_{SKj}	0	10	5.3
F_{SRi}	C_{SRi}	SKj / SRi	SK1	SK2	WW
2.86	0	FW	1.2	1.66	
0.8	0	SR1			0.8
5	14	SR2		4.14	0.86
5.9	25	SR3			5.9
1.4	34	SR4			1.4

FIGURE 9.8
Network structure for RCN of AN case in Example 9.5 (all flowrate terms in kg/s; concentration in ppm).

9.5 Connections and Flowrate Trade-off with Multiobjective Optimization

When there exist numerous sinks and sources in an RCN, its piping connections can be substantial. One may make use of multiobjective optimization techniques to identify the trade-off between piping connections and water saving (or cost). One of the established multiobjective optimization techniques is the *constraint method* (Cohon, 1978). The latter has found the main steps, given as follows:

1. Set up a *pay-off table*. This step consists of the following sub-steps:

   a. Several single-objective optimization problems are formulated and solved, with their optimal solution is obtained for each objective. The optimal solution for the objective u. The optimal solution for the objective u is denoted as:

   $$\mathbf{x}_u = \mathbf{x}_{1,u}, \mathbf{x}_{2,u}, \cdots, \mathbf{x}_{I,u} \qquad (9.23)$$

   b. For each of the U optimal solutions, the objective values are calculated, i.e., $G_1(\mathbf{x}_u), G_2(\mathbf{x}_u), \cdots, G_U(\mathbf{x}_u), u = 1, 2, \cdots, U$.

TABLE 9.5

Pay-off Table for a Multiobjective Problem

	$G_1(\mathbf{x}_u)$	$G_2(\mathbf{x}_u)$	$\cdots$	$G_U(\mathbf{x}_u)$
$\mathbf{x}_1$	$G_1(\mathbf{x}_1)$	$G_2(\mathbf{x}_1)$	$\cdots$	$G_U(\mathbf{x}_1)$
$\mathbf{x}_2$	$G_1(\mathbf{x}_1)$	$G_2(\mathbf{x}_1)$	$\cdots$	$G_U(\mathbf{x}_1)$
$\cdots$	$\cdots$	$\cdots$	$\cdots$	$\cdots$
$\mathbf{x}_U$	$G_1(\mathbf{x}_U)$	$G_2(\mathbf{x}_U)$	$\cdots$	$G_U(\mathbf{x}_U)$

c. Construct a pay-off table (Table 9.5), in which the rows correspond to $\mathbf{x}_1, \mathbf{x}_2, \cdots, \mathbf{x}_U$, and the columns correspond to objectives $G_1(\mathbf{x}_u), G_2(\mathbf{x}_u), \cdots, G_U(\mathbf{x}_u)$.

d. Identify the largest (L_u) and smallest values (S_u) in all column u.

2. Set up constraints.

Convert the multiobjective optimization problem given in Equation 9.24 into its corresponding constrained problem, as given in Equations 9.25 and 9.26.

$$\min G(\mathbf{x}) = \left[G_1(\mathbf{x}), G_2(\mathbf{x}), \cdots, G_Q(\mathbf{x}) \right] \tag{9.24}$$

$$\min G_h(\mathbf{x}) \text{ subject to, } r \tag{9.25}$$

$$G_u(\mathbf{x}) \le \epsilon_u \tag{9.26}$$

where objective h is arbitrarily chosen, and all other objectives are converted into constraints.

3. Set the right-hand side coefficients. For objective u, the lower and upper limits are represented by S_u and L_u, respectively, i.e., $S_u \le \epsilon_u \le L_u$. Choose the number of different values of ϵ_u and denote it by z.

4. Solve the optimization problem. To generate a range of non-inferior solutions, solve the constrained problem in step 2 for every combination of values for ϵ_u ($u = 1, 2, \ldots, U$), as given by Equation 9.27.

$$\epsilon_u = S_u + \frac{n}{z-1}(L_u - S_u); n = 0, 1, 2, \ldots, z-1 \tag{9.27}$$

Example 9.6 Trade-Off Analysis for Complex Network

The hypothetical example of Polley and Polley (2000) in Example 8.1 is revisited, with limiting data given in Table 9.6 (reproduced from Table 8.1). Identify the trade-off between its minimum freshwater flowrate with number of connections with multiobjective optimization.

TABLE 9.6

Limiting Water Data for Example 9.6

Sinks, SK$_j$	F_{SKj} (t/h)	C_{SKj} (ppm)	Sources, SR$_i$	F_{SRi} (t/h)	C_{SRi} (ppm)
SK1	50	20	SR1	50	50
SK2	100	50	SR2	100	100
SK3	80	100	SR3	70	150
SK4	70	200	SR4	60	250

Source: Polley and Polley (2000).

SOLUTION

The basic constraints for this RCN are first discussed, before the constraint method is used for analyzing the water network. The basic constraints are given as follows.

Equation 9.9 is used describe the flowrate balances for process sinks:

$$F_{SR1,SK1} + F_{SR2,SK1} + F_{SR3,SK1} + F_{SR4,SK1} + F_{FW,SK1} = 50$$

$$F_{SR1,SK2} + F_{SR2,SK2} + F_{SR3,SK2} + F_{SR4,SK2} + F_{FW,SK2} = 100$$

$$F_{SR1,SK3} + F_{SR2,SK3} + F_{SR3,SK3} + F_{SR4,SK3} + F_{FW,SK3} = 80$$

$$F_{SR1,SK4} + F_{SR2,SK4} + F_{SR3,SK4} + F_{SR4,SK4} + F_{FW,SK4} = 70$$

The quality requirement of the sinks is described by Equation 9.10:

$$F_{SR1,SK1}(50) + F_{SR2,SK1}(100) + F_{SR3,SK1}(150) + F_{SR4,SK1}(250) + F_{FW,SK1}(0) \leq 1.2(20)$$

$$F_{SR1,SK2}(50) + F_{SR2,SK2}(100) + F_{SR3,SK2}(150) + F_{SR4,SK2}(250) + F_{FW,SK2}(0) \leq 5.8(50)$$

$$F_{SR1,SK3}(50) + F_{SR2,SK3}(100) + F_{SR3,SK3}(150) + F_{SR4,SK3}(250) + F_{FW,SK3}(0) = 80(100)$$

$$F_{SR1,SK4}(50) + F_{SR2,SK4}(100) + F_{SR3,SK4}(150) + F_{SR4,SK4}(250) + F_{FW,SK4}(0) = 70(200)$$

Equation 9.11 describes the flowrate balance of the process sources.

$$F_{SR1,SK1} + F_{SR1,SK2} + F_{SR1,SK3} + F_{SR1,SK4} + F_{SR1,WW} = 50$$

$$F_{SR2,SK1} + F_{SR2,SK2} + F_{SR2,SK3} + F_{SR2,SK4} + F_{SR2,WW} = 100$$

$$F_{SR3,SK1} + F_{SR3,SK2} + F_{SR3,SK3} + F_{SR3,SK4} + F_{SR3,WW} = 70$$

$$F_{SR4,SK1} + F_{SR4,SK2} + F_{SR4,SK3} + F_{SR4,SK4} + F_{SR4,WW} = 60$$

Equation 9.12 indicates that all flowrate variables take non-negative values.

$$F_{SR1,SK1}, F_{SR2,SK1}, F_{SR3,SK1}, F_{SR4,SK1} \geq 0$$

$$F_{SR1,SK2}, F_{SR2,SK2}, F_{SR3,SK2}, F_{SR4,SK2} \geq 0$$

$$F_{SR1,SK3}, F_{SR2,SK3}, F_{SR3,SK3}, F_{SR4,SK3} \geq 0$$

$$F_{SR1,SK4}, F_{SR2,SK4}, F_{SR3,SK4}, F_{SR4,SK4} \geq 0$$

$$F_{FW,SK1}, F_{FW,SK2}, F_{FW,SK3}, F_{FW,SK4} \geq 0$$

$$F_{SR1,WW}, F_{SR2,WW}, F_{SR3,WW}, F_{SR4,WW} \geq 0$$

It is convenient to determine the total wastewater flowrate (F_{WW}) with Equation 9.13:

$$F_{WW} = F_{SR1,WW} + F_{SR2,WW} + F_{SR3,WW} + F_{SR4,WW}$$

Next, binary term [0, 1] is used to define the presence of a material stream between SR_i and SK_j ($B_{SRi,SKj}$), freshwater and SKj, as well as SRi with WW The binary term is activated with Equation 9.18

$$F_{SR1,SK1} \leq 1000 B_{SR1,SK1}; \ F_{SR2,SK1} \leq 1000 B_{SR2,SK1}$$

$$F_{SR3,SK1} \leq 1000 B_{SR3,SK1}; \ F_{SR4,SK1} \leq 1000 B_{SR4,SK1}$$

$$F_{SR1,SK2} \leq 1000 B_{SR1,SK2}; \ F_{SR2,SK2} \leq 1000 B_{SR2,SK2}$$

$$F_{SR3,SK2} \leq 1000 B_{SR3,SK2}; \ F_{SR4,SK2} \leq 1000 B_{SR4,SK2}$$

$$F_{SR1,SK3} \leq 1000 B_{SR1,SK3} \ ; F_{SR2,SK3} \leq 1000 B_{SR2,SK3}$$

$$F_{SR3,SK3} \leq 1000 B_{SR3,SK3} \ ; F_{SR4,SK3} \leq 1000 B_{SR4,SK3}$$

$$F_{SR1,SK4} \leq 1000 B_{SR1,SK4}; \ F_{SR2,SK4} \leq 1000 B_{SR2,SK4}$$

$$F_{SR3,SK4} \leq 1000 B_{SR3,SK4} \ ; F_{SR4,SK4} \leq 1000 B_{SR4,SK4}$$

$$F_{FW,SK1} \leq 1000 B_{FW,SK1}; F_{FW,SK2} \leq 1000 B_{FW,SK2}$$

$$F_{FW,SK3} \leq 1000 B_{FW,SK3} \ ; \ F_{FW,SK4} \leq 1000 B_{FW,SK4}$$

$$F_{SR1,WW1} \leq 1000 B_{SR1,WW1} \ ; F_{SR2,WW2} \leq 1000 B_{SR2,WW2}$$

$$F_{SR3,WW3} \leq 1000 B_{SR3,WW3} \ ; F_{SR4,WW4} \leq 1000 B_{SR4,WW4}$$

LINGO formulation for above constraints are written as follows:

```
Model:
!Total freshwater flowrate (FW);

FW = FW1 + FW2 + FW3 + FW4;

!Number of interconnection;

INC = BFW1 + BFW2 + BFW3 + BFW4 + B11 + B21 + B31 +
B41 + B12 + B22 + B32 + B42 + B13 + B23 + B33 + B43 +
B14 + B24 + B34 + B44 + BF1W + BF2W + BF3W + BF4W;

!Flowrate balance for process sinks;

F11 + F21 + F31 + F41 + FW1 = 50;
F12 + F22 + F32 + F42 + FW2 = 100;
F13 + F23 + F33 + F43 + FW3 = 80;
F14 + F24 + F34 + F44 + FW4 = 70;

!Impurity constraints for process sinks;

F11*50 + F21*100 + F31*150 + F41*250 + FW1*0 <= 50*20;
F12*50 + F22*100 + F32*150 + F42*250 + FW2*0 <= 100*50;
F13*50 + F23*100 + F33*150 + F43*250 + FW3*0 <= 80*100;
F14*50 + F24*100 + F34*150 + F44*250 + FW4*0 <= 70*200;

!Flowrate balance for process sources;

F11 + F12 + F13 + F14 + F1W = 50;
F21 + F22 + F23 + F24 + F2W = 100;
F31 + F32 + F33 + F34 + F3W = 70;
F41 + F42 + F43 + F44 + F4W = 60;

!Total wastewater flowrate (WW);

WW = F1W + F2W + F3W + F4W;

!All variables should take non-negative values;

F11 >= 0; F21 >= 0; F31 >= 0; F41 >= 0;
F12 >= 0; F22 >= 0; F32 >= 0; F42 >= 0;
F13 >= 0; F23 >= 0; F33 >= 0; F43 >= 0;
F14 >= 0; F24 >= 0; F34 >= 0; F44 >= 0;
FW1 >= 0; FW2 >= 0; FW3 >= 0; FW4 >= 0;
F1W >= 0; F2W >= 0; F3W >= 0; F4W >= 0;

! Defining binary variables;

F11 <= M*B11; F21 <= M*B21; F31 <= M*B31; F41 <= M*B41;
F12 <= M*B12; F22 <= M*B22; F32 <= M*B32; F42 <= M*B42;
F13 <= M*B13; F23 <= M*B23; F33 <= M*B33; F43 <= M*B43;
F14 <= M*B14; F24 <= M*B24; F34 <= M*B34; F44 <= M*B44;
FW1 <= M*BFW1; FW2 <= M*BFW2; FW3 <= M*BFW3; FW4 <= M*BFW4;
F1W <= M*BF1W; F2W <= M*BF2W; F3W <= M*BF3W; F4W <= M*BF4W;

M = 1000; !Arbitrary value;
```

```
!Defining binary variables;

@BIN (B11); @BIN (B21); @BIN (B31); @BIN (B41);
@BIN (B12); @BIN (B22); @BIN (B32); @BIN (B42);
@BIN (B13); @BIN (B23); @BIN (B33); @BIN (B43);
@BIN (B14); @BIN (B24); @BIN (B34); @BIN (B44);
@BIN (BFW1); @BIN (BFW2); @BIN (BFW3); @BIN (BFW4);
@BIN (BF1W); @BIN (BF2W); @BIN (BF3W); @BIN (BF4W);
```

END

We then follow the four-step procedure to solve the multiobjective optimization problem.

1. *Set up a pay-off table.*

   For this case, two single-objective optimization problems exists, i.e., minimum freshwater flowrate ($G_{FW}(x)$) and minimum interconnection ($G_{INC}(x)$). The model is solved for both objectives independently. To minimize freshwater flowrate the objective in Equation 9.8 is solved, while the objective to minimize interconnection is given in Equation 9.28.

$$\text{Minimize } INC = \sum_j B_{FW,SKj} + \sum_i \sum_j B_{SRi,SKj} + \sum_i B_{SRi,WW} \quad (9.28)$$

   LINGO formulation for these constraints are written as follows:

```
min = FW;   !Objective 1;
min = INC;  !Objective 2;
```

   When objective 1 is solved (objective 2 is deactivated), the minimum freshwater flowrate is determined as 70 t/h, with a total of 24 connections (INC). We then move on to solve objective 2 (objective 1 is deactivated). It is then determined that a total of 7 minimum connections is needed, while minimum freshwater is determined as 80 t/h. These results are summarized in the pay-off table in Table 9.7. In other words, for freshwater flowrate, the values of L_u and S_u correspond to 80 and 70 t/h, respectively. On the other hand, the values of L_u and S_u correspond to 24 and 7 for number of connection (*INC*).

TABLE 9.7

Pay-off Table for Multiobjective Optimization Problem in Problem 9.6

Objective u	S_u	L_u
$G_{FW}(x)$	70	80
$G_{INC}(x)$	7	24

Source: Polley and Polley (2000).

TABLE 9.8

Pay-off Table for Multiobjective
Optimization Problem in Problem 9.6

n	ϵ_{FW}	*INC*
0	70.0	10
1	72.5	10
2	75.0	9
3	77.5	8
4	80.0	7

Source: Polley and Polley (2000).

2. *Set up constraints.*

In step 2, the freshwater flowrate minimization (objective 1) is converted into constraint (Equation 9.29), while the number of connection (Equation 9.28) is minimized.

$$G_{FW}(x) \leq \epsilon_{FW} \qquad (9.29)$$

3. *Set the right-hand side coefficients.*
4. *Solve the optimization problem.*

In step 3, the number of ϵ_{FW} is chosen to be 5 (z). We then generate five sets of values for ϵ_{FW} with Equation 9.27, given in Table 9.8. For each values of ϵ_{FW}, the number of connection (Equation 9.28) is minimized, subject to the constraint in Equation 9.29, and other constraints in the model. Results of the solution are summarized in the last column of Table 9.8, and also Figure 9.9.

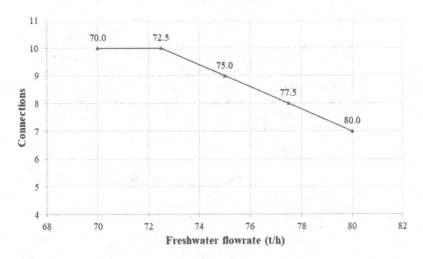

FIGURE 9.9
Trade-off between numbers of connections with freshwater flowrate for Example 9.6.

9.6 Superstructural Model for Material Regeneration Network

To synthesize a material regeneration network, we may make use of the superstructural representation in Figure 9.10 (reproduced from Figure 1.17) or alternatively its matching matrix form in Figure 9.11. Similar to the case in Figures 9.1 and 9.2, all process sources are matched to the process sinks via connections. Besides, the purification unit r acts as both sink and source. When source is sent for regeneration (with flowrate $F_{SRi,REr}$), the inlet of the purification unit r acts as a sink. The outlet stream of the unit r is the regenerated source of higher quality, which will be sent for reuse/recycle in the process sinks (with flowrate $F_{REGr,SKj}$). Note however that no match should be built between the fresh resource and the purification unit r, as well as between the regenerated source r and waste discharge, i.e., $F_{R,REr} = F_{REGr,D} = 0$.

The optimization model for a material regeneration network can be extended from that of the reuse/recycle network, which takes the revised form of Equations 9.30 through 9.34. Note that in each constraint set, a new flowrate term is added for the purification unit (with outlet quality of q_{REGr}),

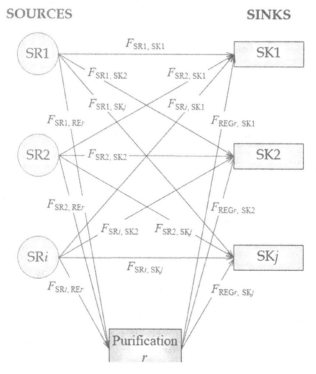

FIGURE 9.10
Superstructural model for material regeneration network.

		F_{SKj}	F_{SK1}	F_{SK2}	F_{SK3}	F_{SKj}	F_{RE}	F_D
		q_{SKj}	q_{SK1}	q_{SK2}	q_{SK3}	q_{SKj}	q_{Rin}	q_D
F_{SRi}	q_{SRi}	SKj \ SRi	SK1	SK2	SK3	SKj	REr	WD
F_R	q_R	FR	$F_{R,SK1}$	$F_{R,SK2}$	$F_{R,SK3}$	$F_{R,SKj}$		
F_{REGr}	q_{REGr}	REGr	$F_{REGr,SK1}$	$F_{REGr,SK2}$	$F_{REGr,SK3}$	$F_{REGr,SKj}$		
F_{SR1}	q_{SR1}	SR1	$F_{SR1,SK1}$	$F_{SR1,SK2}$	$F_{SR1,SK3}$	$F_{SR1,SKj}$	$F_{SR1,REr}$	$F_{SR1,D}$
F_{SR2}	q_{SR2}	SR2	$F_{SR2,SK1}$	$F_{SR2,SK2}$	$F_{SR2,SK3}$	$F_{SR2,SKj}$	$F_{SR2,REr}$	$F_{SR2,D}$
F_{SR3}	q_{SR3}	SR3	$F_{SR3,SK1}$	$F_{SR3,SK2}$	$F_{SR3,SK3}$	$F_{SR3,SKj}$	$F_{SR3,REr}$	$F_{SR3,D}$
F_{SRj}	q_{SRj}	SRi	$F_{SRj,SK1}$	$F_{SRj,SK2}$	$F_{SRj,SK3}$	$F_{SRj,SKj}$	$F_{SRj,REr}$	$F_{SRj,D}$

FIGURE 9.11
Superstructural model for material regeneration network in matching matrix form.

to account for the regenerated source that is sent to process sink ($F_{REGr,SKj}$) or the source to be sent for regeneration ($F_{SRi,REr}$):

$$\sum_i F_{SRi,SKj} + F_{REGr,SKj} + F_{R,SKj} = F_{SKj} \quad \forall j \tag{9.30}$$

$$\sum_i F_{SRi,SKj}q_{SRi} + F_{REGr,SKj}q_{REGr} + F_{R,SKj}q_R \geq F_{SKj}q_{SKj} \quad \forall j \tag{9.31}$$

$$\sum_j F_{SRi,SKj} + F_{SRi,REr} + F_{SKi,D} = F_{SRi} \quad \forall i \tag{9.32}$$

$$F_{SRi,SKj}, F_{R,SKj}, F_{SKi,D}, F_{REGr,SKj}, F_{SRi,REr} \geq 0 \quad \forall i,j \tag{9.33}$$

Note that flowrate balance is needed for purification unit r, given as follows:

$$\sum_i F_{SRi,REr} = \sum_j F_{REGr,SKj} \quad \forall r \tag{9.34}$$

Note also that for purification unit that is characterized by *removal ratio (RR)* of its impurity, its inlet (q_{REr}) and outlet (q_{REGr}) quality values are related with Equations 9.35:

$$RR = \frac{q_{REGr} - q_{REr}}{q_{max} - q_{REr}} \quad \forall r \tag{9.35}$$

where q_{max} is the maximum possible value for such quality (e.g., 100%, 1,000,000 ppm).

The objective of the optimization problem can be set to minimize the operating costs (*OC*) or the *TAC* of the material regeneration network. The *OC* may consider the costs of fresh resource(s), material regeneration and/or waste discharge (with Equation 9.36), while the *TAC* of the RCN is given by Equation 9.3, with the inclusion of capital cost of purification unit *r*. As demonstrated earlier, solving for minimum *TAC* may lead to MILP problems.

$$\text{Minimize } OC = \sum_r \sum_j F_{Rr,SKj} CT_{Rr} + F_D CT_D + F_{REGr} CT_{REGr} \qquad (9.36)$$

Alternatively, the objective may also be set to minimize the flowrates of both fresh resource (F_R) and regenerated source ($\Sigma_r F_{REGr}$). However, since there are two objectives to be minimized, a two-stage optimization approach is needed. In the first stage, the fresh resource flowrate is minimized (with Equation 9.8), subject to the constraints in Equations 9.30 through 9.33. Once the minimum fresh resource flowrate is identified, it is then included as a new constraint in the second stage (Equation 9.37), where the total regenerated source flowrate ($\Sigma_r F_{REGr}$) is minimized, with the optimization objective in Equation 9.38. Similar to the earlier case, this is an LP model for which any solution found is globally optimal.

$$F_{R,min} = F_R \qquad (9.37)$$

$$\text{Minimize } \sum_r F_{REGr} = \sum_j \sum_r F_{REGr,SKj} \qquad (9.38)$$

Example 9.7 Water regeneration network with purification of fixed quality

The classical water minimization case (Wang and Smith, 1994) in Example 8.3 is revisited. Use the superstructural approach to synthesize a water regeneration network. Assuming that a purification unit with fixed outlet concentration (C_{REG}) of 5 ppm is used to purify water sources for further water recovery. Pure freshwater (0 ppm) is available for use in this RCN. Limiting data are given in Table 9.9 (reproduced from Table 8.3). Solve for the following cases:

1. Use MS Excel (with Solver add-in) to determine the minimum operating cost (*OC*) of the RCN. The unit costs of freshwater, wastewater, and regenerated sources are given as 1 $/t ($CT_{FW}$), 1 $/t ($CT_{WW}$), and 0.8 $/t ($CT_{RW}$), respectively.
2. Minimum flowrates for both freshwater and regenerated source.
3. Re-solve minimum flowrates problem in Case 2 with LINGO. Additional constraints are set so that each process sink can only accept a maximum of two reuse/recycle connections due to space constraint.

TABLE 9.9

Limiting Water Data for Classical Water Minimization Case (Wang and Smith, 1994) in Example 9.7

Sinks, SK_j	F_{SKj} (t/h)	C_{SKj} (ppm)	Sources, SR_i	F_{SRi} (t/h)	C_{SRi} (ppm)
SK1	20	0	SR1	20	100
SK2	100	50	SR2	100	100
SK3	40	50	SR3	40	800
SK4	10	400	SR4	10	800
$\sum_i F_{SKj}$	170		$\sum_i F_{SRi}$	170	

SOLUTION

1. Case 1—Use MS Excel to determine the minimum OC of the RCN.

   A superstructure model is constructed in MS Excel interface, as shown in Figure 9.10. All sources including freshwater (FW) and regenerated source (REG) are arranged in rows 6–11, while sinks including regeneration inlet (RE) and wastewater discharge (WW) are arranged in columns E–I.

   Figure 9.12 also shows the constraints described by Equations 9.30–9.33 for this LP model. The objective is set to minimize the OC (cell K20), which is contributed by the flowrates of freshwater, wastewater and regenerated flowrates (cells K17–K19).

   Figure 9.13 shows the results of the LP model, where the OC is determined as \$98.95/h. The model also determined that the freshwater, wastewater and regenerated flowrates are determined as 20, 20 and 73.68 t/h, respectively. Note that regeneration-recycle scheme is carried out here, as water from SR2 and SR3 are

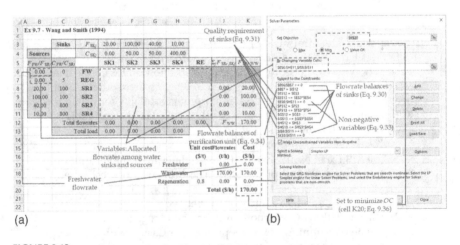

(a)   (b)

FIGURE 9.12

(a) Setting of Excel spreadsheet and (b) Solver for water regeneration network in Example 9.7 (Case 1).

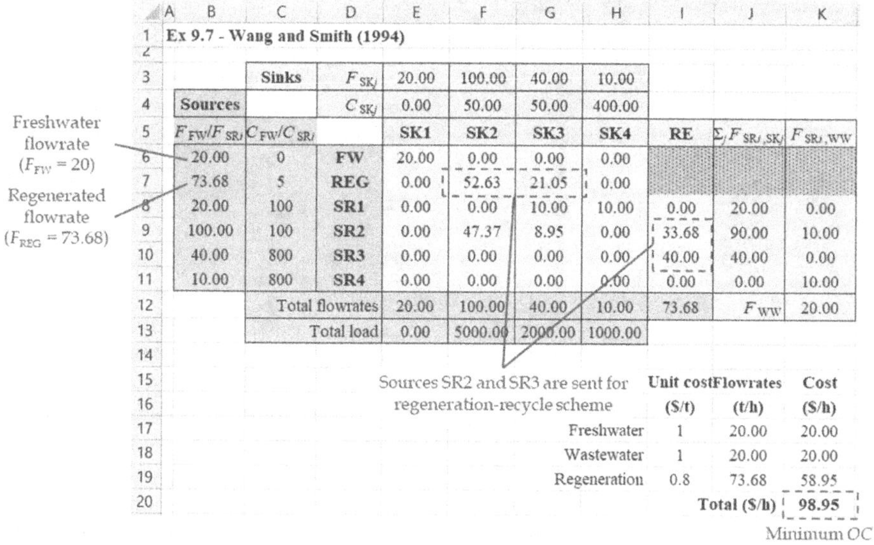

	A	B	C	D	E	F	G	H	I	J	K
1	Ex 9.7 - Wang and Smith (1994)										
2											
3			Sinks	F_{SKj}	20.00	100.00	40.00	10.00			
4		Sources		C_{SKj}	0.00	50.00	50.00	400.00			
5		F_{FW}/F_{SRj}	C_{FW}/C_{SRj}		SK1	SK2	SK3	SK4	RE	$\Sigma_j F_{SRj,SKj}$	$F_{SRj,WW}$
6		20.00	0	FW	20.00	0.00	0.00	0.00			
7		73.68	5	REG	0.00	52.63	21.05	0.00			
8		20.00	100	SR1	0.00	0.00	10.00	10.00	0.00	20.00	0.00
9		100.00	100	SR2	0.00	47.37	8.95	0.00	33.68	90.00	10.00
10		40.00	800	SR3	0.00	0.00	0.00	0.00	40.00	40.00	0.00
11		10.00	800	SR4	0.00	0.00	0.00	0.00	0.00	0.00	10.00
12				Total flowrates	20.00	100.00	40.00	10.00	73.68	F_{WW}	20.00
13				Total load	0.00	5000.00	2000.00	1000.00			
14											
15						Sources SR2 and SR3 are sent for		Unit cost	Flowrates	Cost	
16						regeneration-recycle scheme		(\$/t)	(t/h)	(\$/h)	
17							Freshwater	1	20.00	20.00	
18							Wastewater	1	20.00	20.00	
19							Regeneration	0.8	73.68	58.95	
20									Total (\$/h)	98.95	

Freshwater flowrate (F_{FW} = 20)

Regenerated flowrate (F_{REG} = 73.68)

Minimum OC

FIGURE 9.13
Result for water regeneration network in Example 9.7 (Case 1).

sent to purification unit for regeneration (cells I9–I10), before they are recycled to SK2 and SK3 for recycling (cells F7–G7).

2. Case 2—RCN with minimum flowrates of freshwater and re-generated source

For Case 2 where the flowrates of both freshwater and regenerated source are to be minimized, the two-stage optimization approach is adopted. The setup of Solver for both stages are shown in Figure 9.14. In Stage 1, the optimization objective is set to minimize the freshwater flowrate (F_{FW}) in cell B6 with Equation 9.8 (see Figure 9.14a), while all constraints remain identical as those in Case 1. By solving the model, the result indicates that the minimum freshwater flowrate for this RCN is identified as 20 t/h (solution not shown for brevity).

Once the minimum freshwater flowrate is identified, we move on to solve Stage 2 optimization, where regenerated source flowrate is to be minimized. At this stage, the minimum freshwater flowrate is added as a new constraint to the problem, following Equation 9.39 (see Figure 9.14b).

$$F_{FW} = 20 \qquad (9.39)$$

The optimization objective is then set to minimize the regenerated source flowrate (F_{REG}) in cell B7, which takes the revised form of Equation 9.38:

$$\text{Minimize } F_{REG} = \sum_j F_{REG,SKj} \qquad (9.40)$$

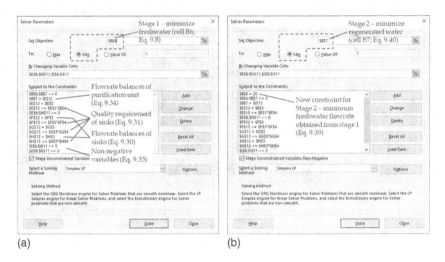

(a)   (b)

FIGURE 9.14
Setup of Solver for two-stage optimization to minimize freshwater and regenerated flowrates (Case 2): (a) Stage 1—minimize freshwater; (b) Stage 2—minimize regenerated flowrate.

Solving the model identify that the minimum regenerated flowrate as 73.68 t/h, with RCN structure identical to that in Case 1 (Figure 9.13), where the objective was set to minimize OC. In other words, both minimum OC and minimum flowrates objectives lead to same RCN design. The final network structure is shown in Figure 9.15.

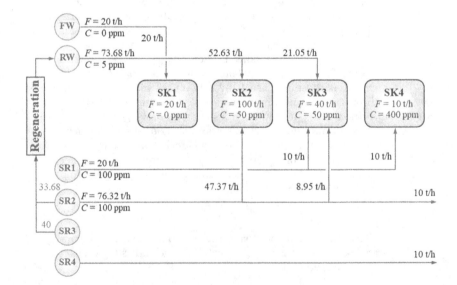

FIGURE 9.15
RCN design in Example 9.7 (Cases 1 and 2) (all flowrate terms in t/h; concentration in ppm).

One may compare the network structure in Figure 9.15 with that designed with *nearest neighbor algorithm* in Figure 8.15. While both networks have the same freshwater and regenerated flowrates fed to the SK2 and SK3, the individual streams for direct reuse/recycle and those for regeneration are different from each other.

3. Case 3—Use LINGO to limit piping connections for Case 2

The basic constrains for the RCN (Equations 9.30–9.33) are given as follows. Equation 9.30 describes the flowrate balances for process sinks:

$$F_{SR1,SK1} + F_{SR2,SK1} + F_{SR3,SK1} + F_{SR4,SK1} + F_{REG,SK1} + F_{FW,SK1} = 20$$

$$F_{SR1,SK2} + F_{SR2,SK2} + F_{SR3,SK2} + F_{SR4,SK2} + F_{REG,SK2} + F_{FW,SK2} = 100$$

$$F_{SR1,SK3} + F_{SR2,SK3} + F_{SR3,SK3} + F_{SR4,SK3} + F_{REG,SK3} + F_{FW,SK3} = 40$$

$$F_{SR1,SK4} + F_{SR2,SK4} + F_{SR3,SK4} + F_{SR4,SK4} + F_{REG,SK4} + F_{FW,SK4} = 10$$

The quality requirement for each sink is described by Equation 9.31. Note that the freshwater term is omitted as it does not contain any impurity (i.e., $C_{FW} = 0$ ppm). Note also that the inequality sign takes a reverse form as that in Equation 9.31, as the quality index is described by impurity concentration.

$$F_{SR1,SK1}(100) + F_{SR2,SK1}(100) + F_{SR3,SK1}(800) + F_{SR4,SK1}(800) + F_{REG,SK1}(5) \leq 20(0)$$

$$F_{SR1,SK2}(100) + F_{SR2,SK2}(100) + F_{SR3,SK2}(800) + F_{SR4,SK2}(800) + F_{REG,SK2}(5) \leq 100(50)$$

$$F_{SR1,SK3}(100) + F_{SR2,SK3}(100) + F_{SR3,SK3}(800) + F_{SR4,SK3}(800) + F_{REG,SK3}(5) \leq 40(50)$$

$$F_{SR1,SK4}(100) + F_{SR2,SK4}(100) + F_{SR3,SK4}(800) + F_{SR4,SK4}(800) + F_{REG,SK4}(5) \leq 10(400)$$

Equation 9.32 next describes the flowrate balance of the process sources. As shown, each process source may be allocated to any process sinks, or the purification unit, while the unutilized source will be sent for wastewater discharge ($F_{SRi,WW}$):

$$F_{SR1,SK1} + F_{SR1,SK2} + F_{SR1,SK3} + F_{SR1,SK4} + F_{SR1,RE} + F_{SR1,WW} = 20$$

$$F_{SR2,SK1} + F_{SR2,SK2} + F_{SR2,SK3} + F_{SR2,SK4} + F_{SR2,RE} + F_{SR2,WW} = 100$$

$$F_{SR3,SK1} + F_{SR3,SK2} + F_{SR3,SK3} + F_{SR3,SK4} + F_{SR3,RE} + F_{SR3,WW} = 40$$

$$F_{SR4,SK1} + F_{SR4,SK2} + F_{SR4,SK3} + F_{SR4,SK4} + F_{SR4,RE} + F_{SR4,WW} = 10$$

Since only one purification unit is used, its flowrate balance is given by Equation 9.34:

$$F_{RE} = F_{REG}$$

Equation 9.33 indicates that all flowrate variables of the optimization model take non-negative values:

$$F_{SR1,SK1}, F_{SR2,SK1}, F_{SR3,SK1}, F_{SR4,SK1} \geq 0$$

$$F_{SR1,SK2}, F_{SR2,SK2}, F_{SR3,SK2}, F_{SR4,SK2} \geq 0$$

$$F_{SR1,SK3}, F_{SR2,SK3}, F_{SR3,SK3}, F_{SR4,SK3} \geq 0$$

$$F_{SR1,SK4}, F_{SR2,SK4}, F_{SR3,SK4}, F_{SR4,SK4} \geq 0$$

$$F_{FW,SK1}, F_{FW,SK2}, F_{FW,SK3}, F_{FW,SK4} \geq 0$$

$$F_{SR1,WW}, F_{SR2,WW}, F_{SR3,WW}, F_{SR4,WW} \geq 0$$

$$F_{SR1,RE}, F_{SR2,RE}, F_{SR3,RE}, F_{SR4,RE} \geq 0$$

$$F_{REG,SK1}, F_{REG,SK2}, F_{REG,SK3}, F_{REG,SK4} \geq 0$$

It is also convenient to relate the total flowrates of the freshwater (F_{FW}), wastewater (F_{WW}), as well as the regenerated sources with their individual flowrate contributions:

$$F_{FW} = F_{FW,SK1} + F_{FW,SK2} + F_{FW,SK3} + F_{FW,SK4}$$

$$F_{WW} = F_{SR1,WW} + F_{SR2,WW} + F_{SR3,WW} + F_{SR4,WW}$$

$$F_{RE} = F_{SR1,RE} + F_{SR2,RE} + F_{SR3,RE} + F_{SR4,RE}$$

$$F_{REG} = F_{REG,SK1} + F_{REG,SK2} + F_{REG,SK3} + F_{REG,SK4}$$

For this case, Equation 9.36 takes the revised form as follows:

$$\text{Minimize } OC(\$/h) = F_{FW}(\$1/\text{ton}) + F_{WW}(\$1/\text{ton}) + F_{REG}(\$0.8/\text{ton})$$

To account for piping connections of the RCN, binary variables are to be used. Similar to Examples 9.4 and 9.5, the presence of a material recovery stream between sources (including regenerated sources) and sinks is to be activated with the binary term* in Equation 9.18, given as follows:

$$F_{SR1,SK1} \leq 1000 B_{SR1,SK1}; \ F_{SR2,SK1} \leq 1000 B_{SR2,SK1}$$

$$F_{SR3,SK1} \leq 1000 B_{SR3,SK1}; \ F_{SR4,SK1} \leq 1000 B_{SR4,SK1}; \ F_{REG,SK1} \leq 1000 B_{REG,SK1}$$

$$F_{SR1,SK2} \leq 1000 B_{SR1,SK2}; \ F_{SR2,SK2} \leq 1000 B_{SR2,SK2}$$

$$F_{SR3,SK2} \leq 1000 B_{SR3,SK2}; \ F_{SR4,SK2} \leq 1000 B_{SR4,SK2}; \ F_{REG,SK2} \leq 1000 B_{REG,SK2}$$

$$F_{SR1,SK3} \leq 1000 B_{SR1,SK3}; \ F_{SR2,SK3} \leq 1000 B_{SR2,SK3}$$

$$F_{SR3,SK3} \leq 1000B_{SR3,SK3}; \ F_{SR4,SK3} \leq 1000B_{SR4,SK3}; \ F_{REG,SK3} \leq 1000B_{REG,SK3}$$

$$F_{SR1,SK4} \leq 1000B_{SR1,SK4}; \ F_{SR2,SK4} \leq 1000B_{SR2,SK4}$$

$$F_{SR3,SK4} \leq 1000B_{SR3,SK4}; \ F_{SR4,SK4} \leq 1000B_{SR4,SK4}; \ F_{REG,SK4} \leq 1000B_{REG,SK4}$$

Since each process sink can accept only a maximum of two reuse/recycle connections, we make use of the revised form of Equation 9.21:

$$B_{SR1,SK1} + B_{SR2,SK1} + B_{SR3,SK1} + B_{SR4,SK1} + B_{REG,SK1} \leq 2$$

$$B_{SR1,SK2} + B_{SR2,SK2} + B_{SR3,SK2} + B_{SR4,SK2} + B_{REG,SK2} \leq 2$$

$$B_{SR1,SK3} + B_{SR2,SK3} + B_{SR3,SK3} + B_{SR4,SK3} + B_{REG,SK3} \leq 2$$

$$B_{SR1,SK4} + B_{SR2,SK4} + B_{SR3,SK4} + B_{SR4,SK4} + B_{REG,SK4} \leq 2$$

Note that the two-stage optimization approach is used, in order to determine the minimum fresh water and regenerated flowrates. LINGO formulation for this MILP model take the following form (only stage 2 objective, i.e., minimum regeneration flowrate is not shown for brevity):

```
Model:
min = REG; !Stage 2;
FW = 20; !Stage 1;

!Flowrate balance for freshwater;

FW = FW1 + FW2 + FW3 + FW4;

!Flowrate balance for purification units;
RE = RE1 + RE2 + RE3 + RE4;
REG = REG1 + REG2 + REG3 + REG4;
RE = REG;

!Flowrate balance for process sinks;

F11 + F21 + F31 + F41 + REG1 + FW1 = 20;
F12 + F22 + F32 + F42 + REG2 + FW2 = 100;
F13 + F23 + F33 + F43 + REG3 + FW3 = 40;
F14 + F24 + F34 + F44 + REG4 + FW4 = 10;

!Impurity constraints for process sinks;

F11*100 + F21*100 + F31*800 + F41*800 + REG1*5 <= 20*0;
F12*100 + F22*100 + F32*800 + F42*800 + REG2*5 <= 100*50;
F13*100 + F23*100 + F33*800 + F43*800 + REG3*5 <= 40*50;
F14*100 + F24*100 + F34*800 + F44*800 + REG4*5 <= 10*400;

!Flowrate balance for process sources;

F11 + F12 + F13 + F14 + RE1 + F1W = 20;
F21 + F22 + F23 + F24 + RE2 + F2W = 100;
F31 + F32 + F33 + F34 + RE3 + F3W = 40;
F41 + F42 + F43 + F44 + RE4 + F4W= 10;
```

```
!Total wastewater flowrate (WW);

WW = F1W + F2W + F3W + F4W;

!All variables should take non-negative values;

F11 >= 0; F21 >= 0; F31 >= 0; F41 >= 0;
F12 >= 0; F22 >= 0; F32 >= 0; F42 >= 0;
F13 >= 0; F23 >= 0; F33 >= 0; F43 >= 0;
F14 >= 0; F24 >= 0; F34 >= 0; F44 >= 0;
FW1 >= 0; FW2 >= 0; FW3 >= 0; FW4 >= 0;
F1W >= 0; F2W >= 0; F3W >= 0; F4W >= 0;

!Binary terms;

F11 <= M*B11; F21 <= M*B21; F31 <= M*B31;
F41 <= M*B41; REG1 <= M*BR1;
F12 <= M*B12; F22 <= M*B22; F32 <= M*B32;
F42 <= M*B42; REG2 <= M*BR2;
F13 <= M*B13; F23 <= M*B23; F33 <= M*B33;
F43 <= M*B43; REG3 <= M*BR3;
F14 <= M*B14; F24 <= M*B24; F34 <= M*B34;
F44 <= M*B44; REG4 <= M*BR4;
M = 1000;
B11 + B21 + B31 + B41 + BR1 <= 2;
B12 + B22 + B32 + B42 + BR2 <= 2;
B13 + B23 + B33 + B43 + BR3 <= 2;
B14 + B24 + B34 + B44 + BR4 <= 2;
@BIN(B11); @BIN(B21); @BIN(B31); @BIN(B41); @BIN(BR1);
@BIN(B12); @BIN(B22); @BIN(B32); @BIN(B42); @BIN(BR2);
@BIN(B13); @BIN(B23); @BIN(B33); @BIN(B43); @BIN(BR3);
@BIN(B14); @BIN(B24); @BIN(B34); @BIN(B44); @BIN(BR4);
```

END

With the two-stage optimization approach, the following is the solution report generated by LINGO, while the RCN structure is given in Figure 9.16:

Global optimal solution found.	
Objective value:	73.68421
Extended solver steps:	0
Total solver iterations:	10
Variable	Value
REG	73.68421
FW	20.00000
FW1	20.00000
FW2	0.000000
FW3	0.000000

(Continued)

Variable	Value
FW4	0.000000
RE1	0.000000
RE2	23.68421
RE3	40.00000
RE4	10.00000
REG1	0.000000
REG2	52.63158
REG3	21.05263
REG4	0.000000
F11	0.000000
F21	0.000000
F31	0.000000
F41	0.000000
F12	0.000000
F22	47.36842
F32	0.000000
F42	0.000000
F13	18.94737
F23	0.000000
F33	0.000000
F43	0.000000
F14	0.000000
F24	10.00000
F34	0.000000
F44	0.000000
F1W	1.052632
F2W	18.94737
F3W	0.000000
F4W	0.000000
WW	20.00000
M	1000.000
B11	0.000000
B21	0.000000
B31	0.000000
B41	0.000000
BR1	0.000000
B12	0.000000

(Continued)

Variable	Value
B22	1.000000
B32	0.000000
B42	0.000000
BR2	1.000000
B13	1.000000
B23	0.000000
B33	0.000000
B43	0.000000
BR3	1.000000
B14	0.000000
B24	1.000000
B34	0.000000
B44	0.000000
BR4	0.000000

From Figure 9.16, it is observed that piping connections of SK3 has been reduced to two, as compared to three connections in Case 2 (see Figure 9.13 or 9.15). This means that a degenerate solution exists for the problem, since the RCNs have the same flowrates, but with different network structure.

		F_{SKj}	20	100	40	10	73.68	20
		C_{SKj}	0	50	50	400		
F_{SRi}	C_{SRi}	SKj \ SRi	SK1	SK2	SK3	SK4	RE	WW
20	0	FR	20					
73.68	5	REG		52.63	21.05			
20	100	SR1			18.95			1.05
100	100	SR2		45.37		10	23.68	18.95
40	400	SR3					40	
10	400	SR4					10	

FIGURE 9.16
Network structure for RCN in Example 9.7 (Case 3) (all flowrates in t/h; concentration in ppm).

Example 9.8 Water regeneration network with purification unit of removal ratio type

Revisit the case study in Example 9.7 (Wang and Smith, 1994). In this case, the water regeneration unit is characterized by RR, described by Equation 9.39, which is the revised form of Equations 9.35. As shown, the RR correlation may be used to determine the outlet concentration of the purification unit, when its inlet concentration is given.

$$RR = \frac{C_{REr} - C_{REGr}}{C_{REr}} \quad \forall r \tag{9.39}$$

Use the superstructural approach to synthesize a water regeneration network with minimum OC, assuming that different purification units are used for sources of different concentrations. Analyze the following cases with different unit costs:

1. Unit costs of freshwater and regenerated sources are given as 1 \$/t (CT_{FW}), and 0.8 \$/t (CT_{RW}), respectively. Unit cost of wastewater is ignored.
2. The unit cost of freshwater is given as 1 \$/t (CT_{FW}), while unit cost of wastewater is ignored. The unit cost of regenerated sources (CT_{RW}) is a function of the impurity load removed by purification unit r (Δm_r), i.e., 0.8 \$/kg, given by Equation 9.40.

$$\Delta m_r = F_{REGr} \left(C_{REr} - C_{REGr} \right) \quad \forall r \tag{9.40}$$

SOLUTION

1. Case 1—With unit costs of 1 \$/t (CT_{FW}) and \$0.8/ton (CT_{RW}). The superstructure model in MS Excel along with its Solver setup are shown in Figure 9.17. Since different purification units

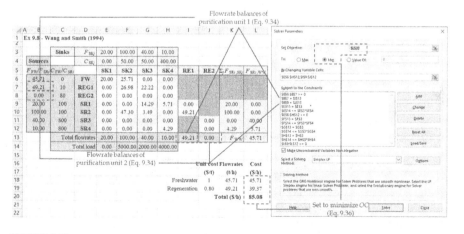

FIGURE 9.17
Water regeneration network for Example 9.8 (Case 1) (all flowrates in t/h; concentration in ppm).

are utilized for sources of different quality, sources SR1 and SR2 (100 ppm) are sent to unit 1, while sources SR3 and SR4 (800 ppm) are sent to unit 2. Hence, flowrate balance in Equation 9.34 is applied to these purification units, respectively, and imposed by constraints in Solver (see Figure 9.17).

Solving the LP model with objective in Equation 9.36 results with an *OC* of 85.08 $/h. As shown in Figure 9.17, the freshwater flowrate is determined as 45.71 t/h, while 49.21 t/h of water sources is regenerated by purification unit 1 for reuse/recycle to the sinks; this purified source is originated from SR2 of 100 ppm (observed from cell I10), which is purified to 10 ppm (C_{Rout}). In other words, purification unit 2 is not needed for this case.

2. Case 2—With unit costs of 1 $/t ($CT_{FW}$) and 0.8 $/kg Δm_r (CT_{RW}).

The setup of MS Excel and Solver for this case is similar to that in Case 1. However, different cost calculation for regeneration water is observed. As shown in Figure 9.18, the load removed by purification units are calculated using Equation 9.40 in cells K19–K20, which contributed to the *OC* in column L (cells C18–C20).

Solving the LP results with an *OC* of 85.08 $/h. As shown in Figure 9.18, the freshwater flowrate is determined as 20 t/h, while 116.67 t/h (= 66.67 + 50 t/h) of water sources is regenerated by both purification units for reuse/recycle to the sinks. Column I of the spreadsheet shows that purification unit 1 will purify 100 t/h of water from SR2 (100 ppm) to 10 ppm, while column J identifies that unit 2 will purify the entire flowrates from SR3 and SR4 (800 ppm) to 80 ppm.

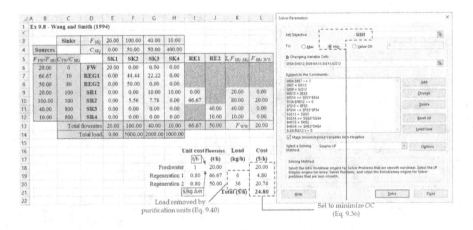

FIGURE 9.18
Water regeneration network for Example 9.8 (Case 2) (all flowrate terms in t/h; concentration in ppm).

9.7 Additional Readings

The superstructural approach discussed in this chapter is the basic model for RCN synthesis, with reuse/recycle and regeneration schemes. The basic model can also be extended to many other variant, e.g., material regeneration networks with partitioning purification units (Tan et al., 2009); total water network (Karuppiah and Grossmann, 2006), etc. Readers may also refer to the subsequent chapters for the synthesis of RCN of batch system and inter-plant integration.

PROBLEMS
Water Minimization Problems

9.1 For the classical water minimization case in Examples 9.7 and 9,8 (Wang and Smith, 1994), solve the following cases:

a. Synthesize a direct reuse/recycle network (i.e., no purification unit) with minimum freshwater flowrate. Compare the network structure with that obtained using the nearest neighbor algorithm (note to readers: solve Problem 8.16 to compare results).

b. Revisit Example 9.8 (Case 2), however with unit costs of 1 $/t ($CT_{FW}$) and 0.8 $/kg Δm_r (CT_{RW}). Compare the results with those in Example 9.8 (Case 2).

9.2 Synthesize an RCN for direct water reuse/recycle scheme for Example 9.6 (Polley and Polley, 2000) for the following cases:

a. RCN with minimum freshwater flowrate, with zero impurity freshwater feed. Compare the resulting network structure with that obtained using the nearest neighbor algorithm (Example 7.1).

b. Redo case (a) when freshwater feed has an impurity of 10 ppm. Compare the network structure with that obtained using the nearest neighbor algorithm (note to readers: solve Problem 7.1 to compare results).

c. Redo case (a) when the number of reuse/recycle streams is limited to three connections.

d. Determine the piping and water costs for the RCN in cases (a) and (c). Distances among the water sources and sinks are given in Table 9.10. Unit costs for freshwater and wastewater are given as $1/ton, each.

e. RCN with minimum TAC. Make use of the cost elements in case (d). The piping cost is annualized to a period of 5 years, with an interest rate of 5%. Assume an annual operation of 8000 h.

f. Perform the trade-off analysis between number of connections and TAC, compare the results with those in Example 9.6.

9.3 Synthesize a minimum freshwater network for the textile plant in Problem 2.4, assuming pure freshwater is used. Is the freshwater and

TABLE 9.10

Distance (m) between Water Sources with Sinks in Problem 9.2

	Sinks			
Sources	SK1	SK2	SK3	SK4
SR1	0	30	150	65
SR2	30	0	45	120
SR3	150	45	0	80
SR4	65	120	80	0

wastewater flowrates identical with that obtained by targeting technique? (Note to readers: solve Problem 3.2 to compare result).

9.4 Synthesize a water reuse/recycle network for the bulk chemical production case in Problem 3.5, assuming pure freshwater is used. The RCN should feature the following characteristics:

a. Minimum freshwater flowrate. Compare the network structure with that obtained using the nearest neighbor algorithm (note to readers: solve Problem 8.6 to compare results).

b. Minimum total annualized cost (*TAC*). Unit cost for freshwater and wastewater are given as $1/ton each. Capital cost of piping is given by Equation 9.20, while the distances between the water sources and sinks are given in Table 9.11. The piping cost is annualized to a period of 5 years, with an interest rate of 5%. Assume an annual operation of 8000 h.

9.5 Revisit the specialty chemical production process in Problems 4.6 and 6.5. Assuming that pure freshwater is used, solve the following cases:

a. Synthesize a water reuse/recycle network with minimum freshwater flowrate. Compare the resulting network structure with that obtained using the nearest neighbor algorithm (solve Problem 8.7 to compare results).

b. Synthesize a water regeneration network that achieves the minimum freshwater and regeneration flowrates. Compare the results

TABLE 9.11

Distance (m) between Water Sources with Sinks for Bulk Chemical Production in Problem 9.4

	Sinks			
Sources	Reactor	Boiler Makeup	Cooling Tower Makeup	Cleaning
Decanter	100	100	100	40
Cooling tower blowdown	100	30	12	50
Boiler blowdown	110	0	40	75

where two purification units are evaluated for use, i.e., one with outlet concentration (C_{Rout}) of 60 ppm, and the other has C_{Rout}) of 200 ppm (solve Problem 6.5 to compare results).

9.6 Synthesize a water reuse/recycle network for the palm oil mill in Problem 4.8, with minimum freshwater flowrate. Assume pure freshwater is used. Compare the network structure with that obtained using the nearest neighbor algorithm (note to readers: solve Problem 8.4 to compare results).

9.7 The water minimization case of the chlor-alkali process in Problem 4.15 is revisited here. Note that two water reuse/recycle networks are to be synthesized as the process consists of acid and base sections. Ion concentration is the major concern for water minimization in this process. Determine the minimum freshwater flowrates needed for these water networks, assuming that the de-ionized water (fresh resource) is free of ion content.

9.8 For the water minimization case in Problems 5.3 and 6.6 (Sorin and Bédard, 1999), synthesize the water networks for the following cases, assuming that pure freshwater is used). Compare the solution with those obtained via nearest neighbor algorithm (solve Problem 8.12a and b to compare results).

a. Direct reuse/recycle network with minimum freshwater (0 ppm) flowrate.

b. Water regeneration network with minimum operating cost. Unit costs for freshwater (CT_{FW}), wastewater (CT_{WW}), and regenerated water (CT_{RW}) are given as \$1/ton, \$0.5/ton, and \$0.5/ton, respectively. The purification unit has a fixed outlet concentration of 10 ppm (C_{Rout}).

9.9 The paper milling process in Problem 4.5 is revisited here. Synthesize a water reuse/recycle network for this case, assuming that pure freshwater is used. The limiting data are given in Table 4.21, while the distances between the water sources and sinks are given in Table 9.12. Unit costs for freshwater and wastewater are given as \$1/ton each. Capital cost

TABLE 9.12

Distance (m) between Water Sources with Sinks of Paper Milling Process in Problem 9.9

Distance	Storage Tank	Pressing Section	Forming Section	De-Inking Pulper (DIP) and Others	DIP
Pressing section	82		5	72	103
Forming section	77	5		67	98
DIP and others	20	67	72		40
DIP	68	98	103	40	
Chemical preparation (CP)	60	53	58	40	85
Approach flow	75	8	8	85	98

of piping is given by Equation 9.20. The piping cost is annualized to a period of 5 years, with an interest rate of 5%. Assume an annual operation of 8000h. Determine the minimum *TAC* for the RCN.

9.10 Use the two-stage optimization approach to synthesize a direct reuse/recycle network for the bioethanol production plant (Liu et al., 2019) in Problem 4.10. The limiting data is given in Table 4.25. Assume that the freshwater feed has impurity (COD) content of 20 mg/L.

9.11 Use the two-stage optimization approach to synthesize a zero water discharge network for the alumina plant in Problem 6.4, assuming a purification unit of (C_{Rout} = 20 ppm) is available, and pure freshwater is used (Note to readers: solve Problem 6.4 to compare result).

9.12 A water minimization case study (Gabriel and El-Halwagi, 2005) is analyzed here, with the limiting water data given in Table 9.13. To enhance water recovery, stream stripping is used to regenerate water sources for further recovery. The unit cost for water regeneration (CT_{RW}) is given as a function of impurity load removal and varies for different removal ratios (RR) for each water source, as shown in Table 9.14. The unit costs for freshwater (CT_{FW}) and wastewater (CT_{WW}) are taken as $0.13/ton and $0.22/ton, respectively. Synthesis a water regeneration network with minimum *OC*.

TABLE 9.13

Limiting Water Data for Problem 9.12

Sinks, SK_j	F_{SKj} (t/h)	C_{SKj} (ppm)	Sources, SR_i	F_{SRi} (t/h)	C_{SRi} (ppm)
SK1	200	20	SR1	150	10
SK2	80	75	SR2	60	50
			SR3	100	85

Source: Gabriel and El-Halwagi (2005).

TABLE 9.14

Cost Data for Purification Unit in Problem 9.12

Sources, SR_i	Removal Ratio (RR)	CT_{RW} ($/kg Impurity Removed)
SR1	0.1	0.68
	0.5	1.46
	0.9	2.96
SR2	0.1	0.54
	0.5	1.16
	0.9	2.36
SR3	0.1	0.45
	0.5	0.97
	0.9	1.97

Source: Gabriel and El-Halwagi (2005).

9.13 For the alkene production process in Problems 2.5, 3.8, 4.14 and 8.21 (Foo, 2019), perform the following tasks:

a. Synthesize a water reuse/recycle network. Compare the solution with that designed using the nearest neighbor algorithm in Problem 8.21.

b. A partitioning purification unit with outlet concentration of 0.8% and recovery factor of 0.8 is used. Determine the minimum flowrates for fresh water and regenerated water with two-stage optimization approach. To simplify the model, the reject stream from purification unit will not be recovered to the sinks. Compare the solution with that designed using the nearest neighbor algorithm in Problem 8.21 (task c).

Multiple-Impurities Problems

9.14 Figure 9.19 shows the water flow diagram of a crude oil refinery, where a wide range of oil products are made, e.g., liquefied petroleum gas, light and heavy naphtha, gasoline, naphtha mixture, etc. (Mughees and Al-Hamad, 2015). In evaluating water minimization potential, two important impurities must be considered, i.e., chemical oxygen demand (COD—impurity A) and water harness (impurity B). The limiting water data for process sinks and sources are given in Table 9.15. Synthesize a water reuse/recycle network, assuming freshwater has zero impurity.

9.15 Figure 9.20 shows the process flow diagram of a corn refinery, where water minimization is to be carried out (Bavar et al., 2018). In evaluating its water minimization potential, two important impurities must be considered, i.e., chemical oxygen demand (COD—impurity A) and total suspended solid (TDS—impurity B). The limiting water data for

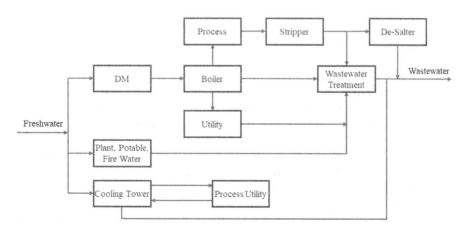

FIGURE 9.19
Water flow diagram of crude oil refinery in Problem 9.14. (Mughees and Al-Hamad, 2015.)

TABLE 9.15

Limiting Water Data of a Crude Oil Refinery in Problem 9.14

j	Sinks, SK_j	F_{SKj} (t/h)	$C_{SKj, A}$ (ppm)	$C_{SKj, B}$ (ppm)
1	Cooling tower	37	1	150
2	Desalter	59	2	12
3	Plant potable, fire water	160	3	400
i	Sources, SR_i	F_{SRi} (t/h)	$C_{SRi, A}$ (ppm)	$C_{SRi, B}$ (ppm)
1	Utility outlet	45	2	0
2	Boiler blowdown	20	4	0
3	Desalter	59	5	160

Source: Mughees and Al-Hamad (2015).

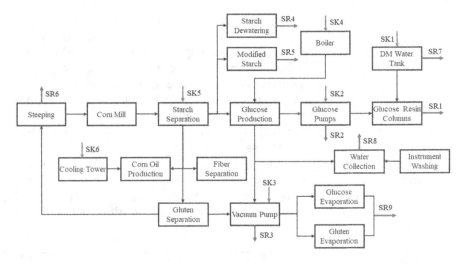

FIGURE 9.20

Process flow diagram of corn refinery in Problem 9.15 (Bavar et al., 2018).

process sinks and sources are given in Table 9.16. Synthesize a water reuse/recycle network, assuming freshwater has zero impurity.

9.16 In a sugarcane refinery (see process flow diagram in Figure 9.21), the water circuit is divided into process water and utility section (Balla et al., 2018). Hence, water minimization will be carried out for these sections independently. In the process water section, total dissolved solids (TDS) is the main impurity that restrict water recovery (i.e., single impurity problem), with limiting data given in Table 9.17. On the other hand, water recovery in the utility section is restricted by the content of biological oxygen demand (BOD, impurity A) and chemical oxygen demand (COD, impurity B), which indicates a multiple

TABLE 9.16

Limiting Water Data of a Corn Refinery in Problem 9.15

Sinks, SK_j	F_{SKj} (t/d)	$C_{SKj, A}$ (ppm)	$C_{SKj, B}$ (ppm)
SK1	445	10	362
SK2	211	162	384
SK3	135	162	384
SK4	200	10	362
SK5	580	10	362
SK6	230	75	500
Sources, SR_i	F_{SRi} (t/d)	$C_{SRi, A}$ (ppm)	$C_{SRi, B}$ (ppm)
SR1	445	434	3780
SR2	210.6	25	372
SR3	135	444	362
SR4	9.68	57,000	2140
SR5	92.24	77,600	12,730
SR6	466.32	43,850	3860
SR7	24.76	10	69,800
SR8	278.88	162	384
SR9	11.04	448	465

Source: Bavar et al. (2018).

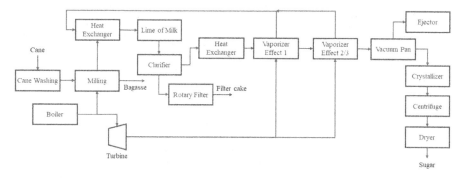

FIGURE 9.21
Process flow diagram of sugarcane refinery in Problem 9.16. (Balla et al., 2018.)

impurity problem. The limiting water data are given in Table 9.18. Perform the following tasks:

a. Synthesize a water reuse/recycle network with minimum fresh-water flowrate for both process water and utility sections.

b. To enhance water reduction potential in the utility section, two purification units are evaluated, with performance given in

TABLE 9.17

Limiting Water Data for Process Water Section of a Sugarcane Refinery in Problem 9.16

SK$_j$	Sinks,	F_{SKj} (t/h)	C_{TDS} (ppm)
SK1	Boiler makeup	110.0	10
SK2	Milk of lime	1.6	40
SK3	Imbibition water	62.0	80
SK4	Filter wash water	9.2	80
SK5	Movement water for pan	4.6	80
SK6	Molasses dilution	4.6	80
SK7	Meltzer	6.9	80
SK8	Mesquite washing and losses	2.3	80
SK9	Floor and equipment washing	2.3	100
SR$_i$	Sources,	F_{SRi} (t/h)	C_{TDS} (ppm)
SR1	Exhaust condensate	110.2	0
SR2	Pan condensate	46.0	23
SR3	Vapor condensate from second effect + heaters 5 and 6	45.2	42
SR4	Vapor condensate from third effect + heaters 3 and 4	37.8	48
SR5	Vapor condensate from fourth effect + heaters 1 and 2	29.7	54

Source: Balla et al. (2018).

TABLE 9.18

Limiting Water Data for Utility Section of a Sugarcane Refinery in Problem 9.16

j	Sinks, SK$_j$	F_{SKj} (t/h)	$C_{SKj, A}$ (ppm)	$C_{SKj, B}$ (ppm)
1	Cane washing	55	25	75
2	Makeup of mill turbine bearing	2.2	25	75
3	Makeup of turbine cooling water	34.7	25	75
4	Makeup of injection water	78	50	75
i	Sources, SR$_i$	F_{SRi} (t/h)	$C_{SRi, A}$ (ppm)	$C_{SRi, B}$ (ppm)
1	Excess water	65.2	150	400
2	Boiler blowdown	5	20	80
3	Purge from turbine cooling water	2.7	20	80

Source: Balla et al. (2018).

Table 9.19. Use the two-stage optimization approach to determine which unit will achieve lower flowrates for fresh water, regenerated water and wastewater.

c. For the two purification units in (b), determine which unit will lead to lower minimum operating cost (*OC*) for the utility section. The unit costs of fresh water and wastewater are both given as

TABLE 9.19

Performance of Purification Unit in Problem 9.16

Purification Unit	Outlet Concentration for BOD (ppm)	Outlet Concentration for COD (ppm)	Unit Cost for Regeneration
A	50	200	0.7 $/t
B	30	50	0.5 $/kg Δm_r

1 $/t, while unit cost for water regeneration is given in the last column of Table 9.19.

9.17 Water minimization is to be carried out for a crude oil refinery, which has large capacity to process sour crude (Sujo-Nova et al., 2009). Due to large consumption of water (see water balance diagram in Figure 9.22), water minimization study is carried out. Two important impurities in considering water recovery are ammonia (impurity A) and hydrogen sulfide (impurity B). The limiting water data for process sinks and sources are given in Table 9.20. Perform the following tasks:

 a. Synthesize a water reuse/recycle network that achieves minimum freshwater flowrate.

 b. Two purification units are evaluated for their water reduction potential, with performance summarized in Table 9.21. Use the two-stage optimization approach to determine which unit will achieve lower flowrates for fresh water, regenerated water and wastewater.

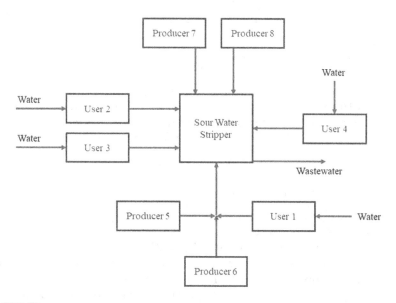

FIGURE 9.22
Water flow diagram of crude oil refinery in Problem 9.17. (Sujo-Nova et al., 2009.)

TABLE 9.20

Limiting Water Data of a Crude Oil Refinery in Problem 9.17

j	Sinks, SK_j	F_{SKj} (kg/s)	$C_{SKj, A}$ (ppm)	$C_{SKj, B}$ (ppm)
1	User 1	1.89	40	10
2	User 2	3.46	1143	1400
3	User 3	3.15	282	284
4	User 4	1.26	235	300
i	Sources, SR_i	F_{SRi} (kg/s)	$C_{SRi, A}$ (ppm)	$C_{SRi, B}$ (ppm)
1	User 1	1.89	142	105
2	User 2	3.51	7,018	9,690
3	User 3	4.82	30,196	23,601
4	User 4	1.26	498	546
5	Producer 5	0.32	100	125
6	Producer 6	3.15	1,247	1,528
7	Producer 7	1.26	235	300
8	Producer 8	0.63	4,500	5,112

Source: Sujo-Nova et al. (2009).

TABLE 9.21

Performance of Purification Unit in Problem 9.17

Purification Unit	Outlet Concentration for Ammonia (ppm)	Outlet Concentration for H_2S (ppm)	Unit Cost for Regeneration
A	100	500	0.7 \$/t
B	50	400	0.5 \$/kg Δm_r

c. Evaluate the two purification units in (b). Determine which unit will lead to lower minimum operating cost (*OC*) of the water regeneration network. The unit costs of fresh water and wastewater are both given as 1 \$/t, while unit cost for water regeneration is given in Table 9.21.

Utility Gas Recovery Problems

9.18 Synthesize an RCN with minimum fresh oxygen supply for oxygen recovery case in Problem 3.9. The fresh oxygen supply has an impurity of 5% (note to readers: solve Problem 8.23 to compare results).

9.19 Synthesize a reuse/recycle network for the refinery hydrogen network in Example 2.4 with minimum fresh hydrogen supply. The fresh hydrogen feed has 1% impurity. Compare the network structure with that obtained via nearest neighbor algorithm (note to readers: solve Problem 8.24 to compare results).

9.20 For the refinery hydrogen network in Problems 2.11, 4.19 and 6.14 (see limiting data in Table 4.31). The fresh hydrogen feed has an impurity content of 5%. Compare the network structure with that obtained using nearest neighbor algorithm (note to readers: solve Problem 8.25 to compare results).

a. A reuse/recycle network with minimum fresh hydrogen. Compare the solution with that obtained in Problem 4.19.
b. A regeneration network with minimum OC. A membrane unit is used as purification unit, with hydrogen recovery factor (RC) of 0.95. The produces a purified stream of 2 mol % impurity (C_{RP}). The unit costs for fresh and regenerated hydrogen are given as 1.6×10^{-3}/mol and 1.0×10^{-3}/mol. Compare the solution with that obtained in Problem 6.14.

9.21 For the aromatic plant in Problems 2.12 and 6.17, synthesize the hydrogen network reuse/recycle for the following cases. Its fresh hydrogen has an impurity of 8.55 mol%.

a. A reuse/recycle network with minimum fresh hydrogen.
b. A regeneration network with minimum fresh hydrogen and regeneration flowrates. A PSA unit is used as purification unit, with a recovery factor of 0.682 (see Equations 6.31 or 6.35), and its product and reject streams have impurity values of 0.10 and 22.00 mol%, respectively.

Property Integration Problems

9.22 Synthesize a direct reuse/recycle network for fiber recovery in the Kraft paper making process in Example 2.6 in order to minimize fresh fiber use. Compare the RCN with that synthesized using the nearest neighbor algorithm (note to readers: solve Problem 8.29 to compare results).

9.23 Synthesize a direct water reuse/recycle network for the microelectronics manufacturing facility in Problem 3.11, in order to minimize its ultrapure water usage. The fresh ultrapure water has a resistivity value of 18,000 kΩ/cm. Compare the solution with the RCN synthesized using the nearest neighbor algorithm (note to readers: solve Problem 8.30 to compare results).

9.24 For the vinyl acetate manufacturing process in Problems 2.14 and 4.22, synthesize direct water reuse/recycle network to minimize the use of the ultrapure water.

9.25 Revisit the wafer fabrication process in Problems 2.13, 4.23, and 6.19, with limiting data given in Table 4.33. Synthesize RCNs for the following cases:

a. Direct reuse/recycle network with minimum freshwater (0 ppm) flowrate, with limiting data in Table 4.33b. Compare the solution with that obtained via nearest neighbor algorithm (Note to reader: solve Problem 8.31b to compare results).

b. Water regeneration network when a purification unit (membrane) is used; the latter has a permeate quality (purified stream) of 15 MΩ m, and a water recovery factor of 0.95 (see Chapter 6 for discussion). The network should have minimum flowrates for both its freshwater and regenerated flowrates. Compare the solution with the RCN synthesized using the nearest neighbor algorithm (note to reader: solve Problem 8.31b to compare results).

9.26 Revisit the biorefinery process in Problem 5.16, with limiting data given in Table 5.8. The fresh ethanol has a purity of 0.996 (fraction). Synthesize an RCN for the following cases:

a. Direct reuse/recycle network with minimum fresh bioethanol. Compare the solution with that obtained via automated targeting model (Note to reader: solve Problem 5.16 to compare results).

b. Regeneration network when a purification unit is used; the latter has a permeate quality of 0.96, and with recovery factor of 0.98.

References

Alves, J. J. and Towler, G. P. 2002. Analysis of refinery hydrogen distribution systems. Industrial and Engineering Chemistry Research, 41, 5759–5769.

Balla, W. H., Rabah, A. A., and Abdallah, B. K. 2018. Pinch analysis of sugarcane refinery water integration, *Sugar Tech*, 20(2), 122–134.

Bavar, M., Sarrafzadeh, M.-H., Asgharnejad, H., Norouzi-Firouz, H. 2018. Water management methods in food industry: Corn refinery as a case study. *Journal of Food Engineering*, 238: 78–84.

Cohon, J. L. ed. 1978. *Multiobjective Programming and Planning*. New York: Academic Press.

El-Halwagi, M. M.1997. *Pollution Prevention Through Process Integration: Systematic Design Tools*. San Diego: Academic Press.

Foo, D. C. Y. 2019. Optimize plant water use from your desk. *Chemical Engineering Progress*, 115: 45–51 (February 2019).

Gabriel, F. B. and El-Halwagi, M. M. 2005. Simultaneous synthesis of waste interception and material reuse networks: Problem reformulation for global optimization, *Environmental Progress*, 24(2), 171–180.

Karuppiah, R. and Grossmann, I. E. 2006. Global optimization for the synthesis of integrated water systems in chemical processes, *Computers and Chemical Engineering*, 30(4), 650–673.

Kim, J.-K. and Smith, R. 2004. Automated design of discontinuous water systems, *Process Safe. Environ. Prot.*, 82, 238–248.

Liu, H., Lijun Ren, L., Zhuo, H., and Fu, S. 2019. water footprint and water pinch analysis in ethanol industrial production for water management, *Water*, 11, 518.

Mughees, W. and Al-Hamad, M. 2015. Application of water pinch technology in minimization of water consumption at a refinery, *Computers and Chemical Engineering*, 73, 34–42.

Polley, G. T. and Polley, H. L. 2000. Design better water networks, *Chemical Engineering Progress*, 96(2), 47–52.

Smith, R.2005. *Chemical Process Design and Integration*. New York: John Wiley & Sons.

Sorin, M. and Bédard, S. 1999. The global pinch point in water reuse networks, *Process Safety and Environmental Protection*, 77, 305–308.

Sujo-Nova, D., Scodari, L. A., Slater, C. S., Dahm, K., and Savelski, M. J. 2009. Retrofit of sour water networks in oil refineries: A case study, *Chemical Engineering and Processing*, 48, 892–901.

Tan, R. R., Ng, D. K. S., Foo, D. C. Y., and Aviso, K. B. 2009. A superstructure model for the synthesis of single-contaminant water networks with partitioning regenerators, *Process Safety and Environmental Protection*, 87, 197–205.

Wang, Y. P. and Smith, R. 1994. Wastewater minimisation, *Chemical Engineering Science*, 49, 981–1006.

10

Extended Application: Synthesis of Inter-Plant Resource Conservation Network

In previous chapters, various process integration techniques for the synthesis of resource conservation network (RCN) have been presented. However, these chapters mainly focus on *in-plant* RCN, where material recovery activities are carried out within a single process/plant. In this chapter, we will focus on material recovery across different process/plant through *inter-plant resource conservation network* (IPRCN). In practice, there may be individual processes/plants that are located at different sites, but geographically near to each other (e.g. in an eco-industrial park). There could also be cases where a huge process plant that is physically segregated into different sections due to its size. Hence, performing material recovery among different sites/sections/plants will lead to reduction of overall fresh resource and waste discharge. Pinch analysis and mathematical programming techniques introduced in previous chapters are extended for the synthesis of IPRCN in this chapter.

10.1 Direct and Indirect Integration Schemes

In general, IPRCN can be achieved with *direct* and *indirect* integration schemes (Chew et al., 2008). In the former, material source(s) from one process/plant is recovered to sinks in another process/plant directly via cross-plant pipelines (Figure 10.1a). Direct integration scheme is favored when the individual

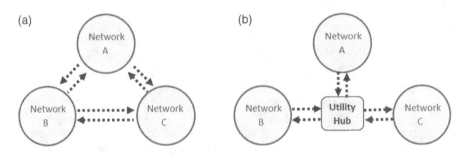

FIGURE 10.1
IPRCN configurations: (a) Direct and (b) indirect integration schemes. (From Chew et al., 2008.)

DOI: 10.1201/9781003173977-11

processes/plants are located near to each other, which allows material recovery to take place easily via cross-plant pipelines. On the other hand, material recovery may also be carried out with indirect integration scheme via a *centralized utility hub*. The latter is commonly found in large industrial areas where it supplies utilities (e.g., industrial gases, steam, electricity, ultrapure water, etc.), or to provide material purification or waste treatment for the nearby processes/plants. Note that direct recovery of material among individual processes/plants is forbidden in indirect integration scheme (Chew et al., 2008).

10.2 Graphical Targeting for Direct Integration Scheme

The philosophy of process integration, i.e., *target ahead of design* is followed. To identify the overall minimum fresh resource and waste discharge targets of an IPRCN, the *material recovery pinch diagram* (MRPD) in Chapter 3 is a useful tool. Note that the MRPD was originally meant for flowrate targeting for in-plant RCN (El-Halwagi et al., 2003; Prakash and Shenoy, 2005). It may be used for IPRCN when all process sinks and sources (of the individual RCNs) are treated as if they belong to a single network (steps 1 and 2 below). The following steps are used to construct the MRPD:

1. Arrange all process sinks (from all individual RCNs) in descending order of quality levels (q_{SKj}). Calculate the maximum load acceptable by the sinks (m_{SKj}) following Equations 10.1, which are given as the product of sink flowrate (F_{SKj}) with its quality level:

$$m_{SK_j} = F_{SK_j} q_{SK_j} \qquad (10.1)$$

2. Arrange all process sources (from all individual RCNs) in descending order of quality levels (q_{SRi}). Calculate the load possessed by the sources (m_{SRi}) following Equation 10.2, which is given as the product of the source flowrate (F_{SRi}) with its quality level:

$$m_{SR_i} = F_{SR_i} q_{SR_i} \qquad (10.2)$$

3. All process sinks are plotted in descending order of their quality levels on a load-versus-flowrate diagram to form the *sink composite curve*. The individual sink segments are connected by linking the tail of a later to the arrowhead of an earlier segment (Figure 10.2).

4. Step 3 is repeated for all process sources to form the *source composite curve* on the same load-versus-flowrate diagram. For a feasible

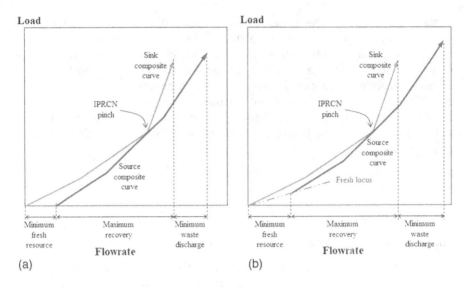

FIGURE 10.2
The MRPD for targeting IPRCN for (a) pure fresh resource and (b) impure fresh resource. (From El-Halwagi et al., 2003; Prakash and Shenoy, 2005.)

MRPD, the entire sink composite curve has to stay above and to the left of the source composite curve. Otherwise, the MRPD is considered infeasible, and step 5 is next followed.

5. For the IPRCN with a single pure fresh resource of superior quality (highest purity among all sources), the source composite curve is moved horizontally to the right until it just touches the sink composite curve, with the latter lying above and to the left of the source composite curve (Figure 10.2a). For an IPRCN with a single impure fresh resource, a fresh locus is to be plotted on the MRPD, with its slope corresponding to its quality level (q_R). The source composite curve is slid on the locus until it just touches the sink composite curve and stays entirely on the right and below the latter (Figure 10.2b).

Note that the MRPDs in Figure 10.2 are essentially identical to those used for targeting for in-plant RCNs (see Figure 3.1). Several useful *IPRCN targets* may be identified from the MRPD. The overall minimum fresh resource target of the IPRCN is given by the opening on the left side of the MRPD, while the opening on the right side represents its overall minimum waste generation target. The overlapping region of the two composite curves indicates the maximum recovery of material resources of the IPRCN. The point where the two composite curves touch is the IPRCN *pinch*, which may be viewed as the overall bottleneck of the IPRCN. The latter restricts the recovery of material among sources and sinks within the IPRCN.

Example 10.1 Inter-plant water network

The inter-plant water network example from Bandyopadhyay et al. (2010) is analyzed. The example consists of two water networks discussed in earlier chapters, i.e., Examples 6.1 (Network A) and 8.1 (Network B). In this case, apart from in-plant direct reuse/recycle, water recovery may also be carried out between both networks. The limiting data for these networks are given in Table 10.1 (reproduced from Tables 6.1 and 8.1, respectively). The freshwater feed is available at 0 ppm. Use the MRPD to identify the overall minimum freshwater and wastewater flowrates for the inter-plant water network.

SOLUTION

The targeting procedure is followed to identify the various targets for the inter-plant water network:

1. *Arrange all process sinks in descending order of quality levels (q_{SKj}). Calculate the maximum load acceptable by the sinks (m_{SKj}):*
2. *Arrange all process sinks (from all individual RCNs) in descending order of quality levels (q_{SRi}). Calculate the load possessed by the sources (m_{SRi})*

   Water sinks of both networks are treated as if they belong to a single network, and arranged in ascending order of concentrations (see column 3 of Table 10.2). The maximum load acceptable by the sinks (m_{SKj}) is calculated using Equation 10.1 and documented in column 4 of Table 10.2.

   Similarly, water sources of both networks are arranged in ascending order of concentrations. The load possessed by the source (m_{SRi}) is calculated using Equation 10.2 and documented in column 4 of Table 10.3.

TABLE 10.1

Limiting Water Data for Individual RCNs in Example 10.1

Sinks, SK$_j$	F_{SKj} (t/h)	C_{SKj} (ppm)	Sources, SR$_i$	F_{SRi} (t/h)	C_{SRi} (ppm)
Network A					
SK1	20	0	SR1	20	100
SK2	100	50	SR2	100	100
SK3	40	50	SR3	40	800
SK4	10	400	SR4	10	800
Network B					
SK5	50	20	SR5	50	50
SK6	100	50	SR6	100	100
SK7	80	100	SR7	70	150
SK8	70	200	SR8	60	250
$\sum_i F_{SKj}$	470		$\sum_i F_{SRi}$	450	

TABLE 10.2

Data for Plotting Sink Composite Curve in Example 10.1

Sinks, SK_j	F_{SKj} (t/h)	C_{SKj} (ppm)	m_{SKj} (g/h)	Cum. F_{SKj} (t/h)	Cum. m_{SKj} (g/h)
SK1	20	0	0	0	0
SK5	50	20	1,000	20	0
SK2	100	50	5,000	70	1,000
SK3	40	50	2,000	170	6,000
SK6	100	50	5,000	210	8,000
SK7	80	100	8,000	310	13,000
SK8	70	200	14,000	390	21,000
SK4	10	400	4,000	460	35,000
				470	39,000

TABLE 10.3

Data for Plotting Source Composite Curve in Example 10.1

Sources, SR_i	F_{SRi} (t/h)	C_{SRi} (ppm)	m_{SRi} (g/h)	Cum. F_{SRi} (t/h)	Cum. m_{SRi} (g/h)
SR5	50	50	2,500	0	0
SR1	20	100	2,000	50	2,500
SR2	100	100	10,000	70	4,500
SR6	100	100	10,000	170	14,500
SR7	70	150	10,500	270	24,500
SR8	60	250	15,000	340	35,000
SR3	40	800	32,000	400	50,000
SR4	10	800	8,000	440	82,000
				450	90,000

3. *All process sinks are plotted in descending order of their quality levels on a load-versus-flowrate diagram to form the sink composite curve.*
4. *Step 3 is repeated for all process sources to form the source composite curve on the same load-versus-flowrate diagram.*

As the flowrate and load values of composite curves in the MRPD are cumulative in nature, the cumulative flowrates (Cum. F_{SKj}) and cumulative loads (Cum. m_{SKj}) for all sinks are first calculated, given in the last two columns of Table 10.1. Similarly, the cumulative flowrates (Cum. F_{SRi}) and cumulative loads (Cum. m_{SRi}) for all sources are calculated and given in the last two columns of Table 10.2. These values are then used to plot the sink and source composite curves on the load-versus-flowrate diagram, as shown in Figure 10.3. Note that the slope of each segment corresponds to their respective concentration values. As shown in Figure 10.3, the sink composite curve indicates that a total flowrate of 470 t/h of water flowrate are needed

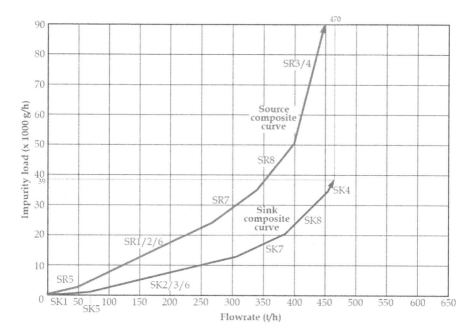

FIGURE 10.3
Infeasible MRPD for inter-plant water network in Example 10.1.

by all sinks in the IPRCN, and these sinks can only accept a total impurity load of 39,000 g/h. On the other hand, the source composite curve indicates that the sources have a total water flowrate of 450 t/h, and possess 90,000 g/h of total impurity load.

It is then observed that the source composite curve stay on the left and above the sink composite curve, indicating that the MRPD is infeasible. Figure 10.3 shows that the sinks do not receive sufficient water flowrate (a total of 470 t/h is required) and is supplied with total impurity load (90,000 g/h) that is higher than its maximum acceptable limit (39,000 g/h).

5. *For the IPRCN with a single pure fresh resource of superior quality, the source composite curve is moved horizontally to the right until it just touches the sink composite curve, with the latter lying above and to the left of the source composite curve.*

As the freshwater supply has no impurity, the source composite curve is moved horizontally to the right until it just touches the sink composite curve and lies completely below and to the right of the latter. This results in a feasible MRPD as shown in Figure 10.4. In this case, both flowrate and load constraints of the sink composite curve are observed.

Several IPRCN targets may be identified from the feasible MRPD in Figure 10.4. The opening on the left side of the MRPD indicates the overall minimum fresh resource target (F_{FW}) of the IPRCN is identified as 155 t/h, while the opening on the

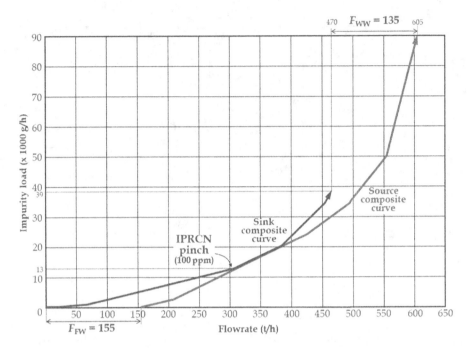

FIGURE 10.4
Feasible MRPD for inter-plant water network in Example 10.1.

right side indicates its overall minimum wastewater generation (F_{WW}) is 135 t/h. The overlapping region of the two composite curves indicates that a total of 315 t/h (= 470 – 155 t/h) of water resources are recovered among the sinks and sources of the IPRCN. The IPRCN pinch is identified as 100 ppm.

10.3 Mathematical Programming for Direct Integration Problems

The synthesis of IPRCNs with mathematical programming technique has been discussed in Chapter 9. In this section, we shall make use of such technique to synthesize an IPRCN. Figure 10.5 (reproduced from Figure 1.19a) shows a superstructural representation of an IPRCN with *direct integration scheme* between two individual RCNs. As shown, the superstructural representation is similar to that of in-plant direct reuse/recycle scheme (Figure 9.1), except that sinks in plant p can now accept both process sources within the same plant p (i.e., with flowrate $F_{SRi,SKj,p}$), as well as cross-plant stream connections from sources of plant p' (with flowrate $F_{SRip',SKjp'}$). The latter are

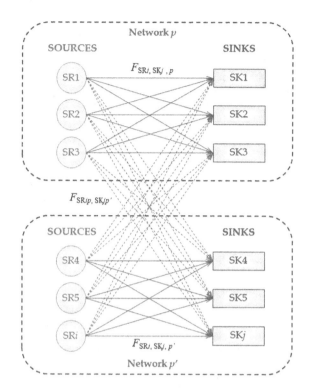

FIGURE 10.5
Superstructural model for direct integration scheme. (From Chew et al. 2008. Reproduced with permission. Copyright 2008 American Chemical Society.)

indicated by the gray area in the *matching matrix* in Figure 10.6. Note that no connection should exist between fresh resource(s) with waste discharge in all networks.

The optimization model for the IPRCN can be extended from that of the in-plant reuse/recycle network (Chapter 9), with constraints given as in Equations 10.3 through 10.6. Note that in each constraint set, a new flowrate term is added for the cross-plant streams ($F_{SRip,SKjp'}$).

$$\sum_i F_{SRi,SKj,p} + F_{R,SKj,p} + \sum_i F_{SRip',SKjp} = F_{SKj,p} \quad \forall j \ \forall p \tag{10.3}$$

$$\sum_i F_{SRi,SKj,p} q_{SRi,c} + F_{R,SKj,p} q_{R,c} + \sum_i F_{SRip',SKjp} q_{SRi,c} \geq F_{SKj,p} q_{SKj,c} \quad \forall j \ \forall p \ \forall c \tag{10.4}$$

$$\sum_i F_{SRi,SKj,p} + F_{SRi,D,p} + \sum_i F_{SRip,SKjp'} = F_{SRi,p} \quad \forall i \ \forall p \tag{10.5}$$

$$F_{SRi,SKj,p} , F_{SRip',SKjp} , F_{R,SKj,p} , F_{SKi,D,p} \geq 0 \ \forall i, j, p \tag{10.6}$$

			F_{SKj}	F_{SK1}	F_{SK2}	F_{SK3}	F_{SKj}	F_D
			C_{SKj}	$q_{SK1,p}$	$q_{SK2,p}$	$q_{SK3,p'}$	$q_{SKj,p'}$	q_D
F_{SRi}	C_{SRi}	SKj / SRi	SK1p	SK2p	SK3p'	SKjp'	WD
F_{Rp}	q_R	FRp	$F_{R\,SK1,p}$	$F_{R\,SK2,p}$			
$F_{SR1,p}$	q_{SR1}	SR1p	$F_{SR1,SK1,p}$	$F_{SR1,SK2,p}$	$F_{SR1p,SK3p'}$	$F_{SR1p,SKjp'}$	$F_{SR1,D,p}$
$F_{SR2,p}$	q_{SR2}	SR2p	$F_{SR2,SK1,p}$	$F_{SR2,SK2,p}$	$F_{SR2p,SK3}$	$F_{SR2p,SKjp'}$	$F_{SR2,D,p}$
$F_{Rp'}$	q_R	FRp'			$F_{Rp,SK3,p'}$	$F_{Rp,SKj,p'}$	
$F_{SR3,p'}$	q_{SR3}	SR3p'	$F_{SR3p',SK1p}$	$F_{SR3p',SK2p}$	$F_{SR3,SK3,p'}$	$F_{SR3,SKj,p'}$	$F_{SR3,D,p'}$
$F_{SRi,p'}$	q_{SRj}	SRip'	$F_{SRip',SK1p}$	$F_{SRip',SK2p}$	$F_{SRi,SK3,p'}$	$F_{SRi,SKj,p'}$	$F_{SRi,D,p'}$

FIGURE 10.6
Superstructural model for direct integration in matrix form (cross-plant connections are indicated by the gray area).

The constraint in Equation 10.3 describes that within each plant p, the flowrate requirement of the sink ($F_{SKj,p}$) is fulfilled by in-plant recovery from various process sources ($F_{SRi,SKj,p}$), fresh resource r ($F_{Rr,SKj,p}$), as well as cross-plant flowrate ($F_{SRip,SKjp'}$). Equation 10.4 describes that the in-plant and inter-plant process sources, as well as fresh resource feed, must fulfill the minimum quality requirement of the process sinks. For cases involving multiple-impurities, the equation should be revised to cater for individual impurities (see following section). The flowrate balance in Equation 10.5 indicates that the flowrate of each process ($F_{SRi,p}$) source may be allocated to sinks within plant p, or to those in plant p' (through cross-plant flowrate), while the unutilized source will be sent for waste discharge ($F_{SRi,D,\,p}$). Equation 10.6 indicates that all flowrate variables of the superstructural model should take non-negative values.

The objective of the optimization model may be set to minimize the cost of the IPRCN. The latter may be contributed by operating cost (*OC*) or *total annualized cost* (*TAC*) of the IPRCN. The *OC* is contributed by the overall fresh resources and waste discharge flowrates; while the *TAC* may include the capital costs of in-plant and cross-plant piping connections[1]. Alternatively, one may also set optimization objective to minimize the overall minimum fresh resource (F_R) and the total flowrate of all cross-plant streams ($F_{SRip,SKjp'}$) with the two-stage optimization approach introduced in Chapter 9. In stage 1, the fresh resource flowrate may be determined with

---

[1] See detailed cost correlation in Chapter 9.

objective in Equation 10.7, or alternatively using the graphical technique of MRPD introduced in an earlier section. The identified overall minimum fresh resource flowrate ($F_{R,min}$) is then included as a new constraint (Equation 10.8) in stage 2, while the cross-plant stream flowrate (F_{CP}) is minimized using the optimization objective in Equation 10.9.

$$\text{Minimize } F_R = \sum_p \sum_j F_{R,SKj,p} \tag{10.7}$$

$$F_{R,min} = F_R \tag{10.8}$$

$$\text{Minimize } F_{CP} = \sum_i F_{SRip,SKjp'} \tag{10.9}$$

As all constraints and objective of the model are linear, this is a linear program (LP) model for which any solution found is globally optimal.

Example 10.2 Example 10.1 is revisited. The overall IPRCN freshwater and wastewater flowrates have been identified as 155 and 135 t/h, respectively. Determine the freshwater and wastewater flowrates of the individual network, as well as the minimum cross-plant flowrate using MS Excel.

SOLUTION

A superstructure for the water network is constructed in the matching matrix form using the MS Excel spreadsheet in Figure 10.7a. Sources and sinks of Network A are arranged as located in cell arrays E6:H10, while

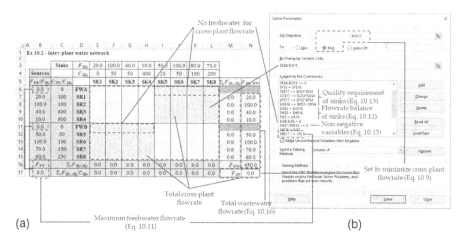

(a)

(b)

FIGURE 10.7

(a) Superstructural model for direct integration in MS Excel; (b) setting of constraints in Solver for Example 10.2 (all flowrate terms in t/h; concentration in ppm).

those of Network B are in cell arrays I11:L15 of the spreadsheet. In other words, these arrays are the regions where in-plant water recovery will take place. On the other hand, arrays I6:L10 and E11:H15 are regions where inter-plant water recovery may take place.

To formulate an optimization model to minimize the overall freshwater flowrate (F_{FW}) of the inter-plant water network, the objective in Equation 10.7 takes the revised form of Equation 10.10:

$$\text{Minimize } F_{FW} = \sum_{p}\sum_{j} F_{FW,SKj,p} \tag{10.10}$$

where $F_{FW,SKj,p}$ is the freshwater flowrate allocated to sink j in plant p. Since this overall freshwater flowrate has been identified using the MRPD as 155 t/h in Example 10.1, the minimum flowrate ($F_{FW,min}$) constraint in Equation 10.11 is added for cell B12 in Solver (see setting in Figure 10.7b).

$$F_{FW,min} = 155 \text{ t/h} \tag{10.11}$$

All other generic constraints in Equations 10.3–10.6 take their revised form as in Equation 10.12–10.15. Equation 10.12 describes the flowrate balances for process sinks. One can derive these flowrate balances by observing columns E–L in Figure 10.7:

$$\sum_{i} F_{SRi,SKj,p} + F_{R,SKj,p} + \sum_{i} F_{SRip',SKjp} = F_{SKj,p} \quad \forall j \; \forall p \tag{10.12}$$

The quality requirement of the sinks is described by Equation 10.13. Note however that the inequality sign takes a reverse form as that in Equation 10.11, as the quality index is described by impurity concentration. In other words, impurity load sent from process sources, freshwater (with concentration C_{FW}), and cross-plant flowrate must not exceed the load limit of the process sinks; the latter is given as the product of its flowrate ($F_{SKj,p}$) and impurity concentration (C_{SKj}):

$$\sum_{i} F_{SRi,SKj,p}C_{SRi} + F_{FW,SKj,p}C_{FW} + \sum_{i} F_{SRip',SKjp}C_{SRi} \leq F_{SKj,p}C_{SKj} \quad \forall j \; \forall p \tag{10.13}$$

Equation 10.14 describes the flowrate balance of the process sources. As shown, each process source may be allocated to any process sinks within the same plant p ($F_{SRi,SKj,p}$), or be sent as cross-plant flowrate ($F_{SRip,SKjp'}$), while the unutilized source will be sent for wastewater discharge ($F_{SRi,WW,p}$). One can derive these flowrate balances by observing rows 6–15 in Figure 10.7.

$$\sum_{i} F_{SRi,SKj,p} + F_{SRi,WW,p} + \sum_{i} F_{SRip,SKjp'} = F_{SRi,p} \quad \forall i \; \forall p \tag{10.14}$$

Equation 12.12 indicates that all flowrate variables take non-negative values. These include the flowrate terms for the matches between source i ($F_{SRi,SKj}$) and freshwater to sink j ($F_{FW,SKj}$), cross-plant flowrate ($F_{SRip',SKjp}$), as well as the wastewater stream from source i ($F_{SRi,WW}$):

$$F_{SRi,SKj,p}, F_{SRip',SKjp}, F_{FW,SKj,p}, F_{SRi,WW,p} \geq 0 \; \forall i, j, p \qquad (10.15)$$

It is also convenient to determine the total flowrate of the wastewater streams (F_{WW}), given by the last column of the matching matrix (array N6:N16):

$$F_{WW} = \sum_p \sum_i F_{SRi,WW,p} \qquad (10.16)$$

Solving the optimization model with MS Excel Solver yield the results in Figure 10.8, and an inter-plant water network in Figure 10.9. The latter shows one cross-plant flowrate (cell K8) exists for this IPRCN, connecting between source SR2 of Network A with sink SK7 of Network B. Besides, the model also determines that freshwater flowrates for Networks A and B are determined as 90 t/h (cell B6, $F_{FW,A}$) and 65 t/h (cell B11, $F_{FW,B}$). The total wastewater (F_{WW}) of this inter-plant water network is determined as 155 t/h (cell B17). Both freshwater and wastewater are identical to those determined using MRPD in Figure 10.4. Note also that the total material recovery can be calculated by the summation of individual cells in column M (cells M7:M15), which gives a total value of 315 t/h, identical to that determined using MRPD in Figure 10.4.

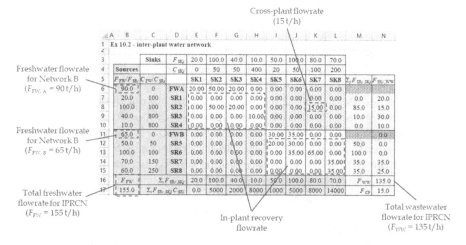

FIGURE 10.8

Results of superstructural model for Example 10.2 (all flowrate terms in t/h; concentration in ppm).

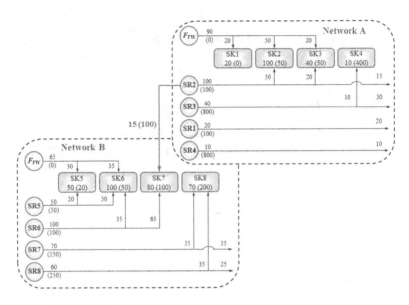

FIGURE 10.9
Final configuration of IPRCN in Example 10.2 (all flowrate terms in t/h; concentration in parenthesis in ppm).

10.4 Industrial Case Study—Integrated Pulp Mill and Bleached Paper Plant

An industrial case study of an integrated pulp mill and bleached paper plant is next analyzed. Figure 10.10 shows the process flow diagram of this process, originally reported by Lovelady et al. (2007). As shown, wood chips and white liquor are fed into a digester unit for pulp cooking operation. The cooked pulp is then sent to brown stock washing in which the pulp is separated from its residual liquor. The pulp is next screened and washed before it is sent for bleaching operation in the bleached paper plant.

On the other hand, the black liquor (residual cooking liquor from brown stock washing) is sent to a series of multiple effect evaporators and concentrator to be concentrated to strong black liquor. The latter is burned in a recovery boiler in order to reduce its oxidized sulfur compounds to become sulfide. Flue gas from the recovery furnace (containing particulate matter) will be sent an electrostatic precipitator (ESP), where the dust particles are collected and be sent to the strong black liquor system. Besides, inorganic chemicals (known as smelt) is recovered from the boiler, and is used to regenerate the cooking liquor in the dissolving tank.

In the dissolving tank, weak wash liquor (from washer/filter) is used to dissolve the smelt in order to form the green liquor. The latter is clarified to remove any undissolved materials, e.g., unburned carbon and inorganic

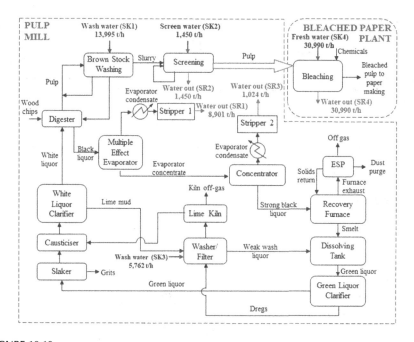

FIGURE 10.10
Process flow diagram of an integrated pulp mill and a bleached paper plant. (From Chew et al., 2008.).

impurities, where are known collectively as dregs. The dregs are washed to remove any sodium compounds and are then discarded. The clarified green liquor is fed to a slaker, where lime and water react to form slaked lime. Any grits (unreactive lime particles and insoluble materials) are removed from the slaker. White liquor is then formed in the causticizer by reacting the slaked lime and sodium carbonate. The white liquor is sent to a clarifier to remove lime mud formed in the causticizing reaction, before it is returned to the digester. Lime mud from the white liquor clarifier is washed to remove entrained alkali, which is then sent to lime kiln where lime mud is converted to reburned lime for use in the slaking reaction.

Figure 10.10 also labels the water sinks and sources of the integrated pulp mill and bleached paper plant. As observed, both plants have high water consumption due to extensive washing operation for the process. As a consequence, the plants have larger flowrate of wastewater discharge. To promote sustainability, water minimization potential is explored. As the bleached paper plant is located next to the pulp mill, direct integration may be carried out for the water sinks and sources of these plants. Table 10.4 summarizes the limiting data for the water sinks and sources (Chew et al., 2008). As shown, water recovery among the water sinks and sources is dictated by the concentration of three ions, i.e., chlorine (Cl^-), potassium (K^+), and sodium (Na^+). Note that concentrations of these ions are reported as 3.7, 1.1, and 3.6 ppm, respectively, in the freshwater resource.

TABLE 10.4

Limiting Water Data for Industrial Case Study

| Sinks, SK$_j$ | F_{SKj} (t/h) | C_{SKj} (ppm) | | | Sources, SRi | F_{SRi} (t/h) | C_{SRi} (ppm) | | |
		Cl⁻	K⁺	Na⁺			Cl⁻	K⁺	Na⁺
Pulp mill									
Brown stock washing (SK1)	13,995	34.4	12.7	89.4	Stripper 1 (SR1)	8,901	0	0	0
Screening (SK2)	1,450	241.0	2.4	241.0	Screening (SR2)	1,450	308.6	115.5	840.3
Washer/filter (SK3)	5,762	0	0	0	Stripper 2 (SR3)	1,024	0	0	0
Bleached paper plant									
Bleaching (SK4)	30,990	3.7	1.1	3.6	Bleaching (SR4)	30,990	500.0	5.0	500.0

Source: Chew et al. (2008).

Note that for a multiple-contaminant case, mathematical programming is more reliable. All constraints used for Example 10.2 are applicable here, except that Equation 10.13 has to be revised to cater for three ions in this case (multiple-impurities), given as in Equation 10.17. As shown, the load of impurity m received from process sources, freshwater (with concentration $C_{FW,\,m}$), and cross-plant flowrate by sink j must not exceed its maximum limit; the latter is given as the product of its flowrate ($F_{SKj,p}$) and concentration of impurity m ($C_{SKj,m}$):

$$\sum_i F_{SRi,SKj,p} C_{SRi,m} + F_{FW,SKj,p} C_{FW,m} + \sum_i F_{SRip',SKjp} C_{SRi,m} \le F_{SKj,p} C_{SKj,m} \quad \forall j \; \forall p \; \forall m$$

(10.17)

We next proceed to use the two-stage optimization approach to minimize both freshwater and cross-plant flowrates of the IPRCN. In stage 1, the objective is set to minimize the overall freshwater flowrate (cell B14) with Equation 10.10, subject to the constraints in Equations 10.12, 10.14–10.17. The model determines that the freshwater flowrate to be 40,126 t/h (result is not shown for brevity). The freshwater flowrate is added as a constraint in stage 2 optimization (see the spreadsheet and Solver setting in Figure 10.11), with the constraint in Equation 10.17. Solving the model again using objective in Equation 10.9, subject to constraints in 10.12, 10.14–10.17, as well as the new constraint in Equation 10.18 (freshwater flowrate limit) yields the minimum cross-plant flowrate (cell K14) of 697 t/h, with results shown in Figure 10.11.

$$F_{FW,min} = 40,126 \text{ t/h} \tag{10.18}$$

The IPRCN results in Figure 10.11 is analyzed. It is observed that two cross-plant flowrates involving SR4 are needed for this case, i.e., SR4–SK1 (cell G13) and SR4–SK2 (cell H13). The former has a small flowrate of 1 t/h,

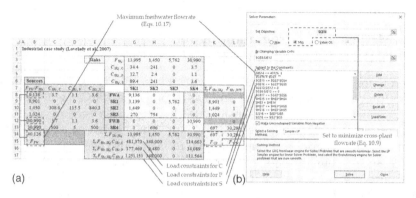

(a)

(b)

FIGURE 10.11
(a) Superstructural model for direct integration in MS Excel; (b) setting of constraints in Solver for industrial case study (all flowrate terms in t/h; concentration in ppm).

which may not be justified due to practical consideration or physical plant constraint. If this cross-plant flowrate will be removed, the flowrate requirement of SK1 can be fulfilled higher flowrate of freshwater, i.e., 9,137 t/h; hence a slightly higher total flowrate of freshwater is resulted at 40,127 t/h (= 9,137 + 30,990 t/h). Such IPRCN structure is given in Figure 10.12.

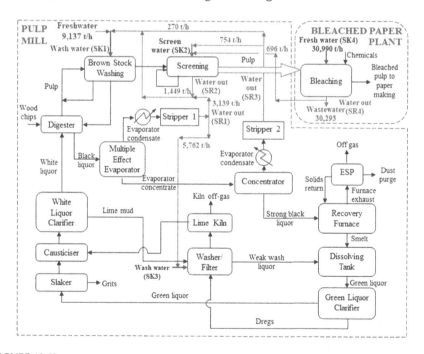

FIGURE 10.12
Water Recovery Scheme for Industrial Case Study (note: no cross-plant flowrate for SR4–SK1; flowrate requirement of SK1 is fulfilled by freshwater).

10.5 Mathematical Programming for Indirect Integration Problems

When there exist several plants/processes in an IPRCN, or the individual plants/processes are located far from each other, an IPRCN with indirect integration scheme is more sensible. Such integration scheme is shown in a superstructural in Figures 10.13 (Chew et al., 2008). As shown, apart from in-plant material recovery, material recovery also occur among plants (e.g., p and p') but through a centralized utility hub. In the latter, material sources may be mixed in storage tank(s), or be purified before they are sent for further reuse/recycle to the process sinks in network p and p'. Comparing the superstructural representations in Figure 10.13 with that of direct integration scheme (Figure 10.5), it is observed that cross-plant connection does not exist between sinks and sources of individual networks for indirect integration scheme. One may also represent the superstructure in a matching matrix form (Figure 10.14). Note that no connection should exist between fresh resource(s) with waste discharge in all networks, as well as

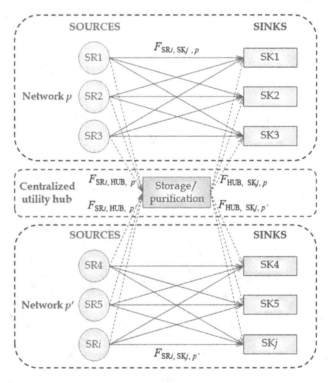

FIGURE 10.13
Superstructural model for indirect integration scheme. (Adapted from Chew et al., 2008.)

		F_{SKj}	F_{SK1}	F_{SK2}	F_{SK3}	F_{SKj}	F_{HUB}	F_D
		C_{SKj}	$q_{SK1,p}$	$q_{SK2,p}$	$q_{SK3,p'}$	$q_{SKj,p'}$		q_D
F_{SRi}	C_{SRi}	SKj \ SRi	**SK1p**	**SK2p**	**SK3p'**	**SKjp'**	**HUB**	**WD**
F_{Rp}	q_R	**FRp**	$F_{R,SK1,p}$	$F_{R,SK2,p}$				
F_{HUBp}	q_H	**HUBp**	$F_{HUB,SK1,p}$	$F_{HUB,SK2,p}$				
$F_{SR1,p}$	q_{SR1}	**SR1p**	$F_{SR1,SK1,p}$	$F_{SR1,SK2,p}$			$F_{SR1,HUB,p}$	$F_{SR1,D,p}$
$F_{SR2,p}$	q_{SR2}	**SR2p**	$F_{SR2,SK1,p}$	$F_{SR2,SK2,p}$			$F_{SR2,HUB,p}$	$F_{SR2,D,p}$
$F_{Rp'}$	q_R	**FRp'**			$F_{Rp,SK3,p'}$	$F_{Rp,SKj,p'}$		
$F_{HUBp'}$	q_H	**HUBp'**			$F_{HUB,SK3,p'}$	$F_{HUB,SKj,p'}$		
$F_{SR3,p'}$	q_{SR3}	**SR3p'**			$F_{SR3,SK3,p'}$	$F_{SR3,SKj,p'}$	$F_{SR3,HUB,p'}$	$F_{SR3,D,p'}$
$F_{SRj,p'}$	q_{SRj}	**SRip'**			$F_{SRj,SK3,p'}$	$F_{SRj,SKj,p'}$	$F_{SRj,HUB,p'}$	$F_{SRj,D,p'}$

FIGURE 10.14
Superstructural model for indirect integration in matrix form (gray area indicates no direct integration among sinks and sources of different networks).

those between sinks and sources of the individual networks (gray section in Figures 10.14).

The optimization model for indirect integration can be extended from that of direct integration, with revised constraints given as in Equations 10.19 through 10.23. Note that in each constraint, no direct cross-plant streams are present between sinks and sources of different network. Instead, the sources in each network may send their material to the utility hub (with flowrate $F_{SRip,Hub}$), while the latter sends the material to individual sinks in network p (with flowrate $F_{HUB,SKjp}$).

$$\sum_i F_{SRi,SKj,p} + F_{R,SKj,p} + F_{HUB,SKj,p} = F_{SKj,p} \quad \forall j \ \forall p \tag{10.19}$$

$$\sum_i F_{SRi,SKj,p} q_{SRi} + F_{R,SKj,p} q_R + F_{HUB,SKjp} q_H \geq F_{SKj,p} q_{SKj} \quad \forall j \ \forall p \tag{10.20}$$

$$\sum_i F_{SRi,SKj,p} + F_{SRi,D,p} + F_{SRip,HUB} = F_{SRi,p} \quad \forall i \ \forall p \tag{10.21}$$

$$\sum_p \sum_i F_{SRip,HUB} = \sum_p \sum_j F_{HUB,SKjp} \tag{10.22}$$

$$F_{SRi,SKj,p}, F_{HUB,SKj,p}, F_{SRip,HUB}, F_{R,SKj,p}, F_{SKi,D,p} \geq 0 \ \forall i, j, p \tag{10.23}$$

The constraint in Equation 10.19 describes that within each plant p, the flowrate requirement of the sink ($F_{SKj,p}$) is fulfilled by in-plant recovery from various process sources ($F_{SRi,SKj,p}$), fresh resource r ($F_{Rr,SKj,p}$), as well as purified flowrate from the utility hub ($F_{HUB,SKjp'}$). Equation 10.20 describes that the material recovered from in-plant sources and those through the utility hub (with quality q_H), as well as the fresh resource must fulfill the minimum quality requirement of the process sinks. Note that in some cases, the utility hub may be equipped purification unit(s), in which the material sources can be regenerated for better quality (i.e., $q_H \geq q_{SRi}$) before they are sent to the sinks[2] The flowrate balance in Equation 10.21 indicates that each process source may be allocated to sinks within plant p, or be sent to utility hub, while the unutilized source(s) will be discharged ($F_{SRi,D, p}$). The flowrate balance of the utility hub is given by Equation 10.22. Equation 10.23 indicates that all flowrate variables of the superstructural model should take non-negative values. The main assumption made in this model is that, material sources of different quality will not be mixed in the utility hub, so to keep the basic model to be an LP model.

Similar as before, the objective of the optimization model may be set to minimize the cost of the IPRCN, which may take the form as the operating cost (*OC*) or *total annualized cost* (*TAC*) of the IPRCN. The *OC* is contributed by the overall fresh resources and waste discharge flowrates; while the *TAC* may include the capital costs of in-plant and cross-plant piping connections (from utility hub)[3]. Alternatively, one may also set the objective to minimize the overall minimum fresh resource (F_R) and the total flowrate sent from the utility hub ($F_{SRip,RE}$) with the two-stage optimization approach introduced in Chapter 9. For this case, the fresh resource flowrate is minimized using objective in Equation 10.7 in stage 1. The identified overall minimum fresh resource flowrate ($F_{R,min}$) is then included as a new constraint (Equation 10.8) in stage 2, while the inter-plant flowrate through the utility hub (F_{HUB}) is minimized using the objective in Equation 10.22.

$$\text{Minimize } F_R = \sum_p \sum_j F_{R,SKj,p} \tag{10.7}$$

$$F_{R,min} = F_R \tag{10.8}$$

$$\text{Minimize } F_{HUB} = \sum_p \sum_i F_{SRip,HUB} \tag{10.24}$$

As all constraints and objective of the model are linear, this is an LP model for which any solution found is globally optimal.

---

[2] See details of modelling of purification units in Chapter 6.
[3] See details on cost correlation in Chapter 9.

Example 10.3 Example 10.1 is revisited, where indirect integration is carried out for this case. A centralized utility hub is available, in which a purification unit is used to perform regeneration of water sources. The impurity concentration (C_H) of the purified source from the hub is given as 10 ppm. Determine the minimum fresh water and cross-plant flowrates with the two-stage optimization approach.

SOLUTION

Figure 10.15 shows the superstructure for the water network is constructed in the matching matrix using MS Excel spreadsheet, similar to that in Figure 10.7 (Example 10.2; direct integration scheme), except that no direct water recovery will take place among sinks and sources of different networks. Besides, column M is added to account for flowrates to be sent from individual sources to the hub, while rows 7 and 13 are added to allow the purified sources from the hub to be recovered to sinks in Network A (cells E7–H7; HUBA) and Network B (cells I13:L13; HUBB).

We next proceed to minimize the fresh water and regeneration flowrates with the two-stage optimization approach. In stage 1, the objective is set to minimize the overall freshwater flowrate (F_{FW}) of the inter-plant water network, with objective in Equation 10.10 (same as Example 10.2):

$$\text{Minimize } F_{FW} = \sum_p \sum_j F_{FW,SKj,p} \qquad (10.10)$$

All other generic constraints in Equations 10.19–10.23 take their revised form as in Equation 10.25–10.28 (note that equation 10.20 remains identical). Equation 10.25 describes the water flowrate balances for process

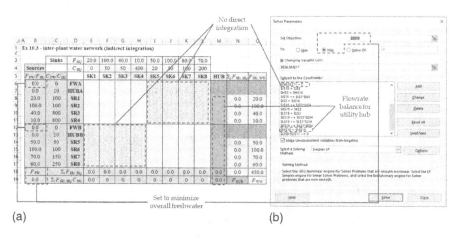

(a)   (b)

FIGURE 10.15
Superstructural model for indirect integration for Example 10.3: (a) setting in MS Excel spreadsheet; (b) setting of constraints in Solver (all flowrate terms in t/h; concentration in ppm).

sinks. One can derive these flowrate balances by observing columns E–L in Figure 10.15:

$$\sum_i F_{SRi,SKj,p} + F_{FW,SKj,p} + \sum_i F_{HUB,SKjp} = F_{SKj,p} \quad \forall j \; \forall p \quad (10.25)$$

The quality requirement of the sinks is described by Equation 10.26. Note however that the inequality sign takes a reverse form as that in Equation 10.20, as the quality index is described by impurity concentration. In other words, impurity load sent from process sources, freshwater (with concentration C_{FW}), and utility hub with concentration C_H) must not exceed the load limit of the process sinks; the latter is given as the product of its flowrate ($F_{SKj,p}$) and impurity concentration (C_{SKj}). Note that the equation has to be written for individual impurities for multiple-impurities case (see Equation 10.17 for example).

$$\sum_i F_{SRi,SKj,p}C_{SRi} + F_{FW,SKj,p}C_{FW} + F_{HUB,SKjp}C_H \leq F_{SKj,p}C_{SKj} \quad \forall j \; \forall p \quad (10.26)$$

Equation 10.27 describes the flowrate balance of the process sources. As shown, each process source may be allocated to any process sinks within the same plant p ($F_{SRi,SKj,p}$), or be sent to utility hub for purification ($F_{SRip,HUB}$), while the unutilized source will be sent for wastewater discharge ($F_{SRi,WW,p}$). One can derive these flowrate balances by observing rows 6–17 in Figure 10.15.

$$\sum_i F_{SRi,SKj,p} + F_{SRi,WW,p} + F_{SRip,HUB} = F_{SRi,p} \quad \forall i \; \forall p \quad (10.27)$$

Equation 10.28 indicates that all flowrate variables take non-negative values, include those of freshwater to sink j ($F_{FW,SKj}$), as well as the wastewater stream from source i (F_{SRiWW}):

$$F_{SRi,SKj,p}, F_{HUB',SKjp}, F_{SRip,HUB}, F_{FW,SKj,p}, F_{SRi,WW,p} \geq 0 \quad \forall i, j, p \quad (10.28)$$

It is also convenient to determine the total flowrate of the wastewater streams (F_{WW}), given by the last column of the matching matrix (cells O8:O17):

$$F_{WW} = \sum_p \sum_i F_{SRi,WW,p} \quad (10.16)$$

Solving the stage 1 optimization model with MS Excel Solver yields the minimum freshwater flowrate of 20 t/h (results not shown for brevity). We next proceed to stage 2 optimization to minimize the regeneration flowrate at the utility hub, with Equation 10.24. The minimum freshwater of 20 t/h is added as a constraint in Equation 10.29.

$$F_{FW,min} = 20 \text{ t/h} \quad (10.29)$$

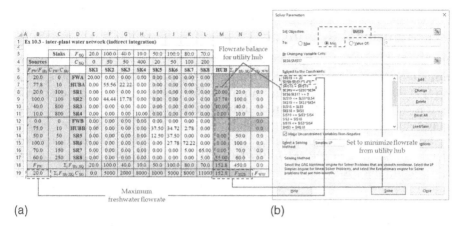

(a)   (b)

FIGURE 10.16
Results for Stage 2 optimization of Example 10.3: (a) setting in MS Excel spreadsheet; (b) setting of constraints in Solver (all flowrate terms in t/h; concentration in ppm).

Solving the model with the same set of constraints (and Equation 10.29) yields the solution in Figure 10.16. The latter shows several sources are sent for purification in the utility hub from both networks, i.e., SR1–SR3 of network A, as well as SR8 of Network B (observed from column M). Upon purification, 77.8 t/h of regenerated water sources are sent to Network A (SK2 and SK3; row 7) and 75 t/h to Network B (SK5–SK7; row 13). Besides, the model also determines that 20 t/h of freshwater is sent for Network A (cell B6, $F_{FW,A}$), while no freshwater is needed for Network B (cell B11, $F_{FW,B}$). Note that a *zero-discharge* IPRCN is achieved, as no wastewater (F_{WW}) is generated from both water networks (cell O18). The final structure of the inter-plant water network is given in Figure 10.17.

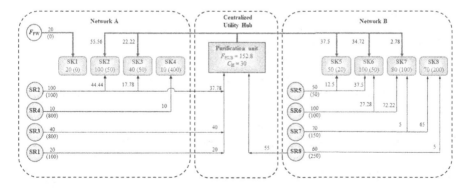

FIGURE 10.17
Results of superstructural model for Example 10.3 (all flowrate terms in t/h; concentration in parenthesis in ppm).

10.6 Additional Readings

In this chapter, graphical pinch method (MRPD) and mathematical programming technique (based on superstructural model) are used for synthesizing an IPRCN. Mathematical programming technique is given bigger emphasis as it is capable in handling direct and indirect integration schemes. Readers may refer to the first edition of this book (Foo, 2012) to have better understanding on the use graphical, algebraic and automated techniques for detailed targeting of IPRCN. In the latter, the method may be categorized as *assisted* and *unassisted integration* schemes (Chew et al., 2010a, 2010b).

PROBLEMS
Water Minimization Problems

10.1. Revisit Examples 10.1 and 10.2. Solve the following tasks:

a. Use *material cascade analysis* (refer to Chapter 4 for its detailed procedure) to identify the minimum flowrate targets for freshwater and wastewater of the overall IPRCN.

b. Use mathematical model to Identify degenerate solution(s) for this IPRCN (hint: refer to Chapter 9 for detailed procedure).

10.2. Revisit Examples 10.3. Synthesize an IPRCN with indirect integration scheme, where a centralized utility hub is present. However, it is assumed that no purification unit is present in the utility hub. In other words, the water quality remains identical as their quality (i.e., no regeneration takes place). Compare the solution with that obtained in Example 10.3.

10.3. Water minimization study is carried out in a catalyst plant (Feng et al., 2006). The plant consists of two separate sections, i.e., synthesis and preparation departments. Limiting water data for these sections is given in Table 10.5. Note that sodium ion (Na^+) content is identified as the main impurity for water recovery (Feng et al., 2006). Freshwater supply for the plant is free of sodium ion. Perform the following tasks:

a. Use MRPD or *material cascade analysis* (refer to Chapter 4 for its detailed procedure) to identify the minimum flowrate targets for freshwater and wastewater of the overall IPRCN, assuming direct integration scheme is used.

b. Use mathematical programming method to synthesize an IPRCN for the plant. Determine the minimum cross-plant flowrate for a direct integration scheme.

c. Assuming that the plant is located in a centralized utility hub that is equipped with a regeneration unit. The latter will purify water

TABLE 10.5

Limiting Water Data for a catalyst plant (Feng et al., 2006) in Problem 10.4

Sinks, SK_j	F_{SK_j} (t/d)	C_{SK_j} (ppm)	Sources, SR_i	F_{SR_i} (t/d)	C_{SR_i} (ppm)
Synthesis Department					
Paste formation	34.6	1300	Gel recovery	28.3	4000
Ion exchange and filtration	48.2	1300	Washing and drying	2022.8	1300
Washing and drying	3114.5	800	Pump cooling	24.4	62
Pump cooling	24.4	60			
Preparation Department					
Paste formation	119.3	400	Pump cooling	525.4	62
Paste formation and filtration	271.7	400	Sedimentation	1380	500
Washing, filtration, and drying	1851	80			
Pump cooling	525.4	60			

sources for further reuse/recycle for sinks in both sections (with C_H = 30 ppm). Determine the minimum flowrates for its freshwater and regenerated water from the hub.

10.4. An inter-plant water network that involves three water networks (Olesen and Polley, 1996) is analyzed, with data shown in Table 10.6. Solve the following cases:

a. Perform flowrate targeting with MRPD or *material cascade analysis* (refer to Chapter 4 for its detailed procedure) for a direct reuse/recycle scheme.

b. Use mathematical model to design an inter-plant water network that achieves the flowrate targets.

c. An purification unit of 10 ppm (C_H) is used for water regeneration in the centralized utility hub. Identify the minimum flowrate of freshwater and the purified source from the utility hub.

Utility Gas Recovery Problems

10.5. Assume that the two refinery hydrogen networks in Problems 4.20 (Network A) and 4.19 (Network B) form an IPRCN with direct integration scheme (with direct reuse/recycle of hydrogen sources). Use a two-stage optimization approach to identify the minimum fresh hydrogen flowrate for the overall IPRCN, as well as the minimum cross-plant stream flowrate(s). The limiting data for these networks are given in Tables 4.31 (Network B) and 4.32 (Network A). Fresh hydrogen feeds

TABLE 10.6

Limiting Water Data for Individual RCNs (Olesen and Polley, 1996) in Problem 10.4

Sinks, SK$_j$	F_{SKj} (t/h)	C_{SKj} (ppm)	Sources, SR$_i$	F_{SRi} (t/h)	C_{SRi} (ppm)
Network A					
SK1	20.00	0	SR1	20.00	100
SK2	66.67	50	SR2	66.67	80
SK3	100.00	50	SR3	100.00	100
SK4	41.67	80	SR4	41.67	800
SK5	10.00	400	SR5	10.00	800
Network B					
SK6	20.00	0	SR6	20.00	100
SK7	66.67	50	SR7	66.67	80
SK8	15.63	80	SR8	15.63	400
SK9	42.86	100	SR9	42.86	800
SK10	6.67	400	SR10	6.67	1000
Network C					
SK11	20.00	0	SR11	20.00	100
SK12	80.00	25	SR12	80.00	50
SK13	50.00	25	SR13	50.00	125
SK14	40.00	50	SR14	40.00	800
SK15	300.00	100	SR15	300.00	150
$\sum_j F_{SKj}$	880.17		$\sum_i F_{SRi}$	880.17	

for the individual networks are available at 0.1 (Network A) and 5% (Network B), respectively.

10.6. IPRCN will be synthesized between two hydrogen networks (Deng et al., 2017) with direct integration scheme, with data given in Table 10.7 Fresh hydrogen feeds for both networks are available at 5%. Use MRPD to identify its minimum fresh hydrogen flowrate and mathematical programming approach it identify the minimum inter-plant flowrate.

Property Integration Problems

10.7. Inter-plant water recovery (Lovelady et al., 2009) is to be carried out between two wafer fabrication plants (originated from Problems 2.13 and 3.13) in an eco-industrial park. To evaluate water recovery potential, the main property in concern is resistivity (R), which reflects the total ionic content in the aqueous streams. Limiting water data of both wafer fabrication plants are summarized in Table 10.8. Ultrapure water (UPW; resistivity of 18 MΩ cm) is used to supplement process sources

TABLE 10.7

Limiting Water Data for Hydrogen Networks (Deng et al., 2017) in Problem 10.6

Sinks, SK$_j$	F_{SKj} (t/h)	C_{SKj} (ppm)	Sources, SR$_i$	F_{SRi} (t/h)	C_{SRi} (ppm)
Network A					
HCU	93,306	13.29	CRU	17,303	20.00
GOHT	82,656	16.42	HCU	60,678	20.00
RHT	39,164	17.43	GOHT	55,281	25.00
DHT	12,472	25.13	RHT	25,870	25.00
NHT	5,726	27.35	DHT	8,004	30.00
			NHT	3,840	35.00
Network B					
HCU	201,197	19.39	SRU	50,303	7.00
NHT	14,531	21.15	CRU	33,530	20.00
DHT	44,707	22.43	HCU	145,305	25.00
CNHT	58,117	24.86	NHT	11,177	25.00
			DHT	27,942	27.00
			CNHT	36,885	30.00
$\sum_j F_{SKj}$	551,876		$\sum_i F_{SRi}$	476,118	

TABLE 10.8

Limiting Data for Inter-plant Water Recovery (Lovelady et al., 2009) in Problem 10.7

j	Sinks, SK$_j$	F_{SKj} (t/h)	Ψ_{SKj} ([MΩ cm]$^{-1}$)	i	Sources, SR$_i$	F_{SRi} (t/h)	Ψ_{SRi} ([MΩ cm]$^{-1}$)
	Network A						
1	Wet	500	0.143	1	Wet I	250	1
2	Lithography	450	0.125	2	Wet II	200	0.5
3	CMP	700	0.100	3	Lithography	350	0.333
4	Etc.	350	0.200	4	CMP I	300	10
				5	CMP II	200	0.5
				6	Etc	280	2
	Network B						
5	Wafer Fab	182	0.0625	7	50% Spent rinse	227.12	0.125
6	CMP	159	0.1	8	100% Spent rinse	227.12	0.5

when they are insufficient for use in the sinks. The mixing rule for resistivity is given as follows (reproduced from Equation 2.85):[4]

$$\frac{1}{R_M} = \sum_i \frac{x_i}{R_i} \tag{10.30}$$

---

[4] Resistivity values are given in their corresponding operator values.

Solve the following cases:

a. Assuming that the CUF only supplies UPW for the wafer fabrication plants, use MRPD to identify the minimum flowrate targets for UPW and wastewater of the overall IPRCN, followed by mathematical programming to identify the minimum cross-plant flowrate for a direct reuse/recycle scheme (direct integration).

b. A centralized purification unit with outlet resistivity value of 9 MΩ. cm is installed in the centralized utility hub so to implement water regeneration via indirect integration. Use the two-stage approach to minimize the UPW and regeneration flowrates.

References

Bandyopadhyay, S., Sahu, G. C., Foo, D. C. Y., and Tan, R. R. 2010. Segregated targeting for multiple resource networks using decomposition algorithm, *AIChE Journal*, 56(5), 1235–1248.

Chew, I. M. L., Ng, D. K. S., Foo, D. C. Y., Tan, R. R., Majozi, T., and Gouws, J. 2008. Synthesis of direct and indirect inter-plant water network. *Industrial & Engineering Chemistry Research*, 47, 9485–9496.

Chew, I. M. L., Foo, D. C. Y., Ng, D. K. S., and Tan, R. R. 2010a. Flowrate targeting algorithm for interplant resource conservation network. Part 1—Unassisted integration scheme, *Industrial and Engineering Chemistry Research*, 49(14), 6439–6455.

Chew, I. M. L., Foo, D. C. Y., and Tan, R. R. 2010b. Flowrate targeting algorithm for interplant resource conservation network. Part 2—Assisted integration scheme, *Industrial and Engineering Chemistry Research*, 49(14), 6456–6468.

Deng, C., Zhou, Y., Jiang, W., and Feng, X. (2017). Optimal design of inter-plant hydrogen network with purification reuse/recycle. *International Journal of Hydrogen Energy*, 42, 19984–20002.

El-Halwagi, M. M., Gabriel, F., and Harell, D. 2003. Rigorous graphical targeting for resource conservation via material recycle/reuse networks. *Industrial and Engineering Chemistry Research*, 42, 4319–4328.

Feng, X., Wang, N., and Chen, E. 2006. Water system integration in a catalyst plant, *Transactions of the Institute of Chemical Engineers, Part A*, 84(A8), 645–651.

Foo, D. C. Y. (2012). *Process Integration for Resource Conservation*, 1st ed., Boca Raton, Florida, US: CRC Press.

Lovelady, E. M., El-Halwagi, M. M., and Krishnagopalan, G.A. 2007. An integrated approach to the optimisation of water usage and discharge in pulp and paper plants. *International Journal of Environment and Pollution*, 29(1/2/3), 274–307.

Lovelady, E. M., El-Halwagi, M. M., Chew, I., Ng, D. K. S., Foo, D. C. Y., and Tan, R. R. (2009). A Property-Integration Approach to the Design and Integration of Eco-Industrial Parks, paper presented in the Seventh International Conference on the Foundations of Computer-Aided Process Design (FOCAPD 2009), Colorado, US.

Olesen, S. G. and Polley, G. T. 1996. Dealing with plant geography and piping constraints in water network design, *Process Safety and Environmental Protection*, 74, 273–276.

Prakash, R. and Shenoy, U. V. 2005. Targeting and design of water networks for fixed flowrate and fixed contaminant load operations. *Chemical Engineering Science*, 60(1), 255–268.

11

Extended Application: Synthesis of Batch Resource Conservation Network

In previous chapters, we learnt how resource conservation is carried out for continuous processes. This chapter focuses on resource conservation network (RCN) for batch processes, or in short "batch RCN." Batch processes are commonly encountered in the chemical process industry, particularly in industrial sectors such as food and beverage, polymer, and biochemical product manufacturing. For a batch RCN, apart from quality constraint, time dimension is another important constraint that needs to be considered in synthesizing a batch RCN. In most cases, a mass storage system is commonly employed to enhance material recovery for a batch RCN. In this chapter, pinch analysis and mathematical programming techniques introduced in previous chapters are extended for the synthesis of a batch RCN, which is consistent with the *target ahead of design* philosophy. As such, the pinch-based graphical targeting technique (Chapter 3)[1] is first used to determine the overall minimum fresh resource and waste discharge targets of the batch RCN. Detailed design of the network is then carried out using extended mathematical programming technique introduced in Chapter 9.

11.1 Types of Batch Resource Consumption Units

Before discussing the use of targeting and design techniques for a batch RCN, it is best to understand the characteristics of various batch resource consumption units. Figure 11.1 shows two typical types of batch operation mode. For Process 1 in Figure 11.1a, its actual operation takes place between the duration of t_2 and t_3. However, before the operation commences, fresh resource (e.g., water, gas) is charged to the operation within the duration of $t_1 - t_2$ (as sink). On the other hand, waste is discharged from the process (as source) at a later duration ($t_3 - t_4$), upon the completion of its operation. This type of operation is classified as the *truly batch* process (Gouws et al., 2010). A batch polymerization reaction is a typical example for this type of operation. In most cases, water (resource) is used as a carrier for the polymerization

---

[1] Algebraic targeting technique of *material cascade analysis* (Chapter 4) may also be used to determine the overall minimum fresh resource and waste discharge targets of a batch RCN.

DOI: 10.1201/9781003173977-12

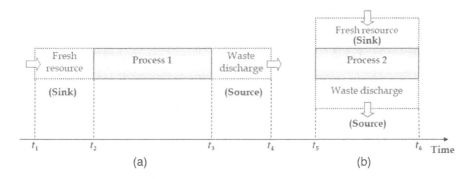

FIGURE 11.1
Types of batch resource consumption units: (a) truly batch and (b) semicontinuous operations. (Adapted with permission from Gouws et al., 2010. Reproduced with permission. Copyright 2010 American Chemical Society.)

reaction. Freshwater is charged to the reactor before the reaction commences, and wastewater is discharged at the end of the polymerization reaction.

Another type of batch process mode is *semicontinuous* operation, as shown in Figure 11.1b. For this case, the charging of fresh resource to the process (sink) and its discharge of waste (source) are both carried out during the cause of operation, i.e., between t_5 and t_6. Vessel washing operation is a typical example of this type. During the washing operation, freshwater is charged to the vessel, while wastewater is withdrawn; both are carried out simultaneously. Note that some operations (e.g., washing) may either be operated in truly batch or semicontinuous operation mode, depending on the process requirement.

Besides, one should also note that both processes in Figure 11.1 serve as process sink and source at the same time. There are also many cases where only process sink or source is found in the operations. Typical example of the former is a batch reactor where a particular resource (e.g., water, gas, etc.) is used as reactant. Being a limiting reactant, the resource is consumed completely upon the completion of the reaction. In other words, the reactor is treated as a process sink for the resource in concern. On the other hand, a chemical reactor may also serve as a process source that produces useful material. Typical example of this operation is fermentation processes, where a significant amount of water is produced. The graphical representation for these types of processes is shown in Figure 11.2 for truly batch or semicontinuous processes.

11.2 Targeting Procedure for Direct Reuse/Recycle in a Batch RCN

To identify the overall minimum fresh resource and waste discharge targets of a batch RCN, the *material recovery pinch diagram* (MRPD) introduced in Chapter 3 is a useful tool. Note that the MRPD was originally meant for

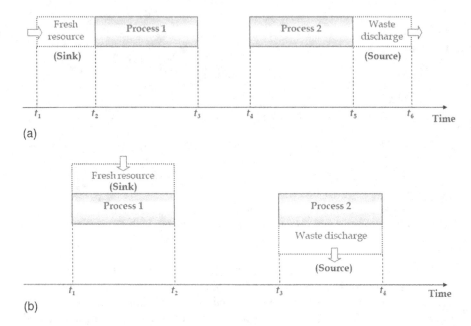

FIGURE 11.2
Batch resource consumption units with either process sink or source that operate in (a) truly batch and (b) semicontinuous mode. (Adapted with permission from Gouws et al., 2010. Reproduced with permission. Copyright 2010 American Chemical Society.)

flowrate targeting for in-plant RCN in continuous processes (El-Halwagi et al., 2003; Prakash and Shenoy, 2005). It may be used for batch RCN with the main assumption that mass storage system is made available for use. With mass storage system, process sources may be recovered to the sinks regardless of their time dimension limitation.

Procedure for flow targeting using MRPD is given as follows:

1. Arrange all process sinks in descending order of quality levels (q_{SKj}). Calculate the maximum load acceptable by the sinks (m_{SKj}) following Equations 11.1, which are given as the product of sink flow (F_{SKj}) with its quality level:

$$m_{SK_j} = F_{SK_j} q_{SK_j} \tag{11.1}$$

2. Arrange all process sources in descending order of quality levels (q_{SRi}). Calculate the load possessed by the sources (m_{SRi}) following Equation 11.2, which is given as the product of the source flow (F_{SRi}) with its quality level:

$$m_{SR_i} = F_{SR_i} q_{SR_i} \tag{11.2}$$

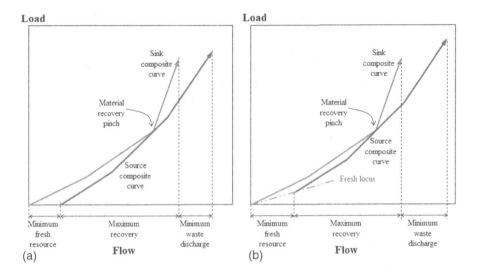

FIGURE 11.3
The MRPD for targeting the batch RCN for (a) pure fresh resource and (b) impure fresh resource. (From El-Halwagi, 2003; Prakash and Shenoy, 2005.)

3. All process sinks are plotted in descending order of their quality levels on a load-versus-flow diagram to form the *sink composite curve*. The individual sink segments are connected by linking the tail of a later segment to the arrowhead of an earlier segment (Figure 11.3).

4. Step 3 is repeated for all process sources to form the *source composite curve* on the same load-versus-flow diagram. For a feasible MRPD, the entire sink composite curve has to stay above and to the left of the source composite curve. Otherwise, the MRPD is considered infeasible, and step 5 is next followed.

5. For a batch RCN with a single pure fresh resource of superior quality (highest purity among all sources), the source composite curve is moved horizontally to the right until it just touches the sink composite curve, with the latter lying above and to the left of the source composite curve (Figure 11.3a). For a batch RCN with a single impure fresh resource, a fresh locus is to be plotted on the MRPD, with its slope corresponding to its quality level (q_R). The source composite curve is slid on the locus until it just touches the sink composite curve and to the right and below the latter (Figure 11.3b).

As described in Chapter 3, the MRPDs are useful in identifying several important *targets* for the batch RCN. First, the overall minimum fresh resource target of the batch RCN is given by the opening on the left side of the MRPD, while opening on the right represents its overall minimum waste generation target. The overlapping region between the two composite curves indicates the maximum recovery target among the sources and the sinks. The point where the two composite curves touch is the mass recovery *pinch*, which may be viewed as the overall bottleneck of the overall network, which restricts the recovery of material among the sources and the sinks. Note that the maximum recovery target is built on the main assumption that the mass storage system is made available for use, so that time constraints can be overcome.

Example 11.1 Flow targeting for a batch water network

A classical *batch water network* example (Wang and Smith, 1995) is analyzed here. The network consists of three water-using processes, each consisting of a pair of water sink and source. The limiting water data of the processes are given in Table 11.1.

Complete the following tasks:

1. Classify the three water-using processes. Are they truly batch or semicontinuous operations?
2. Target the overall minimum freshwater and wastewater flows of this network with MRPD, assuming a water storage system is used. Freshwater supply is assumed to be free of impurity (i.e., 0 ppm).

SOLUTION

1. Classification of water-using processes

    From Figure 11.4, it is observed that the water sinks and sources are present in the same duration as that of the water-using processes. Hence, they are classified as the semicontinuous operations. Note also that the flowrate terms have a unit of t/h.

2. Flow targeting with MRPD

    The five-step procedure is followed to identify the minimum freshwater and wastewater flows of the batch water network:

    i. *Arrange all process sinks in descending order of quality levels (q_{SKj}). Calculate the maximum load acceptable by the sinks (m_{SKj}).*

TABLE 11.1

Limiting Water Data for Example 11.1

SK_j	F_{SKj} (t)	C_{SKj} (ppm)	T^{STT} (h)	T^{END} (h)
SK1	100	100	0.5	1.5
SK2	40	0	0	0.5
SK3	25	100	0.5	1.0
SR_j	F_{SRi} (t)	C_{SRi} (ppm)	T^{STT} (h)	T^{END} (h)
SR1	100	400	0.5	1.5
SR2	40	200	0	0.5
SR3	25	200	0.5	1.0

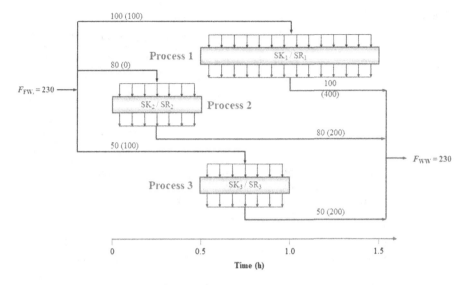

FIGURE 11.4
Batch water network in Example 11.1 (all flow terms in t; concentration in ppm, given in parenthesis). (From Foo et al., 2005, p. 1381. With permission.)

    ii. *Arrange all process sources in descending order of quality levels* *(q_{SRi}). Calculate the load possessed by the sources (m_{SRi}).*
      Water sinks SK1–SK3 are arranged in ascending order of concentrations (see column 3 of Table 10.2). The maximum load acceptable by the sinks (m_{SKj}) is calculated using Equation 11.1 and documented in column 4 of Table 11.2.
      Similarly, water sources of both networks are arranged in ascending order of concentrations. The load possessed by the source (m_{SRi}) is calculated using Equation 11.2 and documented in column 4 of Table 11.3.
  3. *All process sinks are plotted with descending order of their quality levels on a load-versus-flow diagram to form the sink composite curve.*
  4. *Step 3 is repeated for all process sources to form the source composite curve on the same load-versus-flow diagram.*
      To ease the plotting of the MRPD, the cumulative flows (Cum. F_{SKj}) and cumulative loads (Cum. m_{SKj}) for all sinks are

TABLE 11.2

Data for Plotting Sink Composite Curve in Example 11.1 (with Water Storage)

Sinks, SK$_j$	F_{SKj} (t)	C_{SKj} (ppm)	m_{SKj} (g)	Cum. F_{SKj} (t)	Cum. m_{SKj} (g)
SK2	40	0	0	0	0
SK1	100	100	10,000	40	0
SK3	25	100	2,500	140	10,000
				165	12,500

TABLE 11.3

Data for Plotting Source Composite Curve in Example 11.1 (with Water Storage)

Sources, SR_i	F_{SRi} (t)	C_{SRi} (ppm)	m_{SRi} (g)	Cum. F_{SRi} (t)	Cum. m_{SRi} (g)
SR2	40	200	8,000	0	0
SR3	25	200	5,000	40	8,000
SR1	100	400	40,000	65	13,000
				165	53,000

first calculated, and are given in the last two columns of Table 11.1. Similarly, the cumulative flows (Cum. F_{SRi}) and cumulative loads (Cum. m_{SRi}) for all sources are calculated and given in the last two columns of Table 11.2. These values are then used to plot the sink and source composite curves of the MRPD on a load-versus-flow diagram, as shown in Figure 11.5. Note that the slope of each segment corresponds to their respective concentration values.

As shown in Figure 11.5, the sink composite curve indicates that total water flows of 165 t are needed by the sinks, and these sinks can only accept a total impurity load of 12,500 g. On the other hand, the source composite curve indicates that the sources have a total water flow of 165 t, and possess 53,000 g of total impurity load. This means that the MRPD is infeasible, as the source composite curve stays on the left and above the sink composite curve. In other words, if the sources are fully recovered to the sinks, the latter will receive a total impurity load (53,000 g) that are higher than their maximum acceptable limit (12,500 g).

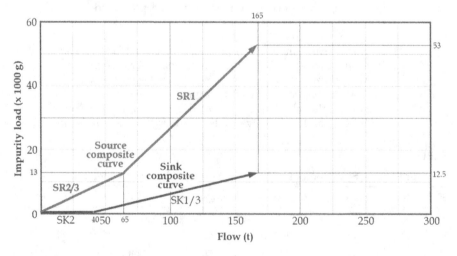

FIGURE 11.5
Infeasible MRPD for Example 11.1.

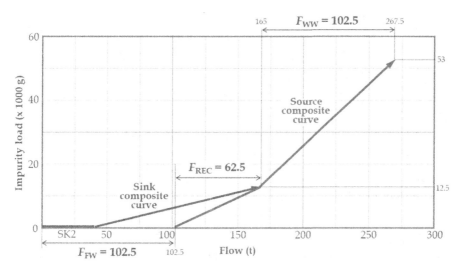

FIGURE 11.6
Feasible for Example 11.1.

5. *For the RCN with a single pure fresh resource of superior quality, the source composite curve is moved horizontally to the right until it just touches the sink composite curve, with the latter lying above and to the left of the source composite curve.*

As the freshwater supply has no impurity, the source composite curve is moved horizontally to the right until it just touches the sink composite curve and lies completely below and to the right of the latter. This results in a feasible MRPD as shown in Figure 11.6. In this case, both flow and load constraints of the sink composite curve are observed.

The feasible MRPD in Figure 11.6 identifies several important network targets. The overall minimum freshwater flow target (F_{FW}) of the batch water network is identified as 102.5 t, given by the opening on the left side of the MRPD, while the opening on the right indicates its overall minimum wastewater flow (F_{WW}) as 102.5 t. The overlapping region of the two composite curves indicates that a total of 62.5 t (= 165 – 102.5 t) of water resources are recovered among the sinks and sources of the batch water network. The overall pinch is identified as 200 ppm, as the two composite curves touch at the segment that is formed by sources SR2 and SR3..

11.3 Design of Batch RCN with Mathematical Programming

Mathematical programming is a convenient tool to perform detailed design of a batch RCN. Apart from the overall fresh resource and waste generation, the model can also determine the fresh resource and waste generation for

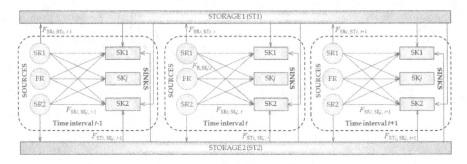

FIGURE 11.7
Superstructure for batch RCN, where storage system is used to store process sources for recovery to sinks across time intervals.

each time interval, as well as the sizing of the mass storage. A general superstructure for a batch RCN is shown in Figure 11.7. As shown, within each time interval t, each source i may be recovered to the sink j directly, or be sent for storage s, where it can be recovered to sinks at a later time interval $t + 1$.

To synthesize a batch RCN a three-step procedure should be followed, which is given as follows.

1. Define the time intervals for the batch RCN based on start and end times of the process sinks and sources.
2. Locate the process sinks and sources in their respective time intervals, and determine their interval flows, based on the ratio of the duration of the time interval to that of the sink/source.
3. Solve the following mathematical model:

The basic constraints for a batch RCN with direct reuse/recycle scheme are given in Equations 11.3–11.7. Note that they are extended from the basic form in Chapter 9, except that the flowrate term in continuous processes is now replaced with flow terms. Besides, each constraint are now added with a new term to indicate the stored source from storage s that is sent to sink j ($F_{STs,SKj,t}$), or flows of source i that is sent to storage s ($F_{SRi,STs,t}$) in each time interval t.

$$\sum_i F_{SRi,SKj,t} + \sum_s F_{STs,SKj,t} + F_{R,SKj,t} = F_{SKj,t} \quad \forall j \; \forall t \tag{11.3}$$

$$\sum_i F_{SRi,SKj,t} q_{SRi} + \sum_s F_{STs,SKj,t} q_{STs} + F_{R,SKj,t} q_R \geq F_{SKj,t} q_{SKj} \quad \forall j \; \forall t \tag{11.4}$$

$$\sum_j F_{SRi,SKj,t} + \sum_s F_{SRi,STs,t} + F_{SRi,D,t} = F_{SRi,t} \quad \forall i \; \forall t \tag{11.5}$$

$$\sum_t \sum_i F_{SRi,STs,t} = \sum_t \sum_j F_{STs,SKj,t} \quad \forall s \tag{11.6}$$

$$F_{SRi,SKj}, F_{STs,SKj}, F_{SRi,STs}, F_{R,SKj}, F_{SKi,D} \geq 0 \;\; \forall i,j \tag{11.7}$$

The constraint in Equation 11.3 describes that the flow requirement of the sink is fulfilled by various process sources ($F_{SRi,SKj,t}$), fresh resource ($F_{R,SKj,t}$), as well as the stored source ($F_{STs,SKj,t}$) available within the time interval t. Equation 11.4 describes that the flows from the process sources, stored sources, and fresh resource feed (with quality of q_{SRi}, q_{STs}, and q_R, respectively) must fulfill the minimum quality requirement of the process sinks. Next, Equation 11.5 indicates that the flow of each process source ($F_{SRi,t}$) may be allocated to sinks, while the unutilized source will be sent for waste discharge ($F_{SRi,D,t}$). The flow balance for storage s is given by Equation 11.6. Equation 11.7 indicates that all flow variables of the superstructural model should take non-negative values.

The objective of the optimization model may be set to minimize the overall fresh resource flow, and the total stored flows of the process sources. Since it involves two objectives, the two-stage optimization approach (introduced in Chapter 9) may be used. For Stage 1, the minimum overall fresh resource flow (across all time intervals) may be determined using the MRPD (see earlier section of this chapter); alternatively, the flow may be minimized using Equation 11.8, subject to the constraints in Equations 11.3 through 11.7.

$$\text{Minimize } F_R = \sum_t \sum_j F_{R,SKj,t} \tag{11.8}$$

Once the overall minimum fresh resource flow ($F_{R,min}$) is identified, it is added as a new constraint in Stage 2 (Equation 11.9), where the total material storage size is minimized with optimization objective in Equation 11.10. Note that additional constraints for sizing the mass storage are necessary, given as in Equations 11.10–11.13 (Foo, 2010).

$$F_{R,min} = F_R \tag{11.9}$$

$$\text{Minimize } F_{ST} = \sum_s F_{ST,s} \tag{11.10}$$

$$\left(CF_{ST,t} + \sum_i F_{SRi,STs,t} - \sum_j F_{STs,SKj,t} \right)_s = \left(CF_{ST,t+1} \right)_s \;\; \forall s \;\; \forall t \tag{11.11}$$

$$F_{ST,s} \geq \left(CF_{ST,t+1} \right)_s \;\; \forall s \;\; \forall t \tag{11.12}$$

$$F_{ST,s}, CF_{ST,t,s} \geq 0 \;\; \forall s \;\; \forall t \tag{11.13}$$

In Equation 11.11, the cumulative size of the mass storage at time interval t ($CF_{ST,t,i}$) is determined by taking into account the material that is stored from process sources i to the storage ($F_{SRi,STs,t}$) as well as the material that is recovered from storage to the sinks ($F_{STs,SKj,t}$). Note that the main assumption made is that, each process source is equipped with an independent storage system. The main rationale for this assumption is to avoid the mixing of sources of different concentrations of impurity (which will lead to nonlinear model). Note that sources of the same impurity concentration may be fed to the same storage, so to reduce the number of mass storage. Equation 11.12 indicates that the size of storage for source i must be bigger than its cumulative size. Lastly, both the size and cumulative size of the mass storage must take non-negative values, as given in Equation 11.13.

For cases when the mass storage cost is insignificant, the objective may be set to minimize the operating cost (OC); the latter is contributed by the overall fresh resource (F_R) and waste discharge (F_{WD}) across all time intervals, given as in Equation 11.14. Note that the overall waste discharge flow is contributed by the individual waste discharge flows of source i among all time interval t ($F_{SRi,D,t}$), given as in Equation 11.15.

$$\text{Minimize } OC = F_R CT_R + F_{WD} CT_D \tag{11.14}$$

$$F_{WD} = \sum_t \sum_i F_{SRi,D,t} \tag{11.15}$$

where CT_R and CT_D are the unit costs of fresh resource and waste discharge, respectively. Similar to the basic variant in Chapter 9, this is an LP model for which any solution found is globally optimal.

Example 11.2 Detailed design of a batch water network

Example 11.1 is revisited. Assuming pure freshwater supply (i.e., 0 ppm) is available. Design the batch water network. Determine also the minimum size of the water storage tank(s).

SOLUTION

The three-step procedure is followed to design the batch water network:

1. *Define the time intervals for the batch RCN based on start and end times of the process sinks and sources.*
2. *Locate the process sinks and sources in their respective time intervals, and determine their interval flows, based on the ratio of the duration of the time interval to that of the sink/source.*

    Based on the start and end time of water sinks and sources, three time intervals may be defined accordingly, i.e., 0–0.5 h (time interval t = A), 0.5–1.0 h (t = B), and 1.0–1.5 h (t = C).

TABLE 11.4

Water Flows for Process Sinks and Sources in Their Respective Time Intervals (All Flow Terms in Ton)

SK$_j$/SR$_i$	Time Interval, t		
	t = A (0–0.5 h)	t = B (0.5–1.0 h)	t = C (1.0–1.5 h)
SK1		$F_{SK1,B} = 50$	$F_{SK1,C} = 50$
SK2	$F_{SK2,A} = 40$		
SK3		$F_{SK3,B} = 25$	
SR1		$F_{SR1,B} = 50$	$F_{SR1,C} = 50$
SR2	$F_{SR2,A} = 40$		
SR3		$F_{SR3,B} = 25$	

Following Step 2 of the procedure, water sinks, and sources are located in their respective time intervals. The interval flows of the water sinks and sources are next determined and are given in Table 11.4.

For instance, sink SK1 that exists between 0.5–1.5 h is located in time intervals B and C. Since both of these intervals have a duration of 0.5 h, the interval flows of SK1 are determined as half of its total flow (interval flows of other sinks and sources are determined similarly), i.e.,

$$F_{SK1,B} = F_{SK1,C} = 100 \text{ ton} \left(\frac{0.5 \text{ h}}{1.0 \text{ h}} \right) = 50 \text{ ton}$$

3. *Solve the mathematical model.*

A superstructure for the water network is constructed in the matching matrix form using the MS Excel spreadsheet in Figure 11.8a. Water sources and sinks, as well as potential storage system of interval A (0 – 0.5 h) are arranged as located in cell arrays E6:H10, while those of interval B (0.5 – 1.0 h) in cell arrays I11:L15 of the spreadsheet, and those of interval C (1.0 – 1.5 h) are located cell arrays M16:O19. These arrays are potential regions where water recovery can take place among the process sources and sinks, directly or indirectly (via storage). Note that sources SR2 and SR3 can share the same storage tank, as they have the same concentration of impurity (i.e., 200 ppm; see Table 11.3), while SR1 has a separate storage tank (400 ppm).

To formulate an optimization model to minimize the overall freshwater flow (F_{FW}) of the batch water network, the objective in Equation 11.8 takes the revised form of Equation 11.16:

$$\text{Minimize } F_{FW} = \sum_t \sum_j F_{FW,SKj,t} \tag{11.16}$$

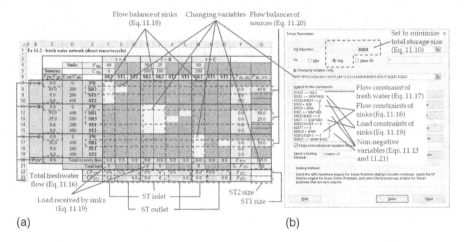

FIGURE 11.8
(a) Superstructure in MS Excel spreadsheet and (b) Solver for batch water network in Example 2 (all flow terms in t; concentration in ppm).

where $F_{FW,SKj,t}$ is the freshwater flow allocated to sink j in time interval t. Since this overall freshwater flow has been identified using the MRPD as 102.5 t in Example 11.1, the objective in Equation 11.12 is omitted. A maximum flow constraint is added for cell C21 in Solver (see setting in Figure 11.8) following Equation 11.13. The model is set to minimize water storage size with Equation 11.10.

$$F_{FW,min} = 102.5 \text{ t} \tag{11.17}$$

All other generic constraints in Equations 10.3–10.6 take their revised form as in Equations 11.18–11.21. Equation 11.18 describes the flow balances for process sinks. These flow balances are implemented in columns F, I, J, and M (row 20) in the spreadsheet, along with the constraint in Solver (see Figure 11.8):

$$\sum_i F_{SRi,SKj,t} + \sum_s F_{STs,SKj,t} + F_{FW,SKj,t} = F_{SKj,t} \quad \forall j \ \forall t \tag{11.22}$$

The quality requirement of the sinks is described by Equation 11.19. Note however that the inequality sign takes a reverse form as that in Equation 11.4, as the quality index is described by impurity concentration. In other words, impurity load sent from process sources, freshwater water (with concentration C_{FW}), and stored water (with concentration C_{STs}) must not exceed the load limit of the process sinks; the latter is given as the product of its flow ($F_{SKj,t}$) and impurity concentration (C_{SKj}). These impurity load constraints are imposed for each sink in columns

F, I, J, and M (row 21) in the spreadsheet, along with the constraint in Solver (see Figure 11.8):

$$\sum_i F_{SRi,SKj,t}C_{SRi} + \sum_s F_{STs,SKj,t}C_{STs} + F_{FW,SKj,t}C_{FW} \le F_{SKj,t}C_{SKj} \ \forall j \ \forall t \qquad (11.19)$$

Equation 11.20 describes the interval flow balances of the process sources. As shown, each process source may be recovered directly to any process sinks within the same time interval t ($F_{SRi,SKj,t}$), or be sent to water storage ($F_{SRi,STs,t}$), while the unutilized source will be sent for wastewater discharge ($F_{SRi,WW,t}$). One can derive these flow balances by observing rows 8, 12, 13, and 17 in Figure 11.8.

$$\sum_j F_{SRi,SKj,t} + \sum_s F_{SRi,STs,t} + F_{SRi,WW,t} = F_{SRi,t} \ \forall i \ \forall t \qquad (11.20)$$

Equation 11.21 indicates that all flow variables take non-negative values. These include the flow terms for the matches between source i ($F_{SRi,SKj,t}$) and freshwater to sink j ($F_{FW,SKj,t}$), inlet and outlet flows of water storage ($F_{SRi,STs,t}$, $F_{STs,SKj,t}$), as well as the wastewater stream from source i ($F_{SRi,WW,t}$):

$$F_{SRi,SKj,t}, F_{STs,SKj,t}, F_{SRi,STs,t}, F_{FW,SKj,t}, F_{SKi,WW,t} \ge 0 \ \forall i, j, p \qquad (11.21)$$

It is also convenient to determine the total flow of the wastewater streams (F_{WW}), given by the summation of the individual wastewater flows (column Q of Figure 11.8):

$$F_{WW} = \sum_t \sum_i F_{SRi,WW,t} \qquad (11.18)$$

Solving the optimization model with objective in Equation 11.10 (minimum water storage) yield the batch water network in Figure 11.9. The latter shows only water storage ST1 is needed for this batch water network, with a size of 37.5 t (indicated by cells Q22). ST1 will stores 37.5 t of water from source SR2 in interval A (cell G8), so for indirect reuse in sink SK1 in intervals B (cell I14; 12.5 t) and C (cell M18; 25 t). Note that ST2 is not necessary, as it is flow is determined as zero (cells Q23). Note that Figure 12.24 also shows that direct water reuse/recycle are observed in time interval B, i.e., between SR3 with SK1 and SK3 (cells I13 and J13). The total direct and indirect recovery of water may be added to 62.5 t, identical to that determined using the MRPD in Figure 11.6. Besides, the model also determines that a total of 102.5 t of freshwater flows (F_{FW}, cell C20) are needed for the batch water network, i.e., 40 t for interval A (cell C7, $F_{FW,A}$), 37.5 t for interval B (cell I11 and J11, $F_{FW,B}$), and 25 t for interval C (cell M16, $F_{FW,C}$). The total wastewater (F_{WW}) is determined as 102.5 t (cell Q20), contributed by SR2 (cell Q8) and SR1 (cell Q12 and Q17). Note that both freshwater and wastewater flows are also identical to those determined using MRPD in Figure 11.6.

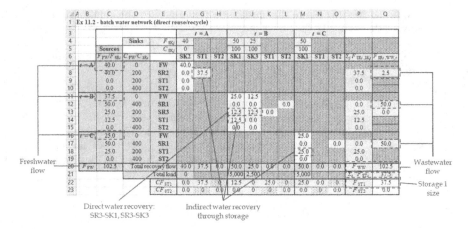

FIGURE 11.9
Results for batch water network in Example 11.2 (all flow terms in t; concentration in ppm).

11.4 Synthesis of Batch Regeneration Network

To further reduce the flows of fresh resource and waste discharge, regeneration-reuse/recycle scheme may be considered for a batch RCN, or in short *batch regeneration network*. As discussed in Chapter 6, purification units may be categorized as single pass and partitioning units; and their performance may be characterized as fixed outlet quality and removal ratio type. In this section, synthesis of batch regeneration network is demonstrated with single pass purification unit of fixed outlet quality type. Readers may refer to Chapter 6 for other types of purification units.

The basic constraints for a batch regeneration network are modified from Equations 11.3 to 11.7, given in Equations 11.23–11.28. Note that in each constraint, the flow terms of purification inlet ($F_{SRi,REr,t}$) and outlet ($F_{REGr,SKj,t}$) are added. It is also assumed that each regeneration unit r is equipped with a storage unit, so that the purified sources can be recovered to the sinks at other time intervals.

$$\sum_i F_{SRi,SKj,t} + \sum_s F_{STs,SKj,t} + \sum_r F_{REGr,SKj,t} + F_{R,SKj,t} = F_{SKj,t} \quad \forall j \; \forall t \quad (11.23)$$

$$\sum_i F_{SRi,SKj,t} q_{SRi} + \sum_s F_{STs,SKj,t} q_{STs} + \sum_r F_{REGr,SKj,t} q_{REGr} + F_{R,SKj,t} q_R \geq F_{SKj,t} q_{SKj} \quad \forall j \; \forall t$$

$$(11.24)$$

$$\sum_j F_{SRi,SKj,t} + \sum_s F_{SRi,STs,t} + \sum_r F_{SRi,REr,t} + F_{SRi,D,t} = F_{SRi,t} \ \forall i \ \forall t \tag{11.25}$$

$$\sum_t \sum_i F_{SRi,STs,t} = \sum_t \sum_j F_{STs,SKj,t} \ \forall s \tag{11.26}$$

$$\sum_t \sum_i F_{SRi,REr,t} = \sum_t \sum_j F_{REGr,SKj,t} \ \forall r \tag{11.27}$$

$$F_{SRi,SKj}, F_{STs,SKj}, F_{SRi,STs}, F_{SRi,REr,t}, F_{REGr,SKj,t}, F_{R,SKj}, F_{SKi,D} \geq 0 \qquad \forall i,j \tag{11.28}$$

The constraint in Equation 11.23 describes that the flow requirement of the sink is fulfilled by various process sources ($F_{SRi,SKj,t}$), stored source ($F_{STs,SKj,t}$), regenerated source from purification unit r ($F_{REGr,SKj,t}$), fresh resource ($F_{R,SKj,t}$). Equation 11.24 describes that the flows from the process sources, stored sources, regenerated source, and fresh resource feed (with quality of q_{SRi}, q_{STs}, q_{REGr}, and q_R, respectively) must fulfill the minimum quality requirement of the process sinks. Next, Equation 11.25 indicates that the flow of each process source ($F_{SRi,t}$) may be allocated to sinks or sent to purification unit r ($F_{SRi,REr,t}$), while the unutilized source will be sent for waste discharge ($F_{SRi,D,t}$). The flow balances for storage s and regeneration unit r are given by Equations 11.26 and 11.27, respectively. Equation 11.27 indicates that all flow variables of the superstructural model should take non-negative values.

As mentioned earlier, all purification unit r are equipped with a storage tank, size of the latter may be modified from Equations 11.11 to 11.13, given as in Equations 11.29–11.31 that follow. Note that the constraints in Equations 11.11–11.13 are still applicable for all storage that are used for indirect recovery of process sources to the sinks across different time intervals.

$$\left(CF_{RST,t} + \sum_i F_{SRi,REr,t} - \sum_j F_{REGr,SKj,t} \right)_r = \left(CF_{RST,t+1} \right)_r \ \forall r \ \forall t \tag{11.29}$$

$$F_{RST,r} \geq \left(CF_{RST,t+1} \right)_r \ \forall r \ \forall t \tag{11.30}$$

$$F_{RST,r}, CF_{RST,t,r} \geq 0 \ \forall r \ \forall t \tag{11.31}$$

For a batch regeneration network, it is sensible to use the two-stage optimization approach to minimize the overall freshwater and regeneration flows. The former has been given in Equation 11.8, i.e.:

$$\text{Minimize } F_R = \sum_t \sum_j F_{R,SKj,t} \tag{11.8}$$

Once the overall minimum fresh resource flow ($F_{R,min}$) is identified, it is added as a new constraint in Stage 2 (using Equation 11.9), where the total regeneration flow is minimized with optimization objective in Equation 11.32.

$$F_{R,min} = F_R \tag{11.9}$$

$$\text{Minimize } F_{REG} = \sum_r \sum_j F_{REG,r\ SKj} \tag{11.32}$$

Example 11.3 Batch water regeneration network

The water minimization case in Examples 11.1 and 11.2 is revisited here. A purification unit with outlet concentration (C_{Rout}) of 20 ppm is evaluated for its potential for further water reduction with regeneration-reuse/recycle scheme. Use the superstructural approach to synthesize a batch water regeneration network with minimum freshwater and regeneration flowrates, assuming pure freshwater supply (0 ppm) is used.

SOLUTION

The three-step procedure outlined earlier is followed to synthesize a batch water regeneration network with minimum freshwater and regeneration flowrates. Note that Steps 1 and 2 are omitted here, as they are the same as in the earlier example.

Solve the following mathematical model.

A superstructure for the water network is constructed in the matching matrix form using the MS Excel spreadsheet in Figure 11.10. Note that it

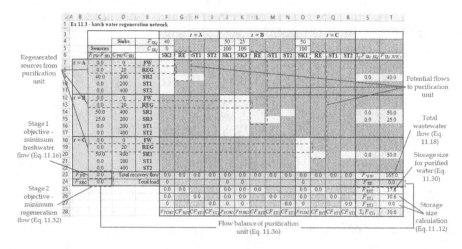

FIGURE 11.10
Superstructure in MS Excel spreadsheet for batch water regeneration network in Example 11.3 (all flow terms in t; concentration in ppm).

is similar to that of reuse/recycle in Figure 11.8a, except that potential flows from sources to sink (columns G, L, and P) and those from regenerated sources (rows 8, 13, and 19) are now added in the spreadsheet. Besides, water sources are also allowed to be recovered directly or indirectly (via storage) to the water sinks, as in Example 11.2. Similarly, SR2 and SR3 will share the same storage tank, as they have the same impurity concentration of 200 ppm), while SR1 has a separate storage tank (400 ppm). Note also that cell T25 is set to calculate the storage needed for the regenerated water, following Equation 11.30 (with constraints set in Solver).

The two-stage optimization approach is used to determine the minimum freshwater and regeneration flow. In Stage 1 of the optimization, the overall freshwater flow (F_{FW}) is minimized, with objective in Equation 11.16:

$$\text{Minimize } F_{FW} = \sum_t \sum_j F_{FW,SKj,t} \tag{11.16}$$

where $F_{FW,SKj,t}$ is the freshwater flow allocated to sink j in time interval t. While most constraints are similar to those in Example 11.2, additional flow variables are added to account for inlet ($F_{SRi,REr,t}$) and outlet flows ($F_{REG,SKj,t}$) of the purification unit. The generic constraints in Equations 11.23–11.25 take their revised form as in Equations 11.33–11.37. Equation 11.29 describes the flow balances for process sinks. As shown, the sink may be fed with water from sources ($F_{SRi,SKj,t}$), storage ($\Sigma_s F_{STs,SKj,t}$), regenerated water ($F_{REG,SKj,t}$) and freshwater ($F_{FW,SKj,t}$). These flow balances are implemented in columns F, J, K, and O in the spreadsheet (see Figure 11.10).

$$\sum_i F_{SRi,SKj,t} + \sum_s F_{STs,SKj,t} + F_{REG,SKj,t} + F_{FW,SKj,t} = F_{SKj,t} \;\; \forall j \; \forall t \tag{11.33}$$

The quality requirement of the sinks is described by Equation 11.30. As shown, the impurity load sent from process sources, regenerated sources (with concentration C_{RE}), freshwater water, and stored water must not exceed the load limit of the process sinks; the latter is given as the product of its flow ($F_{SKj,t}$) and impurity concentration (C_{SKj}).

$$\sum_i F_{SRi,SKj,t} C_{SRi} + \sum_s F_{STs,SKj,t} C_{STs}$$
$$+ \sum_r F_{REG,SKj,t} C_{REG} + F_{FW,SKj,t} C_{FW} \leq F_{SKj,t} C_{SKj} \;\; \forall j \; \forall t \tag{11.34}$$

Equation 11.35 describes the interval flow balances of the process sources. As shown, each process source may be recovered directly to any process sinks within the same time interval t ($F_{SRi,SKj,t}$), or be sent to water storage ($F_{SRi,STs,t}$) or purification unit ($F_{SRi,REr,t}$), while the unutilized source will be sent for wastewater discharge ($F_{SRi,WW,t}$). One can derive these flow balances by observing rows 9, 14, 15, and 20 in Figure 11.10.

$$\sum_j F_{SRi,SKj,t} + \sum_s F_{SRi,STs,t} + \sum_r F_{SRi,RE,t} + F_{SRi,WW,t} = F_{SRi,t} \quad \forall i \; \forall t \qquad (11.35)$$

Since only one purification unit is used, Equation 11.36 describes the flow balance of this unit.

$$\sum_t \sum_i F_{SRi,RE,t} = \sum_t \sum_j F_{REG,SKj,t} \qquad (11.36)$$

Equation 11.37 indicates that all flow variables take non-negative values, including the inlet and outlet flows of the purification unit.

$$F_{SRi,SKj,t}, F_{STs,SKj,t}, F_{SRi,STs,t}, F_{SRi,RE,t}, F_{REG,SKj,t}, F_{FW,SKj,t}, F_{SKi,WW,t} \geq 0 \; \forall i, j, p$$

$$(11.37)$$

The total flow of wastewater (F_{WW}), given by the summation of the individual wastewater flows, given as the earlier Equation 11.18 (column T of Figure 11.10):

$$F_{WW} = \sum_t \sum_i F_{SRi,WW,t} \qquad (11.18)$$

Besides, the constraints for sizing water storages (Equations 11.10–11.13) as well as storage for regenerated water (Equations 11.29–11.31) are also necessary, with their setting for Solver shown in Figure 11.11a.

Solving the above model yields the minimum freshwater flow of 40 t (cell C23), as shown in Figure 11.12. As observed, the overall wastewater flow is calculated as 40 t (cell T23). Figure 11.12 also shows that some sources are directly recovered to the sinks within time interval B, while some are recovered indirectly via water storage in intervals A and C.

In Stage 2 optimization, a maximum flow constraint is added for cell C23 in Solver (see Figure 11.11b) following Equation 11.38, while other constraints remain identical. The model is set to minimize the regenerated water flow with Equation 11.32.

$$F_{FW,min} = 40 \text{ t} \qquad (11.38)$$

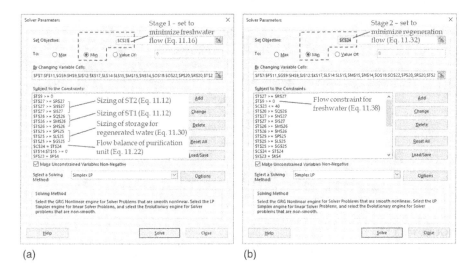

FIGURE 11.11
Setting in MS Excel Solver for batch water regeneration network in Example 11.3: (a) Stage 1 optimization—to minimize freshwater (b) Stage 2 optimization—to minimize regeneration flow.

Solving the model again yields the results in Figure 11.13, where the regeneration flow (cell C24) is minimized to 69.4 t. For this case, the storage for storing 200 ppm water is determined as 30.6 t (cell T26), while that for storing the regenerated water is 17.8 t (cell T25). Note that some water sources are regenerated and recovered within interval B, i.e., without passing through the water storage tank (see Figure 11.13). The final water network is shown in Figure 11.14.

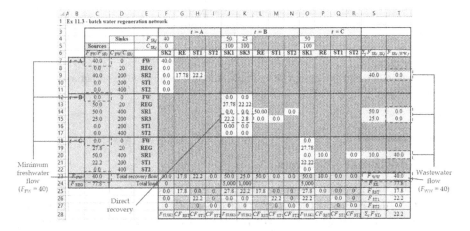

FIGURE 11.12
Results for Stage 1 optimization in Example 11.3 (all flow terms in t; concentration in ppm).

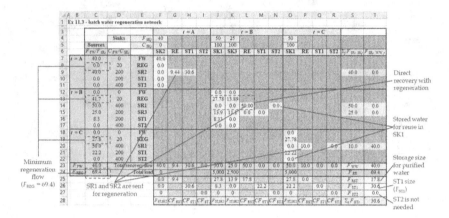

FIGURE 11.13
Results for Stage 2 optimization in Example 11.3 (all flow terms in t; concentration in ppm).

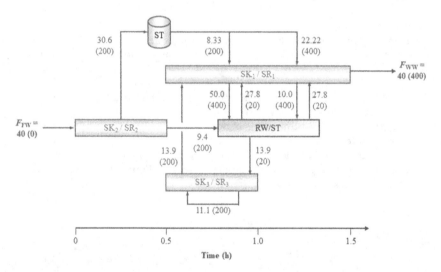

FIGURE 11.14
Water regeneration network design for Example 11.3 (all flow terms in t; concentration in ppm in parenthesis).

11.7 Additional Readings

In this chapter, graphical pinch method (MRPD) and mathematical programming technique (based on superstructural model) are used for the synthesis of a batch RCN. Mathematical programming technique is given bigger emphasis as it is more robust for batch process systems where time factor is an important system constraint. Readers may refer to the first edition of this book (Foo, 2012) to have better understanding on the use graphical, algebraic, and automated techniques for advanced targeting of a batch RCN. Readers may also find a good comparison of various targeting and mathematical techniques for batch RCN design in the review by Gouws et al. (2010).

PROBLEMS
Water Minimization Problems

11.1 A batch water network that consists of four semicontinuous water-using processes (Kim and Smith, 2004) is analyzed for its water recovery potential. Limiting water data for this case study are given in Table 11.5. Freshwater feed is assumed to be free of impurity. Complete the following tasks:

a. Use MRPD to target the minimum freshwater and wastewater flows for direct reuse/recycle scheme for the network, assuming that water storage tank is present (hint: beware of the water flow-rates of the semicontinuous processes).

b. Synthesize a batch water network that achieves the minimum flow of freshwater as determined in (a). Determine the minimum size of the water storage(s).

c. An purification unit with outlet concentration (C_{Rout}) of 10 ppm is used to regenerate water sources for further recovery. Synthesize a batch regeneration network with minimum flows of freshwater, and regenerated water.

11.2 Table 11.6 shows the limiting water data for a batch water network that consists of four truly batch water-using processes (Chen and Lee, 2008). Freshwater feed is assumed to be free of impurity. Complete the following tasks:

a. Use MRPD to target the minimum freshwater and wastewater flows of the water network, assuming water storage tanks are available.

TABLE 11.5

Limiting Water Data for Problem 11.1 (Kim and Smith, 2004)

SK_j	F_{SKj} (t/h)	C_{SKj} (ppm)	T^{STT} (h)	T^{END} (h)
SK1	20	0	0	1
SK2	100	50	1	3.5
SK3	40	50	3	5
SK4	10	400	1	3
SR_i	F_{SRi} (t/h)	C_{SRi} (ppm)	T^{STT} (h)	T^{END} (h)
SR1	20	100	0	1
SR2	100	100	1	3.5
SR3	40	800	3	5
SR4	10	800	1	3

TABLE 11.6

Limiting Water Data for Problem 11.2 (Chen and Lee, 2008)

SK_j	F_{SKj} (ton)	C_{SKj} (kg/kg)	T^{STT} (h)	T^{END} (h)
SK1	20	0	0	1.0
SK2	16	0.25	0	0.5
SK3	15	0.1	5.0	6.5
SK4	24	0.25	2.0	2.5
SK5	15	0.1	7.0	8.5
SR_i	F_{SRi} (ton)	C_{SRi} (kg/kg)	T^{STT} (h)	T^{END} (h)
SR1	20	0.2	4.0	5.0
SR2	16	0.5	4.5	5.0
SR3	15	0.1	5.0	6.5
SR4	24	0.4	6.5	7.0
SR5	15	0.12	7.0	8.5

b. An purification unit with outlet concentration (C_{Rout}) of 0.1 kg/kg
   is used to regeneration water sources. Synthesize a batch regeneration network with minimum freshwater, and regenerated water.

11.3 Figure 11.15 shows five water-using processes for an agrochemical manufacturing plant, where water is used extensively as a washing agent
   and reaction solvent to produce three types of products, i.e., A, B, and C

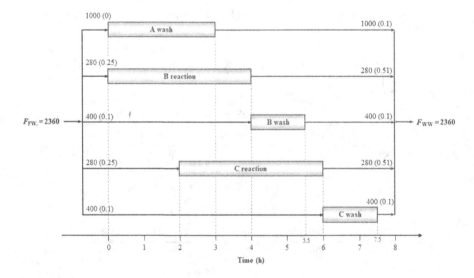

FIGURE 11.15
Water-using processes for an agrochemical plant in Problem 11.3 (all flow terms in kg; values in
parenthesis represent limiting concentration in kg salt/kg water).

(Majozi et al., 2006). To reduce the freshwater and wastewater flows, a water recovery study is to be conducted. However, during the manufacturing process, sodium chloride (NaCl) is formed as a by-product, which is the main impurity of concern for the water recovery study.

Product A was produced in an organic solvent that is immiscible in water. Water is hence used for washing the NaCl content in product A. For products B and C, water is used for both reaction solvent as well as the washing agent. However, since the reaction solvent water has removed most of the NaCl content, the wash water for products B and C does not remove any NaCl content. The feed water and wastewater flows for the five water-using processes are given in Figure 11.15.

Besides, the water-using processes have the following characteristics. The feed water of the processes can only be charged prior to the start of their actual operation, while wastewater is discharged upon the completion of the operation. In other words, no water flow is charged to and discharged from the processes during the operation. Note that the exact duration of water charge and wastewater discharge is insignificant as compared with that of the actual operation.

Perform the following tasks:

a. Should the water-using processes be classified as truly batch or semicontinuous operations? Identify the limiting data (which should include the info on time dimension) for a water minimization study.

b. Use MRPD to target the minimum freshwater and wastewater flows for direct reuse/recycle scheme, assuming water storage tanks are available.

c. Use superstructural model to design the batch water network that achieves the flow targets in (b).

11.4 Figure 11.16 shows the process flow diagram for a fruit juice production plant. As shown, fruits are first sent for washing, before they are pressed. Effluent from the pressing unit is then sent for a series of

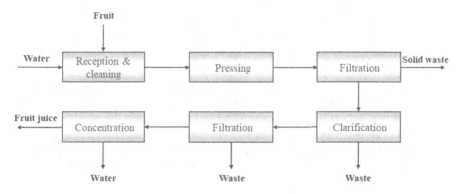

FIGURE 11.16
Process flow diagram for fruit juice production plant in Problem 11.4.

TABLE 11.7

Limiting Water Data for Fruit Juice Production Plant
in Problem 11.4 (Chen and Lee, 2010; Foo, 2010)

SK_j	F_{SK_j} (t)	C_{SK_j} (ppm)	T^{STT} (h)	T^{END} (h)
SK1	20	0	0.5	2.5
SK2	20	6	5	7
SK3	20	15	9.5	11.5
SK4	16	5	17	19
SK5	20	7	6	8

SR_i	F_{SR_i} (t)	C_{SR_i} (ppm)	T^{STT} (h)	T^{END} (h)
SR1	20	5	2.5	4.5
SR2	20	14	7	9
SR3	20	20	11.5	13.5
SR4	8	25	17	19
SR5	16	10	10.5	14.5

filtration, clarification, and concentration, where the fruit juice is
produced. Due to the significance of water use, water minimization
study is to be carried out for the process. Table 11.7 shows the limiting
water data for the water minimization study (Almató et al., 1999; Li
and Chang, 2006; Chen and Lee, 2010; Foo, 2010). Solve the following
cases:

a. Determine the minimum freshwater and wastewater flow targets
   using the MRPD.
b. Determine the minimum *OC* for a water regeneration network,
   when a purification unit is used to regenerate process sources. The
   unit has an outlet concentration of 5 ppm. The unit costs for freshwater (CT_{FW}), wastewater (CT_{WW}), and water regeneration (CT_{RW})
   are given as 1, 1, and 0.8 $/t. Identify also the time intervals where
   the purification unit will be in operation.

11.5 The process flow diagram for a batch polyvinyl chloride (PVC) resin
manufacturing plant is shown in Figure 11.17 (Chan et al., 2008). Fresh
vinyl chloride monomer (VCM) is mixed with recycled VCM to be fed
to a series of batch polymerization reactors. A huge amount of water is
used as reaction carrier in the reactor. Besides, toward the end of the
polymerization process, water is added to maintain the reactor temperature. Hence, upon the completion of the polymerization reaction, the
reactor content is in a slurry form that contains PVC resin and water.
The reactor effluent is sent to a blowdown vessel, which serves as a
buffer for the downstream purification processes that are operated in
continuous mode. A steam stripper is used to recover the unconverted
VCM from the reactor effluent. The stripped VCM, along with the VCM

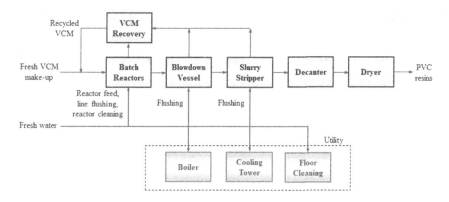

FIGURE 11.17
Process flow diagram for PVC manufacturing in Problem 11.5. (Chan et al., 2008. Springer Science+Business Media. With permission.)

that emits from the reactors and blowdown vessel, are sent to the VCM recovery system. Effluent from the stripper is sent to a decanter where the PVC is separated from its water content. PVC product from the decanter is then forwarded to the dryer to reduce its moisture content before sending to the product storage tank.

Apart from the process water being used in the reactor, a significant amount of utility water is consumed in the plant. Utility water is used to flush the reactors, blowdown vessel, and the stripper as well as for reactor and general plant cleaning. Besides, boiler and cooling tower makeup consume a huge amount of water too. Water recovery potential is explored to reduce the overall water flows of the process. Suspended solid content is the main impurity of concern for water recovery.

Due to the polymerization reactors that are operated in batch mode, water sinks associated with these units are also available in batch mode, as given in Table 11.8 (Chan et al., 2008). Note that since the new batch of the process starts prior to the completion of the previous batch, there are three batches of operation ($N - 1$, N, and $N + 1$) that exist within the time interval of analysis (0–14.78 h). In contrast, the single water source of the process is originated from the decanter that is operated in continuous mode, with its limiting flowrate given in Table 11.9.

Complete the following tasks:

a. Identify the interval flows for the water sinks and sources for the process (hint: the water source can be modeled as a semicontinuous stream).
b. Use MRPD to target the minimum freshwater and wastewater flows for direct reuse/recycle scheme, assuming water storage tanks are available, and freshwater feed is free from suspended solid content.
c. Use superstructural model to design the batch water network. Determine also the minimum size of the water storage.

TABLE 11.8

Limiting Data for Water Sinks in Problem 11.5 (Chan et al., 2008)

Batch	SK_j	Operation	F_{SKj} (kg)	C_{SKj} (ppm)	T^{STT} (h)	T^{END} (h)
	2a	Line flushing	12,500	10	0	4
$N-1$	5a	Reactor cleaning	5,400	20	0	4.5
	6a	Cooling tower	28,350	50	0	4.5
	7a	Flushing	8,000	200	0	4
	1b	Reactor feed	105,000	10	0	3.33
	2b	Line flushing	15,625	10	9.28	14.28
	3b	Boiler makeup 1	3,300	10	0	0.25
	4b	Boiler makeup 2	3,300	10	8.00	8.25
N	5b	Reactor cleaning	6,000	20	9.78	14.78
	6b	Cooling tower	93,114	50	0	14.78
	7b	Flushing	10,000	200	9.28	14.28
	8b	Floor cleaning 1	1,500	500	0	0.25
	9b	Floor cleaning 2	1,500	500	8.00	8.25
	1c	Reactor feed	105,000	10	10.28	13.61
$N+1$	3c	Boiler makeup 1	3,300	10	10.28	10.53
	6c	Cooling tower	28,350	50	10.28	14.78
	8c	Floor cleaning 1	1,500	500	10.28	10.53

TABLE 11.9

Limiting Data for Water Source in Problem 11.5
(Chan et al., 2008)

SR_i	Operation	Flowrate (kg/h)	C_{SRi} (ppm)
1	Decanter	9244	50

Property Integration Problems

11.6 Table 11.10 shows the limiting data for a batch property integration problem (Foo, 2010), where two water sources are to be recovered to two water sinks. Both water sources and sinks are operated in semicontinuous mode. In considering water recovery, pH is the main stream quality of concern. To supplement the water sources, two external fresh resources are eligible, i.e., acid (pH = 5) and freshwater (pH = 7). The mixing rule for pH is given as

$$10^{-pH} = \sum_i x_i 10^{-pH_i} \qquad (11.36)$$

TABLE 11.10

Limiting Water Data for Problem 11.6 (Foo, 2010)

SK$_j$	F_{SKj} (kg/s)	pH (SK$_j$)	T^{STT} (h)	T^{END} (h)
SK1	8	7.2	5	6
SK2	3	7.35	7	8
SR$_i$	F_{SRi} (kg/s)	pH (SR$_i$)	T^{STT} (h)	T^{END} (h)
SR1	6	9	0	1
SR2	2	7.7	3	4

Assuming that the process is to be operated with water storage tanks, complete the following tasks:

a. Assuming only freshwater is used as external fresh resource (i.e., acid is not considered), use MRPD to determine the minimum freshwater and wastewater flow for the batch water network.
b. Use superstructural model to synthesize a batch water network that achieves the flow targets in (a).
c. Due to technical reasons, it is determined that the total flow of freshwater to be added to both sinks should not be more than 2.97 kg/s. Use superstructural model to synthesize the batch RCN and to determine the amount of acid needed for this case.

References

Almató, M., Espuña, A., and Puigjaner, L. 1999. Optimisation of water use in batch process industries, *Computers and Chemical Engineering*, 23, 1427–1437.

Chan, J. H., Foo, D. C. Y., Kumaresan, S., Aziz, R. A., and Hassan, M. A. A. 2008. An integrated approach for water minimisation in a PVC manufacturing process, *Clean Technologies and Environmental Policy*, 10(1), 67–79.

Chen, C.-L. and Lee, J.-Y. 2008. A graphical technique for the design of water-using networks in batch processes, *Chemical Engineering Science*, 63, 3740–3754.

Chen, C.-L. and Lee, J.-Y. 2010. On the use of graphical analysis for the design of batch water networks, *Clean Technologies and Environmental Policy*, 12, 117–123.

El-Halwagi, M. M., Gabriel, F., and Harell, D. 2003. Rigorous graphical targeting for resource conservation via material recycle/reuse networks. *Industrial and Engineering Chemistry Research*, 42, 4319–4328.

Foo, D. C. Y. 2010. Automated targeting technique for batch process integration, *Industrial and Engineering Chemistry Research*, 49(20), 9899–9916.

Foo, D. C. Y. (2012). *Process Integration for Resource Conservation*, 1st ed. Boca Raton, Florida, US: CRC Press.

Foo, D. C. Y., Lee, J.-Y., and Ng, D. K. S., and Chen, C.-L. Targeting and design for batch regeneration and total networks. *Clean Technologies and Environmental Policy* (in review).

Foo, D. C. Y., Manan, Z. A., and Tan, Y. L. 2005. Synthesis of maximum water recovery network for batch process systems, *Journal of Cleaner Production*, 13(15), 1381–1394.

Gouws, J., Majozi, T., Foo, D. C. Y., Chen, C. L., and Lee, J.-Y. 2010. Water minimisation techniques for batch processes, *Industrial and Engineering Chemistry Research*, 49(19), 8877–8893.

Kim, J.-K. and Smith, R. 2004. Automated design of discontinuous water systems, *Process Safety and Environmental Protection*, 82(B3), 238–248.

Li, B. H. and Chang, C. T., 2006, A mathematical programming model for discontinuous water-reuse system design, *Industrial and Engineering Chemistry Research*, 45:5027–5036.

Majozi, T., Brouckaert, C. J., and Buckley, C. A. 2006. A graphical technique for wastewater minimization in batch processes, *Journal of Environmental Management*, 78, 317–329.

Prakash, R. and Shenoy, U. V. 2005. Targeting and design of water networks for fixed flowrate and fixed contaminant load operations. *Chemical Engineering Science*, 60(1), 255–268.

Wang, Y. P. and Smith, R. 1995. Time pinch analysis, *Chemical Engineering Research and Design*, 73, 905–914.

12

Extended Application: Retrofit of Resource Conservation Network

Process integration techniques presented in previous chapters are mainly dedicated to *grassroots* design, i.e., new processes that are yet to be built. However, in many cases, engineers will have to deal with existing processes/plants that have implemented some material recovery activities, i.e., some material recovery streams exist in the process. Note however that these processes may not have explored their full potential in maximizing material recovery. Hence, these processes should undergo *retrofit* exercises in order to enhance their material recovery potential. Some techniques in previous chapters are extended for use here. The retrofit procedure follows the work reported in Foo (2018).

12.1 MRD and Its Insights

To identify the actual material recovery potential for an RCN, the *material recovery pinch diagram* (MRPD) introduced in Chapter 3 is a useful tool. Note that the MRPD (El-Halwagi et al., 2003; Prakash and Shenoy, 2005) was originally meant for flowrate targeting for grassroots design problem. In performing RCN retrofit, flowrate targeting with MRPD may be used to as a benchmark for comparison on existing RCNs. The following steps are used to construct the MRPD (reproduced from Chapter 3):

1. Arrange all process sinks in ascending order of their impurity concentration levels (C_{SKj}). Calculate the maximum load acceptable by the sinks (m_{SKj}) following Equations 12.1, which are given as the product of sink flowrate (F_{SKj}) with its concentration level (the equations may be extended for property integration problems with the use of property operators[1] instead of impurity concentrations):

$$m_{SK_j} = F_{SK_j} C_{SK_j} \qquad (12.1)$$

2. Arrange all process sources (from all individual RCNs) in ascending order of their impurity concentration levels (C_{SRi}). Calculate the load possessed by the sources (m_{SRi}) following Equation 10.2, which

---

[1] See details in Chapter 2.

is given as the product of the source flowrate (F_{SRi}) with its concentration level (the equations may be extended for property integration problems with the use of property operators[1]):

$$m_{SR_i} = F_{SR_i} C_{SR_i} \qquad (10.2)$$

3. All process sinks are plotted in ascending order of their concentration levels on a load-versus-flowrate diagram to form the *sink composite curve*. The individual sink segments are connected by linking the tail of a later to the arrowhead of an earlier segment.

4. Step 3 is repeated for all process sources to form the *source composite curve* on the same load-versus-flowrate diagram. For a feasible MRPD, the entire sink composite curve has to stay above and to the left of the source composite curve. Otherwise, the MRPD is considered infeasible, and step 5 is next followed.

5. For the RCN with a single pure fresh resource of superior quality (highest purity among all sources), the source composite curve is moved horizontally to the right until it just touches the sink composite curve, with the latter lying above and to the left of the source composite curve. Figure 12.1a shows a feasible MRPD with a single pure fresh resource (reproduced from Figure 3.1)[2].

Several useful *RCN targets* may be identified from the MRPD. As shown in Figure 12.1a, the overlapping region between the sink and source composite

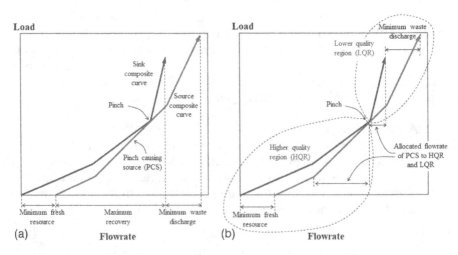

FIGURE 12.1
The MRPD and the allocated flowrates of the PCS.

---

[2] For RCN with a single impure fresh resource, a fresh locus is plotted on the MRPD. The source composite curve is slid on the locus until it just touches the sink composite curve, and stays on the right and below the latter (see Chapter 3 for details).

curves determines the recovery flowrate among the process sources and the sinks; while the opening at the left and right sides of the MRPD represent the minimum flowrates of fresh resource and waste discharge of the RCN, respectively. An important characteristic of the MRPD is the *pinch* point, as it represents the bottleneck of material recovery within the RCN. At the pinch, the maximum tolerable load of the sink composite curve is matched with that of the source composite curve. Hence, the following rules may be devised based on the MRPD:

1. The pinch divides the overall RCN into *higher* (HQR) and *lower quality regions* (LQR). Process sinks and sources that have higher quality (or lower impurity concentration) than the pinch are located in the HQR, while the others of lower quality (higher impurity concentration) in the LQR. This also means that process sources should only be sent to the sinks in the same region, and not to those in different regions.

2. The *pinch-causing source* (PCS) is exceptional to rule 1, as it belongs to both HQR and LQR. One may determine its *allocated flowrates* to the HQR and LQR by inspecting its horizontal distances in these regions (see Figure 12.1b).

If an RCN was designed following the above rules, it will achieve the maximum material recovery potential, with minimum fresh resource and waste discharge targets. In other words, process sinks in the HQR are fed with maximum tolerable load (i.e., which leads to the pinch), such as that shown in Figure 12.1. Note however that this is often not the case when an RCN was designed with other ad-hoc decision, or by intuitive decision based on experiences, etc. For such cases, the RCN will have higher flowrate of fresh resource than its actual minimum value, which in turn leads to higher waste discharge.

12.2 Load–Flowrate Diagram

In order to better analyze the existing RCN, the *load–flowrate diagram* (Foo, 2018) is a useful tool. As shown in Figure 12.2, the load–flowrate diagram displays the maximum load acceptable by the sinks (m_{SKj}), as well as its existing loads ($m_{SKj,ext}$). These values are calculated using Equations 12.1 and 12.3, respectively.

$$m_{SKj,ext} = \sum_i F_{SRi,SKj,ext}\, C_{SRi} \quad \forall j \tag{12.3}$$

where $F_{SRi,SKj,ext}$ is the existing flowrate received by sink j (SKj) from source i (SRi).

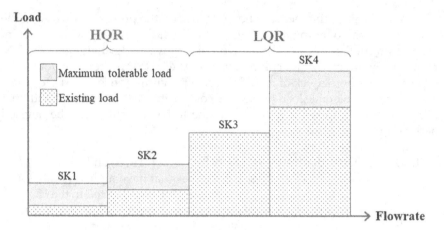

FIGURE 12.2
The load–flowrate diagram. (From Foo, 2018.)

For the case in Figure 12.2, among the four sinks, only SK3 in the LQR is fed with the maximum acceptable load, while other process sinks (SK1, SK2, and SK4) are fed with loads that are below their maximum limits. Hence, these sinks are identified as the candidates for retrofit, as they may receive more impurity load from the process sources.

Example 12.1 Illustrative example (Foo, 2018)

Figure 12.3 shows an RCN where source SR1 is partially recycled to sink SK2, while the remaining of SR1 and SR2 are sent for waste discharge. Hence, it is observed that the RCN has a fresh resource (FR) flowrate of 50 t/h, and waste discharge of 45 t/h (= 20 + 25 t/h).

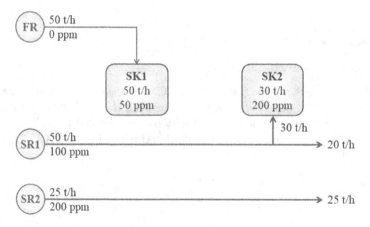

FIGURE 12.3
An existing RCN for Example 12.1. (From Foo, 2018.)

Perform the following tasks:

1. Determine the minimum fresh resource and waste discharge flowrates of the RCN, for a grassroots design problem.
2. Use the load–flowrate diagram to analyze if the sinks are fed with maximum acceptable loads.

SOLUTION

1. Minimum fresh resource and waste discharge flowrates for grassroots design.

   The MRPD in Figure 12.4a shows that in grassroots design, the RCN only requires 25 t/h of fresh resource (F_R), and will discharge 20 t/h of waste (F_D). SR1 is identified as the PCS where it belongs to both HQR and LQR. Figure 12.4a also shows that 25 t/h (= 50 – 25 t/h) of this RCS should be allocated to the HQR (F_{HQR}), while 25 t/h (= 50 – 25 t/h) to the LQR (F_{LQR}).

2. Use the load-flowrate diagram to analyze the RCN.

   The existing RCN in Figure 12.4a shows that the RCN consumes 50 t/h of fresh resource and discharges 45 t/h of waste, where both are 25 t/h higher than the minimum flowrates identified in task 1. Next, the load–flowrate diagram in Figure 12.4b) is constructed, using Equations 12.1 and 12.2. It is observed that both SK1 and SK2 do not receive their maximum acceptable loads. In fact, SK1 does not receive any load as it is fed with fresh resource completely. On the other hand, SK2 in the LQR receives 3 kg/s from the PCS (SR1), which is half of its maximum acceptable loads of 6 kg/s. Hence, the RCN may be retrofitted in order to reduce its fresh resource and waste discharge; this can be done by recovering more sources to both SK1 and SK2, which are now discharged as waste.

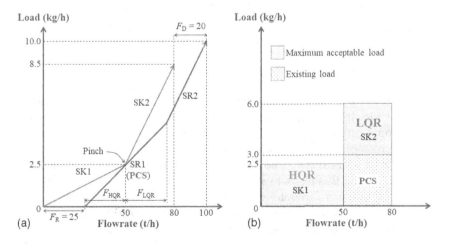

FIGURE 12.4
(a) MRPD for grassroots design; (b) load–flowrate diagram for Example 12.1 (From Foo, 2018).

12.3 Retrofit Procedure

For an existing process, it is best to incur minimum changes to process where retrofit is performed. Hence, the following assumptions are made (Foo, 2018):

1. Existing recycle connections of the RCN (if any) are to be kept as much as possible. In other words, changes (if any) will be kept to the minimal when retrofit is to be conducted on the RCN.

2. Fresh resource flowrates for the RCN may be reduced, or completely removed. In most cases, there are control valve(s) that allow fresh resource flowrate to be manipulated.

The procedure for retrofitting an RCN consists of the following three steps (Foo, 2018):

1. Identification of grassroots design targets. The latter consist of the minimum fresh resource and waste discharge flowrates, allocated flowrates of the PCS to the HQR and LQR, as well as the pinch concentration. These targets may be done using the MRPD, or the algebraic targeting technique[3].

2. Perform load analysis of the process sinks with a load–flowrate diagram. This step allows the identification of sinks that are capable of receiving recycle connection(s) from the process sources. The latter are usually discharged as waste.

3. Revamping of piping connections. Two main strategies may be used to identify process source(s) to be recycled to the process sink(s) identified in Step 2:

   i. Strategy 1: recovery of sources that are discharged as waste.

   ii. Strategy 2: re-piping of cross-pinch sink-sources matches, along with the recovery of waste discharge.

   Among the two options, Strategy 1 will incur minimum changes to process operation as it involves the recovery of waste streams. Hence, the revamp of piping connections on existing RCN is kept to the minimal. On the other hand, Strategy 2 usually incurs a higher degree of changes on existing RCN, as a revamp will take place on some existing cross-pinch piping connections. In most cases, Strategy 2 involves the revamp of the PCS so that it is better utilized in both HQR and LQR. Doing this leads to a higher extent of material recovery, which results in lower flowrates of fresh resource and waste discharge.

   For both strategies, the existing RCN will be retrofitted by reducing its fresh resource flowrate, when the identified sinks (in step 2) will be fed with process source(s) (that is originally discharged as waste).

---

[3] See Chapter 2 for details.

These strategies are carried out using a revised nearest neighbor algorithm (NNA)[4], given as follows.

i. The identified sink in step 2 is designated as h. In some cases, more than one sink are identified in step 2. For such cases, sink of highest quality requirement (i.e., lowest impurity concentration) will first be analyzed.

ii. A source is to be identified (following Strategy 1 or 2). If the identified source has an equal or lower concentration than sink h, the source is fed to the sink directly. Flowrate of the fresh resource that is originally fed to the sink will be reduced.

iii. If the identified source has a higher concentration (than sink h), it will be mixed with fresh resource to be fed to sink h. These are the *neighbor candidates* which are designated as N1 and N2, respectively. Before the flowrates of these neighbor candidates are determined, one will have to first determine the associated *retrofit flowrate* ($F_{SKh,ret}$) and *retrofit load* ($m_{SKh,ret}$) of sink h, these are calculated using Equations 12.4 and 12.5, respectively.

$$F_{SKh,ret} = F_{SKh} - \sum_i F_{SRi,SKh} \tag{12.4}$$

$$m_{SKh,ret} = m_{SKh} - m_{SKh,ext} \tag{12.5}$$

where $F_{SRi,SKh}$ is flowrate of existing matches between source i with sink h. As shown in Equation 12.4, the retrofit flowrate of sink h is determined from the difference between its required flowrate and the total flow rates of its existing matches. Note that the two load terms on the right side of Equation 12.5 have been given in Equations 12.1 and 12.3 (note that $m_{SKj,ext}$ is the same as $m_{SKh,ext}$ for this case).

With the determination of retrofit flowrate and retrofit load, the allocated flowrates of the neighbor candidates N1 and N2 (F_{N1} and F_{N2}) can be determined using the revised NNA equations that follow.

$$F_{SKh,ret} = F_{N1} + F_{N2} \tag{12.6}$$

$$m_{SKh,ret} = F_{N1}C_{N1} + F_{N2}C_{N2} \tag{12.7}$$

Similar to the original NNA procedure, when the calculated flowrates of the neighbor candidates (N1 and/or N2) are higher than their available flowrates, the limited flowrate of the source is used completely, and a new pair of neighbors is used to fulfill the remaining flowrate and load requirements as determined by

---

[4] See Chapter 8 for the original NNA for grassroots design of RCN

Equations 12.6 and 12.7. Note also that when there are more than one source candidate, the candidate with the quality closest to the sink should be prioritized, following the original NNA principle[5].

iv. Move on to Step i to analyze the process sink of higher concentration.

Example 12.2 Illustrative example revisited

For the existing RCN in Example 12.1, use the two retrofit strategies to enhance its material recovery.

SOLUTION

Example 12.1 shows that both sink SK1 and SK2 do not receive their maximum acceptable loads. Hence, the sources that are now discharged as waste may be recovered to both SK1 and SK2. Both strategies are analyzed next:

 1. Retrofit with Strategy 1—recovery of sources that are discharged as waste

    Strategy 1 is first evaluated for the retrofit of RCN. For this case, both existing waste streams may be considered for recovery to the sinks. Next, SK1 is identified as sink h that will undergo retrofit, as it has a lower concentration among the two sinks (step i). Since none of the sources has an equal or lower concentration than that of SK1 (step ii), step iii is next followed so that a pair of neighbor candidates are identified to be fed to SK1. We then identify the fresh resource (FR) as N1, while SR1 as N2 for SK1 (Figure 12.5a).

    As no source is recovered to SK1 in the existing RCN, Equations 12.4 and 12.5 determine that the retrofit flowrate ($F_{SK1,ret}$) and load ($m_{SK1,ret}$) of SK1 as 50 t/h and 2.5 kg/h. Next, Equations 12.5 and 12.7 are solved to determine the allocated flowrates of N1 (FW) and N2 (SR1) to SK1 as 25 t/h (F_{N1}) and 25 t/h (F_{N2}), respectively. As the waste flowrate from SR1 (20 t/h) is smaller than its calculated value of F_{N2}, the entire SR1 is recovered to SK1, and a new pair of neighbor candidates is required. The fresh resource (FR) is paired with SR2 as the new neighbor candidates for SK1 (Figure 12.5b). A new set of retrofit flowrate ($F_{SK1,ret}$) and load ($m_{SK1,ret}$) need to be recalculated, as SK1 has received flowrate and load from SR1. Equations 12.4 and 12.5 next determine that the new $F_{SK1,ret}$ and $m_{SK1,ret}$ values correspond to 30 t/h (= 50 − 20 t/h) and 0.5 kg/h (= 2.5 − (20 × 100) kg/h), respectively. Hence, the allocated flowrates of N1 (FW) and N2 (SR2) to SK1 are determined using Equations 12.6 and 12.7 as 27.5 t/h (F_{N1}) and 2.5 t/h (F_{N2}), respectively. This results in an improved RCN in Figure 12.6a, with reduced fresh resource and waste discharge of 27.5 and 22.5 t/h, respectively. Note that the waste stream from SR1 is completely removed.

---

[5] See Chapter 8 for details.

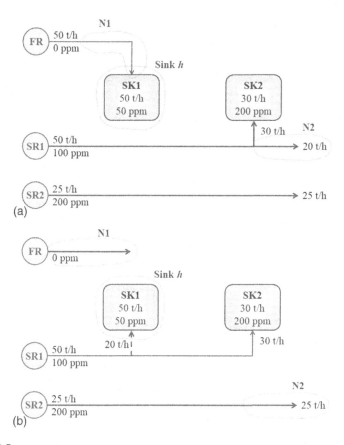

FIGURE 12.5
RCN retrofit with Strategy 1 for Example 12.2: (a) SR1 and FR as neighbor candidates; (b) SR2 and FR as neighbor candidates.

2. Retrofit with Strategy 2—re-piping of cross-pinch sink-sources matches, along with the recovery of waste discharge.

In Strategy 2, any cross-pinch sources-sink match is to be identified and re-piped. For this case, the MRPD in Example 12.1 identifies that SR1 is the PCS, which should be fed to both HQR and LQR, with flowrates of 25 t/h each (Figure 12.4a). In other words, SR1 should be allocated to SK1 in the HQR with flowrate of 25 t/h. However, SR1 is now allocated fully to SK2 in the LQR (Figure 12.5a). Hence, the SR1–SK2 match is to be revamped. Step iii is next where fresh resource (N1) and SR1 (N2) are identified as the neighbor candidates for SK1. The earlier calculation in Strategy 1 indicated that allocated flowrates of these neighbor candidates correspond to 25 (N1) and 25 t/h (N2), respectively. Hence, apart from recovering 20 t/h of waste stream from SR1, the remaining 5 t/h (= 25 – 20 t/h) of the allocated flowrate should come from the existing SR1–SK2 match. Note however

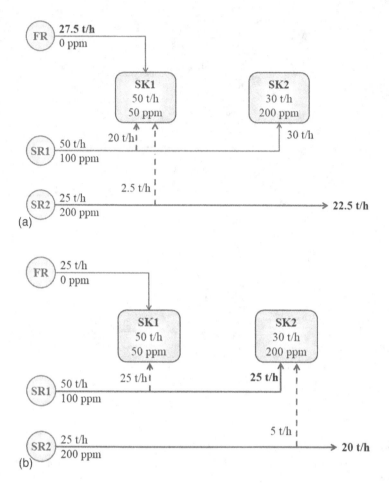

FIGURE 12.6
Improved RCN design with (a) Strategy 1; (b) Strategy 2 for Example 12.2. (From Foo, 2018.)

that doing this leads to flowrate deficit of 5 t/h with SK2. As SK2 is located in the LQR, no fresh resource should be used to supplement its flowrate deficit. One may recover additional waste stream from SR2 to SK2, since the former has lower impurity concentration (i.e., cleaner). Figure 12.6b shows the retrofitted RCN, with much reduced fresh resource and waste discharge flowrates, identical to the minimum targets as identified by the MRPD for grassroots design (Figure 12.4a).

Comparing the RCN designed using two strategies in Figure 12.6, one note that the RCN in Figure 12.6b involves connections with both process sinks, while that in Figure 12.6a only involves SK1. In other words, network designed using Strategy 2 involves more changes with process operation, though can achieve a larger extent of material recovery.

Example 12.3 Water minimization (Polley and Polley, 2000)

Table 12.1 shows the limiting water data for a classical water minimization example from Polley and Polley (2000). One of its water network designs is given in Figure 12.7. Use the three-stage procedure to retrofit the network for better water recovery, assuming freshwater of zero impurity is used.

SOLUTION

The three-step procedure is followed to perform the retrofit analysis.

1. Identification of grassroots design targets.
   The MRPD is used to identify the RCN targets, i.e., freshwater and waste water flowrates, as well as the allocation flowrates of

TABLE 12.1

Limiting Water Data for Example 12.3

j	Sinks, SK$_j$	F_{SKj} (t/h)	C_{SKj} (ppm)	i	Sources, SR$_i$	F_{SRi} (t/h)	C_{SRi} (ppm)
1	SK1	50	20	1	SR1	50	50
2	SK2	100	50	2	SR2	100	100
3	SK3	80	100	3	SR3	70	150
4	SK4	70	200	4	SR4	60	250

Source: Polley and Polley (2000).

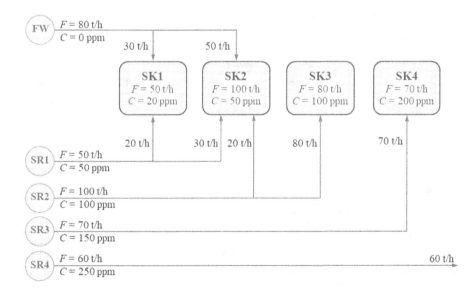

FIGURE 12.7
Existing water network for Example 12.3.

the PCS to the HQR and LQR. The MRPD is generated with the following steps.

  i. *Arrange all process sinks in ascending order of their impurity concentration levels (C_{SKj}). Calculate the maximum load acceptable by the sinks (m_{SKj}).*

 ii. *Arrange all process sources (from all individual RCNs) in ascending order of their impurity concentration levels (q_{SRi}). Calculate the load possessed by the sources (m_{SRi}).*

Water sinks are arranged in ascending order of concentrations (see column 3 of Table 12.2). The maximum load acceptable by the sinks (m_{SKj}) is calculated using Equation 12.1 and located in column 4 of Table 12.2. On the other hand, water sources are also arranged in ascending order of concentrations. Their loads (m_{SRi}) are calculated using Equation 12.2 and documented in column 4 of Table 12.3.

iii. *All process sinks are plotted with ascending order of their concentration levels on a load-versus-flowrate diagram to form the sink composite curve.*

 iv. *Step 3 is repeated for all process sources to form the source composite curve on the same load-versus-flowrate diagram.*

To plot the MRPD, the cumulative flowrates (Cum. F_{SKj}) and loads (Cum. m_{SKj}) for all sinks are first calculated, given in the last two columns of Table 12.2. Similar calculations were also done for the cumulative flowrates (Cum. F_{SRi}) and loads (Cum. m_{SRi}) for all sources, and are given in the last two columns of Table 12.3. These values are used to plot the sink and source composite curves, respectively, on the load-versus-flowrate diagram, given in Figure 12.8. Note that the slope of each segment corresponds to their respective concentration values.

TABLE 12.2

Data for Plotting Sink Composite Curve in Example 12.3

Sinks, SK_j	F_{SKj} (t/h)	C_{SKj} (ppm)	m_{SKj} (g/h)	Cum. F_{SKj} (t/h)	Cum. m_{SKj} (g/h)
SK1	50	20	1,000	50	1,000
SK2	100	50	5,000	150	6,000
SK3	80	100	8,000	230	14,000
SK4	70	200	14,000	300	28,000

TABLE 12.3

Data for Plotting Source Composite Curve in Example 12.3

Sources, SR_i	F_{SRi} (t/h)	C_{SRi} (ppm)	m_{SRi} (g/h)	Cum. F_{SRi} (t/h)	Cum. m_{SRi} (g/h)
SR1	50	50	2,500	50	2,500
SR2	100	100	10,000	150	12,500
SR3	70	150	10,500	220	23,000
SR4	60	250	15,000	280	38,000

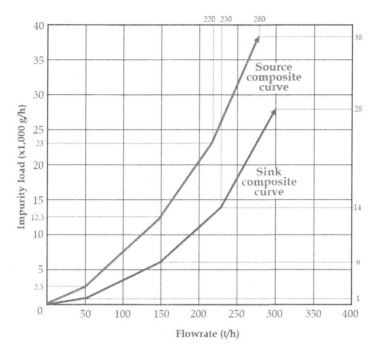

FIGURE 12.8
Infeasible MRPD for Example 12.3.

As shown in Figure 12.8, the sink composite curve indicates that the water sinks require a total flowrate of 300 t/h, and these sinks can only accept a total impurity load of 28,000 g/h. On the other hand, the source composite curve indicates that the water sources have a total water flowrate of 280 t/h, and possess 38,000 g/h of total impurity load. It is also observed that the source composite curve stays on the left and above the sink composite curve, indicating that the MRPD in Figure 12.8 is infeasible. The MRPD also indicates that the sinks do not receive sufficient water flowrate, as it requires a total flowrate of 300 t/h but is supplied with 280 t/h of water sources. Furthermore, the sinks can accept a maximum total impurity load of 28,000 g/h, but are fed with 38,000 g/h of impurity load. Both flowrate and load infeasibility are to be resolved with freshwater feed.

v. *For the RCN with a single pure fresh resource of superior quality (highest purity among all sources), the source composite curve is moved horizontally to the right until it just touches the sink composite curve, with the latter lying above and to the left of the source composite curve.*

As the freshwater supply is free of impurity, the source composite curve is moved horizontally to the right of the sink composite curve, until it stays completely below and

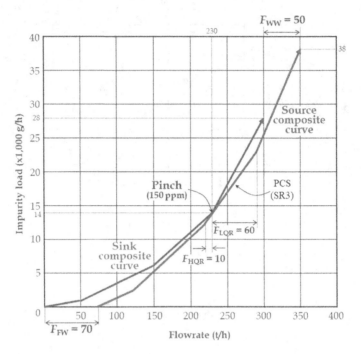

FIGURE 12.9
Feasible MRPD for Example 12.3. (From Foo, 2018.)

to the right of the latter, resulted in a feasible MRPD as shown in Figure 12.9. In this case, both flowrate and load constraints of the sink composite curve are observed.

The MRPD in Figure 12.9 identifies several RCN targets. The opening on the left of the MRPD indicates that this RCN has a minimum freshwater flowrate (F_{FW}) of 70 t/h, while opening on the right indicates its minimum wastewater flowrate (F_{WW}) is 50 t/h. The pinch concentration is identified as 150 ppm, formed by the PCS of SR3. The latter has its allocated flowrates of 10 t/h (F_{HQR}) and 60 t/h (F_{LQR}) to the HQR and LQR, respectively.

2. *Load analysis of process sinks with load–flowrate diagram.*

The load–flowrate diagram in Figure 12.10 is constructed using Equations 12.1 and 12.2. It is observed that both SK2 in the HQR and SK4 in the LQR do not receive their maximum acceptable loads. As shown in Figure 12.7, SK2 is fed with freshwater (FW), SR1 (30 t/h, 50 ppm), and SR2 (20 t/h, 100 ppm). Hence, it receives a load of 3,500 g/h (= 30 × 50 + 50 × 100 g/h), while its limit is 5,000 g/h (see Table 12.2). On the other hand, SK4 is fed with 70 t/h of SR3 (150 ppm), which receives 10,500 g/h (= 70 × 150 g/h) of load, while it has a maximum limit of 14,000 g/h (see Table 12.2). Hence, SK2 and SK4 may undergo retrofit in order to reduce their freshwater and/or wastewater flowrates.

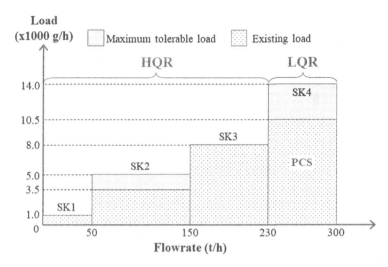

FIGURE 12.10
Load–flowrate diagram for Example 12.3. (From Foo, 2018.)

3. *Revamping of piping connections.*

Retrofit with Strategy 1—recovery of sources that are discharged as waste

In Strategy 1, the wastewater stream from SR4 is considered
for recovery to SK2 (sink *h*), as it has a lower concentration
among the two sinks (step (i)). Since SR4 has a higher concentration than SK2, step (ii) is skipped, while step (iii) is followed so
that a pair of neighbor candidates are identified for recovery. For
this case, fresh water was identified as N1, while SR4 as N2 (see
Figure 12.11).

Equations 12.4 and 12.5 are used to determine the retrofit
flowrate ($F_{SK2,ret}$) and load ($m_{SK2,ret}$) for SK2 as 50 t/h (= 100 – 30 –
20 t/h) and 1,500 g/h (= 5,000 – 3,500 g/h), respectively. Next,
solving Equations 12.6 and 12.7 determines the allocated flowrates of the two neighbors (i.e., FW and SR4) to SK2 as 44 t/h
(F_{N1}) and 6 t/h (F_{N2}), respectively. In other words, the freshwater
flowrate to SK2 is reduced from 50 t/h (Figure 12.7) to 44 t/h.
This results in an improved RCN with lower freshwater (F_{FW} =
74 t/h) and wastewater (F_{WW} = 54 t/h) flowrates, as shown in
Figure 12.12.

Retrofit with Strategy 2—re-piping of cross-pinch sink-
sources matches, along with the recovery of waste discharge.

In Strategy 2, re-piping of cross-pinch water sink-sources
matches is explored, along with the recovery of wastewater
(from SR4). The MRPD in Figure 12.9 shows that the PCS (SR3)
should have an allocated flowrates of 10 t/h sent to the HQR
(F_{HQR}), while 60 t/h to the and LQR (F_{LQR}). Figure 12.7 however
shows that the PCS is utilized entirely for SK4 in the LQR (i.e.,
no allocated flowrate sent to the HQR). Hence, 10 t/h of the PCS

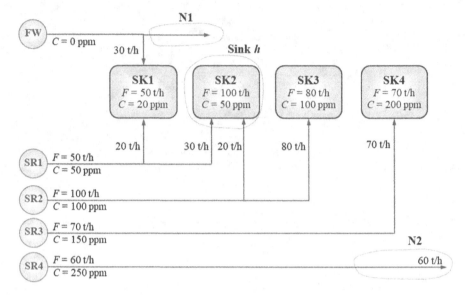

FIGURE 12.11
Neighbor candidates for SK2 following Strategy 1.

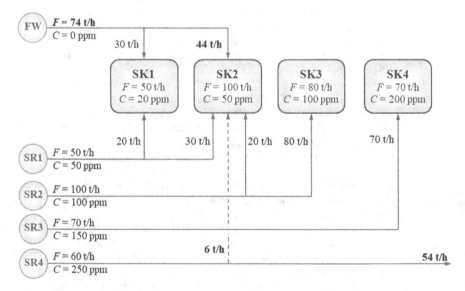

FIGURE 12.12
Improved RCN with Strategy 1 for Example 12.3 (new matches are shown in dashed lines and revised flowrates in bold font).

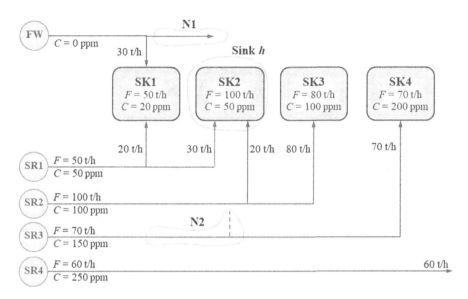

FIGURE 12.13
Neighbor candidates for SK2 following Strategy 2.

is considered as the cross-pinch flowrates to be utilized for SK2. We then consider using fresh water (N1) and SR3 (N2) as neighbor candidates for SK2 following step (iii) of the revised NNA procedure (see Figure 12.13).

Similar to Strategy 1, the retrofit flowrate ($F_{SK2,ret}$) and load ($m_{SK2,ret}$) of SK2 are calculated using Equations 12.4 and 12.5 to be 50 t/h (= 100 – 30 – 20 t/h) and 1,500 g/h (= 5,000 – 3,500 g/h), respectively. Next, Equations 12.6 and 12.7 are solved simultaneously to obtain the flowrates of the neighbor candidates of freshwater and SR3, i.e., 40 (F_{N1}) and 10 t/h (F_{N2}), respectively. Hence, freshwater flowrate for SK2 is reduced to 40 t/h (from 50 t/h), while 10 t/h of water from SR3 (that is currently sent to SK4) is also re-channeled to SK2. However, doing this leads to flowrate deficit experienced by SK4. Since SK4 is located in the LQR, it should not be fed with freshwater. One option to supplement the flowrate requirement of SK4 is to recover the wastewater from SR4. Performing a load analysis with Equation 12.3 indicates that the total load sent to this sink is now increased to 11,500 g/h, but yet to reach its maximum acceptable limit of 14,000 g/h (see Table 12.2). Doing so leads to the improved RCN design in Figure 12.14. As shown, both freshwater and wastewater flowrates of the RCN are now reduced to those of the grassroots design (Figure 12.9).

A comparison on flowrate reduction among the base case network with the two strategies is given in Figure 12.15. It is observed that greater flowrate reduction of freshwater and

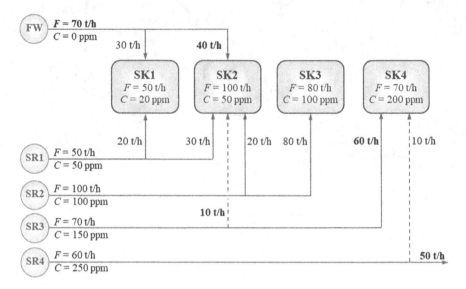

FIGURE 12.14
Improved RCN with Strategy 2 for Example 12.3 (new matches are shown in dashed lines and revised flowrates in bold font).

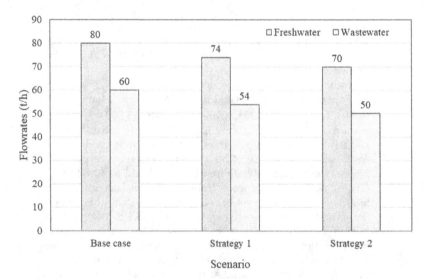

FIGURE 12.15
Comparison on freshwater and wastewater flowrate among base case and retrofit strategies.

wastewater is observed in Strategy 2. However, the latter involves the retrofit of more piping connection, and water sinks. Hence, it can be concluded that trade-off exists between flowrate reduction and piping connections in performing retrofit on existing RCN.

12.4 Additional Readings

In this chapter, only pinch-based procedures for the retrofit of RCN have been described. Readers may refer to the work of Lee et al. (2021) for extended mathematical programming in retrofitting RCN. Essentially, the work of Lee et al. (2021) considers the addition of recycling pipeline, as well as the removal of existing pipelines that are inappropriately placed; both of these have an impact on the overall cost of the RCN.

PROBLEMS
Water Minimization Problems

12.1 Revisit Example 12.2. Another alternative water network of this example is given in Figure 12.16. Use the three-stage procedure to retrofit the network for better water recovery.

12.2 Figure 12.17 (reproduced from Figure 4.4) shows an existing water network in a paper milling process (Foo et al., 2006). Its detailed process description is found in Problem 4.5. To evaluate the possibility of further enhancing water recovery potential, retrofit analysis is to be carried out. Table 12.4 summarizes the limiting water data for the water network, where total suspended solid (TSS) has been identified as the most significant water quality factor in considering water minimization

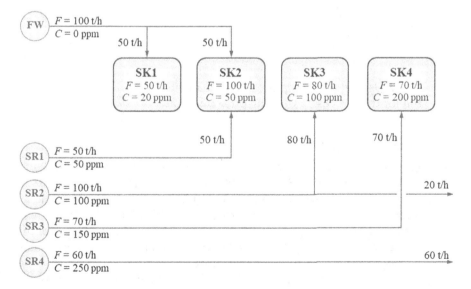

FIGURE 12.16
Alternative design of water network in Problem 12.1.

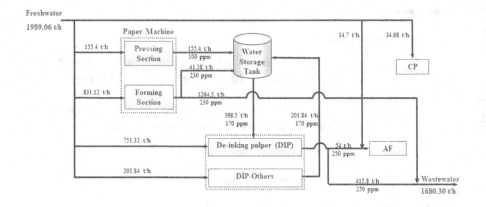

FIGURE 12.17
Existing water network of a paper mill in Problem 12.2. (From Foo et al., 2006. Reproduced with permission. Copyright 2006 American Institute of Chemical Engineers [AIChE].)

TABLE 12.4

Limiting Water Data for Paper Milling Process in Problem 12.2

j	Sinks, SK_j	F_{SKj} (t/h)	C_{SKj} (ppm)	i	Sources, SR_i	F_{SKi} (t/h)	C_{SKi} (ppm)
1	Pressing section	155.40	20	1	Pressing section	155.40	100
2	Forming section	831.12	80	2	Forming section	1305.78	230
3	DIP-Others	201.84	100	3	DIP-Others	201.84	170
4	DIP	1149.84	200	4	DIP	469.80	250
5	Chemical preparation (CP)	34.68	20				
6	Approach flow (AF)	68.70	200				

Source: Foo et al. (2006).

study. Use the three-step procedure to identify the water-saving potential for this process, assuming pure freshwater is used.

Utility Gas Recovery Problems

12.3 For the hydrogen network in Problems 2.10 and 4.18 (Umana et al., 2014), its PFD is shown in Figure 12.18 (reproduced from Figure 2.22). As shown, the network consists of an external hydrogen (H_2) plant, and an internal source, i.e., catalytic reformer (CCR). There are four hydrotreaters in the network, i.e., naphtha hydrotreater (NHT), cracked naphtha hydrotreater (CNHT), diesel hydrotreater (DHT), and vacuum gas oil hydrocracker (VGOHC). Their limiting data are given in Table 4.18a. Use the three-step retrofit procedure to enhance hydrogen recovery for this process. The fresh hydrogen feed (with an impurity content of 7.44%) is reported to have the minimum flowrate of 6.36 t/h (solution for Problem 4.18).

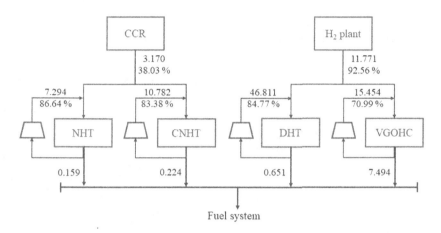

FIGURE 12.18
Hydrogen network for Problem 12.3; numbers represent the total gas flowrate in t/h and hydrogen purity concentration in % (Umana et al., 2014.)

12.4 For the refinery hydrogen network in Problem 4.19 (Alves and Towler, 2002), its existing hydrogen network is given in Figure 12.19, and limiting data are found in Table 4.31. The minimum fresh hydrogen feed is reported as 268.82 mol/s (with an impurity content of 5%; solution for Problems 4.19 and 5.13). Use the three-step retrofit procedure to enhance hydrogen recovery for this process.

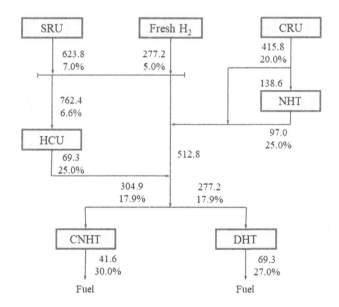

FIGURE 12.19
Refinery hydrogen network in Problem 12.4 (numbers represent the total gas flowrate in mol/s and impurity concentration in mol%). (Reprinted with permission from Alves and Towler, 2002. Reproduced with permission. Copyright American Chemical Society.)

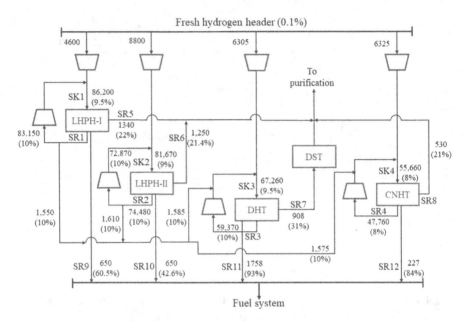

FIGURE 12.20
Refinery hydrogen network in Problem 12.5 (Deng et al., 2020).

12.5 For the refinery hydrogen network in Problem 5.12 (Deng et al., 2020), with its existing hydrogen network given in Figure 12.20. As shown, the network contains lubricating oil hydrogenation processes I (LHPH-I) and II (LHPH-II), diesel hydrogenation (DHT) and coking naphtha hydrogenation (CNHT), and desulfurization tower (DST). There are 12 sources (SR1–SR12) and 4 sinks (SK1–SK4) in the hydrogen network, while the outlet from the desulfurization tower may be ignored, with limiting data given in Table 5.7. The fresh hydrogen of the network is supplied by a hydrogen production plant, which has a pressure swing adsorption (PSA) unit. The fresh hydrogen supply has an impurity concentration of 0.1%. Note that both the hydrogen production plant and the PSA unit should be excluded from the retrofit analysis. Targeting reveals that the hydrogen network has a fresh hydrogen feed (0.1%) of 22,377.7 Nm³/h (solution for Problem 5.12). Use the three-step retrofit procedure to enhance hydrogen recovery for this process.

References

Alves, J. J. and Towler, G. P. 2002. Analysis of refinery hydrogen distribution systems, *Industrial & Engineering Chemistry Research*, 41, 5759–5769.

Deng, C., Zhu, M., Liu, J., and Xiao, F. 2020. Systematic retrofit method for refinery hydrogen network with light hydrocarbons recovery, *International Journal of Hydrogen Energy*, 45, 19391–19404.

El-Halwagi, M. M., Gabriel, F., and Harell, D. 2003. Rigorous graphical targeting for resource conservation via material recycle/reuse networks. *Industrial and Engineering Chemistry Research*, 42, 4319–4328.

Foo, D. C. Y., Manan, Z. A., and Tan, Y. L. 2006. *Use Cascade analysis to optimize water networks*, *Chemical Engineering Progress*, 102(7), 45–52.

Foo, D. C. Y. 2018. Systematic retrofit procedure for resource conservation network based on pinch analysis technique, *Clean Technologies and Environmental Policy*, 20(10), 2255–2273.

Lee, J.-Y., Tsai, C.-H., and Foo, D. C. Y. 2021. Single and multi-objective optimisation for the retrofit of process water networks, *Journal of the Taiwan Institute of Chemical Engineers*, 117, 39–47.

Polley, G. T. and Polley, H. L. 2000. Design better water networks. *Chemical Engineering Progress*, 96(2), 47–52.

Prakash, R. and Shenoy, U. V. 2005. Targeting and design of water networks for fixed flowrate and fixed contaminant load operations. *Chemical Engineering Science*, 60(1), 255–268.

Umana, B., Shoaib, A., Zhang, N., and Smith, R. 2014. Integrating hydroprocessors in refinery hydrogen network optimisation, *Applied Energy*, 133, 169–182.

Part II

Heat Integration

In Part II of this book, the readers will learn how to use pinch analysis and mathematical programming techniques for several important topics on heat integration, i.e., *heat exchanger network* (HEN) synthesis, *combined heat and power* schemes, as well as *heat-integrated water network*. During the 70s–80s in the last century, HEN synthesis as well as combined heat and power schemes were the main focus of process integration research, where their systematic synthesis methods were developed. Later, the developed techniques were extended to address other aspects of energy-intensive chemical process design, e.g., integration of dryers, reactor and distillation, heat-integrated water network, etc. The main driving force in addressing energy efficiency topics back then was mainly operating cost reduction.

In the past two decades, emission of greenhouse gases and the awareness of global warming have become the main factor in encouraging energy efficiency efforts. Since the beginning of this century, industrial emissions have been reported to rise by 70%, due to the increased demand of industrial products (IEA, 2024). CO_2 emissions from the chemical sector were reported as 1.5 Gt, equivalent to a total of 18% of industrial CO_2 emissions (World Economic Forum, 2021). As fossil fuels are the main energy sources for these industrial processes, emphasis should be placed on better design of industrial processes, so as to reduce (if not eliminate) the impact of global warming.

The three chapters in this part of the book are certainly too little to cover the broad topics of heat integration. Interested readers may refer to various important good sources of material on heat integration elsewhere, e.g., Smith (2016), Linnhoff et al. (1982). To facilitate self-learning, all calculation files for examples in Chapters 13–15 are made available to the readers on the publisher website (www.routledge.com/9781032003931).

References

IEA 2024. https://www.iea.org/energy-system/industry (accessed January 2024).

Linnhoff, B., Townsend, D. W., Boland, D., Hewitt, G. F., Thomas, B. E. A., Guy, A. R., and Marshall, R. H. 1982. *A User Guide on Process Integration for the Efficient Use of Energy*. Rugby, U.K: IChemE.

Smith, R. 2016. *Chemical Process: Design and Integration*, 2nd Ed. West Sussex, England: John Wiley.

World Economic Forum 2021. www.weforum.org/agenda/2021/10/how-petrochemicals-industry-can-reduce-its-carbon-footprint/ (accessed January 2024).

13

Synthesis of Heat Exchanger Network

When there are more than one process stream where heat recovery may be performed, *heat exchanger network* (HEN) exists for such case. The development of process integration methods for HEN synthesis dated back in 1970s, and become rather mature in 1990s. In this chapter, the basic elements of HEN synthesis are introduced. Essentially, the main philosophy of process integration methods is to identify the benchmark *energy targets* ahead of detailed HEN design. In order to do so, graphical or algebraic targeting techniques may be used to determine the hot and cold utility consumption of the HEN using basic thermodynamic principles.

13.1 Basic Understanding

In any HEN synthesis problem, two types of streams exist, i.e., *hot* and *cold streams*. As shown in Figure 13.1a, a hot stream refers to stream that encounters temperature drop, i.e., from higher *supply temperature* (T_S) to lower *target temperature* (T_T). In other words, cooling utility (e.g., cooling water) may be used in order for the hot stream to achieve its T_T value. On the other hand, cold stream experiences temperature rise, i.e., from lower T_S to higher T_T values, as shown in Figure 13.1b. This can achieved by supplying heat to this stream via hot utility (e.g., steam). A good alternative is to consider process-to-process heat recovery, where energy can be transferred from hot to cold streams; doing this leads reduced hot and cold utilities for the HEN.

When a stream encounters temperature drop/rise, its enthalpy change can be calculated as the product of heat capacity flowrate of the stream (CP) with its temperature change. Enthalpy change of hot and cold streams are calculated using Equations 13.1a and 13.1b, respectively; while CP value is given in Equation 13.2 For instance, the hot stream H1 in Figure 13.1a has a CP value

FIGURE 13.1
Type of steams: (a) hot stream; (b) cold stream (temperatures given in °C, and enthalpy in MW).

DOI: 10.1201/9781003173977-15

of 1 MW/°C (= 100/(150 − 50)). Similarly, *CP* value of cold stream C2 is calculated as 0.5 MW/°C (= 50/(120 − 20)):

$$\Delta H_H = CP_S (T_{TS} - T_{ST}) \tag{13.1a}$$

$$H_C = CP_S (T_T - T_S) \tag{13.1b}$$

$$CP_S = m_s C_p \tag{13.2}$$

where m_S is the mass flowrate of the stream, with constant pressure heat capacity C_p. The *CP* value in Equation 13.1 is usually a constant parameter for process streams that do not experience phase change. For phase change operations, the stream may be represented by segments of different *CP* values (see Smith, 2016 for details).

13.2 Graphical Targeting Method—Heat Recovery Pinch Diagram (HRPD)

To identify the heat recovery potential of a problem, an established technique for use is the graphical tool known as *heat recovery pinch diagram* (HRPD), first introduced by Hohmann (1971). The HRPD consists of *hot composite curve* and *cold composite curve* constructed on a temperature versus enthalpy diagram. Steps to plot the HRPD are illustrated using a simple example consisting of two hot streams (H1 and H2) and two cold streams (C1 and C2), given as follows.

1. Construct the hot composite curve on a temperature versus enthalpy diagram. As shown in Figure 13.2a, hot streams H1 and H2 are located next to each other on a temperature versus enthalpy diagram.

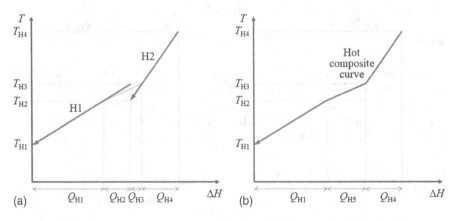

FIGURE 13.2
(a) Two hot streams; (b) Construction of hot composite curve.

H1 is to be cooled from T_{H3} to T_{H1}, while H2 is from T_{H4} to T_{H2}. These supply and target temperatures form three temperature intervals, i.e., $T_{H1}-T_{H2}$, $T_{H2}-T_{H3}$, and $T_{H3}-T_{H4}$.

It is more convenient to begin the analysis on interval of lower temperature, before moving to those of high temperature. Hence, we first analyze the temperature interval of $T_{H1}-T_{H2}$. As shown in Figure 13.2a, only part of stream H1 exists in this interval, with an enthalpy of Q_{H1}. Hence, no additional work is necessary, as this segment becomes part of the hot composite curve. Next, we move to the interval of $T_{H2}-T_{H3}$, where both H1 and H2 exist; their segments have the enthalpy values of Q_{H2} and Q_{H3}. These segments may be merged to become a composite curve segment with an enthalpy value of Q_{H5}, which is the summed value of Q_{H2} and Q_{H3}. Essentially, the joint segment is the result of vector summation of the two individual segments of H1 and H2, as shown in Figure 13.2b. In other words, a total enthalpy of Q_{H5} has to be removed in order for the temperature to drop from T_{H3} to T_{H2} for both streams. In the highest temperature interval, $T_{H3}-T_{H4}$, only H2 exists, with an enthalpy of Q_{H4}. Hence this segment forms the last segment of the hot composite curve. Finally, note that within each temperature interval, the *CP* value of the composite curve segment may be added by the *CP* values of the individual segments that were merged (readers may try to prove it).

2. Construct the cot composite curve on a temperature versus enthalpy diagram, similar to the hot composite curve. As shown in Figure 13.3a. Within the temperature interval of $T_{C1}-T_{C2}$, only part of stream C1 exists, with an enthalpy requirement of Q_{C1}. Within the interval of $T_{C2}-T_{C3}$, both C1 and C2 exist. Vector summation is

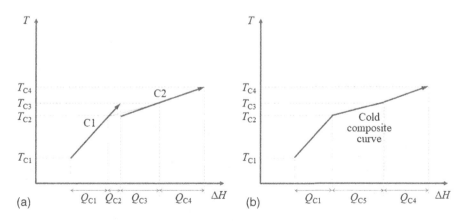

(a)   (b)

FIGURE 13.3
(a) Two cold streams; (b) Construction of cold composite curve.

performed to merge them, in which their enthalpy values of Q_{C2} and Q_{C3} is summed to a total of Q_{C5}, and represented by a composite curve segment, as shown in Figure 13.3b. In other words, a total enthalpy of Q_{C5} has to be supplied to C1 and C2, in order for their temperature to rise from T_{C2} to T_{C3}. In the highest temperature interval, T_{C3}–T_{C4}, only C2 exists, with an enthalpy of Q_{C4}. Hence the C2 segment forms the last segment of the cold composite curve.

3. Overlaying the hot and cold composite curves on the same diagram to form the HRPD, as shown in Figure 13.4a. To ensure a feasible heat transfer, the hot composite curve has to stay above the entire cold composite curve with a *minimum temperature difference* (ΔT_{min}). In order to do so, either of the composite curves may be shifted horizontally. Note that vertical shift is forbidden, as the *y*-axis is the supply/target temperature of the hot and cold streams. The overlapping region between the two composite curves represents the amount of energy recovery among all hot and cold streams (Q_{REC}). The extension on the right of the HRPD gives the *minimum hot utility target* (Q_{Hmin}) of the HEN, while the extension on the left gives the *minimum cold utility target* (Q_{Cmin}). Hence, these utility targets are obtained ahead of the detailed design of the HEN.

There are several important characteristics of the MRPD. As shown in Figure 13.4b, the pinch divides the HEN problems into regions above and below the pinch, and may be viewed as the overall bottleneck of the heat recovery problem. In the region above the pinch, it is observed that the cold composite curve has a larger enthalpy than the hot composite curve (represented by their horizontal distance). In other words, this region experiences energy deficit, which

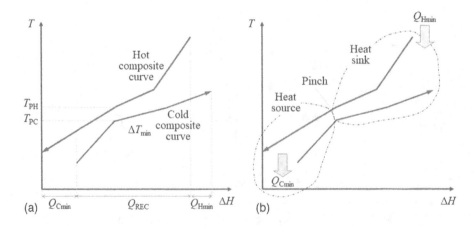

FIGURE 13.4
(a) Heat recovery pinch diagram (MRPD), and the minimum hot and utility targets; (b) MRPD is divided to regions above (heat sink) and below the pinch (heat source).

may be regarded as a *heat sink*. Hence, hot utility (e.g., steam, hot oil, etc.) has to be used in this region so that all cold streams can reach their target temperatures. It is important to note that no cold utility should be used in this region, as doing so will lead to additional usage of hot utility.

On the other hand, the hot composite curve is observed to have a larger enthalpy that the cold composite curve in region below the pinch (see Figure 13.4b), indicating that this region has excess energy. Hence, the excess energy of this *heat source* is to be removed using cold utility (e.g., cooling water), so that all hot streams will reach their target temperature. Since this region experiences excess of energy, no hot utility is to be used in this region, as doing so will lead to additional usage of cold utility. Note also that no hot and cold streams should be paired across the two regions, as doing so would disturb the energy balances of the regions, which leads to excess use of hot and/or cold utilities (Smith, 2016).

Example 13.1 Graphical targeting with HRPD

Figure 13.5 shows the process flow diagram of chemical manufacturing. As shown, a feed stream is to be heated prior to feeding to the reactor. The effluent from the reactor is sent to a flash unit, where the vapor stream leaves at the top. The flash bottom stream is fed to a distillation unit where its bottom product is cooled, before being fed to the storage facilities. Top stream from the distillation will be mixed with another feed stream before being sent to another purification unit, where its bottom waste is discarded. Top stream of the purification unit will undergo a cooling process prior to mixing with another pre-heated stream, to

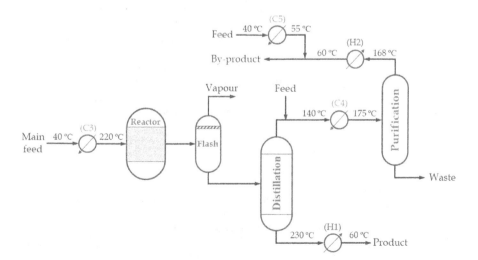

FIGURE 13.5
Process flow diagram for a chemical manufacturing example.

TABLE 13.1

Data for Hot and Cold Streams for Example 13.1

Streams	Supply Temperature T_S (°C)	Target Temperature T_T (°C)	Mass Capacity Flowrate, CP (MW/°C)	Enthalpy Change, ΔH (MW)
H1	230	60	1.5	255
H2	168	60	2.5	270
C3	40	220	2.0	360
C4	140	175	3.0	105
C5	40	55	2.0	30

form a side product. Data for the hot and cold stream of this example is given in Table 13.1. Solve the following tasks:

1. Determine the minimum hot and cold utility targets for a ΔT_{min} value of 10°C.
2. Determine the utility targets when the ΔT_{min} value is increased to 20°C.

SOLUTION

1. Determine the minimum hot and cold utility targets for a ΔT_{min} value of 10°C.

   The three-step procedure is followed for plotting the HRPD.

   i. *Construct the hot composite curve.*

   Both hot streams H1 and H2 are located next to each other in the temperature versus enthalpy diagram. Two temperature intervals are formed based on their start and target temperatures, i.e., 60–168°C and 168–230°C (Figure 13.6). Within the interval of 60–168°C, segments of both hot streams are found, with enthalpy of 270 and 162 MW (= 1.5 × (168 − 60) MW). Hence, vector summation is performed to merge their segments as a composite curve segment, as shown in Figure 13.6. In other words, the composite segment will have a total enthalpy of 432 MW (= 270 + 162 MW). Next, in the interval of 168–230°C, only the remaining segment of H1 is found. Hence it becomes the composite segment automatically. Hence, the construction of the hot composite curve is completed.

   ii. *Construct the cold composite curve.*

   Similar to the hot composite curve, all cold streams C3–C5 are located next to each other in the temperature versus enthalpy diagram (see Figure 13.7). Four temperature intervals are formed based on their start and target temperatures. In the lowest temperature interval, i.e., 40–55°C, C5 (30 MW) and part of C3 (2 × (55 − 40) = 30 MW) exist. Vector summation is performed to merge these segments as a composite segment with a total enthalpy of 60 MW. Next, in the interval 55–140°C, the segment of C3 becomes the composite segment, as it is the only stream that exists here.

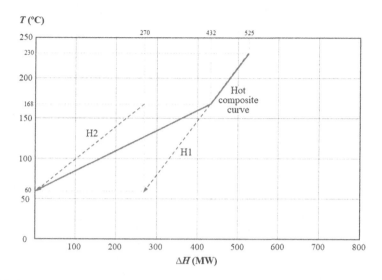

FIGURE 13.6
Construction of hot composite curve of Example 13.1.

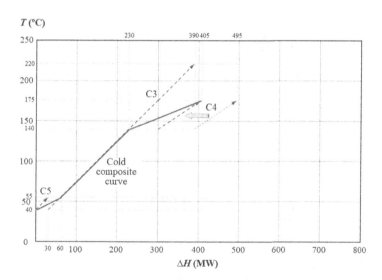

FIGURE 13.7
Construction of cold composite curve of Example 13.1 (for intervals between 40°C and 175°C).

We then move to interval 140–175°C, where C3 segment and C4 (105 MW; see Table 13.1) exist. The enthalpy of C3 segment may be calculated with Equation 13.1b as 70 MW (= 2 x (175 - 140) MW). As there exists a gap between them, C4 is shifted to the left until it stays next to the C3 segment, before vector summation may be performed to merge them as a composite segment of 175 MW (see Figure 13.7).

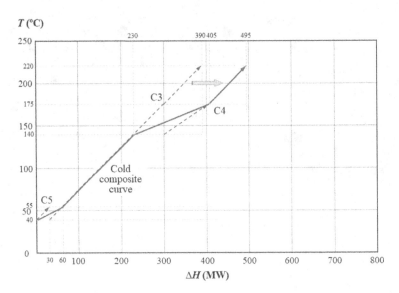

FIGURE 13.8
Construction of cold composite curve of Example 13.1 (for an interval of 175–220°C).

        In the highest interval 175–220°C, only the last segment of C3 (2 × (220 − 175) = 90 MW) exists. It is moved to the most right side of the diagram to form the last segment of the HRPD (see Figure 13.8). Note again that only horizontal shift is permitted for the segments.

  iii. *Overlaying the hot and cold composite curves to form the HRPD.*
      The composite curves are located on the same temperature versus enthalpy diagram. They are shifted horizontally, until the ΔT_{min} value of 10°C exists in between them, with the hot composite curve stays above cold composite curve entirely. As shown in the HRPD in Figure 13.9, the minimum hot utility (Q_{Hmin}) and cold utility (Q_{Cmin}) are identified as 100 and 130 MW, while the amount of energy recovery (Q_{REC}) is identified as 395 MW. The pinch temperatures are identified as 150°C and 140°C for this case.

  2. Determine the utility targets when the ΔT_{min} value is increased to 20°C.

      The cold composite curve is shifted further to the right until the ΔT_{min} value increases to 20°C. As shown in Figure 13.10, the minimum hot utility (Q_{Hmin}) and cold utility (Q_{Cmin}) increase to 140 and 170 MW, respectively. In other words, both utility targets increase by 40 MW, as compared to the earlier case with ΔT_{min} of 10°C (Figure 13.9). On the other hand, the amount of energy recovery (Q_{REC}) reduces by the same magnitude (i.e., 40 MW) to 355 MW. Note also that the new pinch temperatures for this case are identified as 160°C and 140°C (see Figure 13.10).

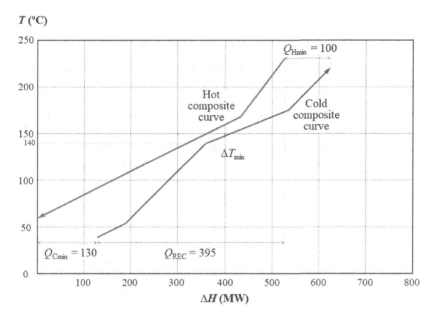

FIGURE 13.9
HRPD for of Example 13.1 with ΔT_{min} of 10°C.

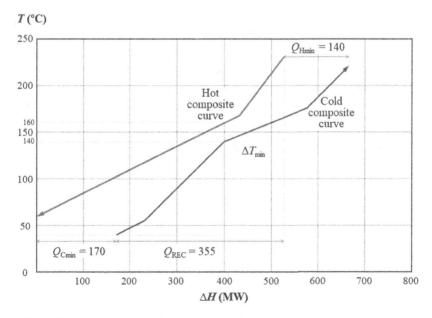

FIGURE 13.10
HRPD shows higher hot and cold utility targets for Example 13.1 with ΔT_{min} of 20°C.

13.3 Algebraic and Automated Targeting Methods

Even though the HRPD is an intuitive tool in setting utility targets ahead of detailed HEN design, its graphical nature makes it cumbersome for construction, especially when the problem involves many hot and cold streams. Besides, all graphical tools suffer from numerical accuracy. Hence, an alternative tool for setting the utility targets is through an algebraic method known as *problem table algorithm* (PTA) developed by Linnhoff and Flower (1978a, 1978b), which followed the pioneering work of Hohmann (1971).

Procedure to carry out the PTA is given as the following steps.

1. Determine the *shifted temperatures* (T^*) of all streams, based on the supply and target temperatures. For hot streams, their shifted temperatures will be colder by half of the ΔT_{min} value, following Equation 13.3. On the other hand, shifted temperatures of the cold streams will be hotter by half of the ΔT_{min} value, following Equation 13.4.

$$T^*_{Hot} = T - \frac{1}{2}\Delta T_{min} \quad \forall \text{hot streams} \tag{13.3}$$

$$T^*_{Cold} = T + \frac{1}{2}\Delta T_{min} \quad \forall \text{cold streams} \tag{13.4}$$

2. Setting up the *shifted temperature scale* by arranging all the shifted temperatures in descending order; all T^* values are placed in the first two columns of the *heat cascade table* (see the generic structure in Table 13.2). Within each interval, we then calculate the temperature difference using Equation 13.5, given in column 3 of the heat cascade table (see Table 13.2). All hot and cold streams are then located

TABLE 13.2

A Generic Heat Cascade Table

k	T^*	ΔT_k^*	H1	H2	C3	C4	ΔH_k	φ_k
1	T_1^*							φ_0
		ΔT_1^*	↓		↑		ΔH_1	
2	T_2^*		↓		↑			φ_1
...	...	ΔT_2^*	↓	↓	↑	↑	ΔH_2	...
$k-1$	T_{k-1}^*	...		↓		↑	...	φ_{k-1}
		ΔT_k^*		↓		↑	ΔH_k	
k	T_k^*							φ_k

within the shifted temperature intervals in the following columns, with hot streams pointing downwards and cold streams upwards.

$$\Delta T_k^* = T_k^* - T_{k+1}^* \quad \forall k \tag{13.5}$$

3. Perform the energy balance within each temperature interval k (ΔH_k). Within temperature interval k, the net CP values ($CP_{Net,k}$) among all hot streams ($CP_{Hot,k}$) and cold streams ($CP_{Cold,k}$) are determined, following Equation 13.6. We then calculate the net enthalpy within interval k following Equation 13.7. Doing this allows us to know if interval k experiences energy surplus or deficit. When an interval has a positive value of ΔH_k, it means that it has excess heat that has to be removed. On the other hand, negative ΔH_k value indicates that additional heat needs to be supplied to this interval to overcome its energy deficit.

$$CP_{Net,k} = \sum_{Hot} CP_{Hot,k} - \sum_{Cold} CP_{Cold,k} \quad \forall k \tag{13.6}$$

$$\Delta H_k = CP_{Net,k} \Delta T_k^* \quad \forall k \tag{13.7}$$

4. Perform *heat cascade* across the temperature intervals. Excess heat at higher temperature intervals may be cascaded (φ_k) to lower temperature interval, in order to remove the energy deficit. This is mathematically described by Equation 13.8. For a feasible heat transfer, the cascaded heat can only take a positive value, as described by Equation 13.9. Hence, a heat cascade is considered infeasible if any of the cascaded heat takes a negative value. To restore feasibility, the absolute value of the largest negative cascaded heat will be placed at the highest level (i.e., as φ_0), which serves as the minimum hot utility (Q_{Hmin}). The heat cascade exercise is then repeated. The last entry of the heat cascade indicates the minimum cold utility (Q_{Cmin}), while a cascaded heat that diminishes ($\varphi_k = 0$) is identified as the pinch of the given problem (T_{Pinch}).

$$\varphi_k = \varphi_{k-1} + \Delta H_k \quad \forall k \tag{13.8}$$

$$\varphi_k \geq 0 \quad \forall k \tag{13.9}$$

Apart from solving the PTA algebraically, one may also solve it via the *automated targeting model* (ATM)[1]. For such a case, the objective in Equation 13.10 is solved to minimize the minimum hot utility (Q_{Hmin}), subject to the constraints in Equations 13.6–13.9. This is a linear problem (LP) which may be solved to obtain a global optimum solution, if it exists.

$$\text{Minimize } \varphi_0 = Q_{Hmin} \tag{13.10}$$

---

[1] This was called the *transhipment model* originally reported by Papoulias and Grossmann (1983).

5. Determine the actual pinch temperatures. The actual pinch temperature of the hot streams ($T_{P,Hot}$) will be hotter by half of ΔT_{min} than the T_{Pinch} value (temperature corresponds to $\varphi_k = 0$, following Equation 13.11. On the other hand, the actual pinch temperature of the cold streams will be colder by half of ΔT_{min} than T_{Pinch}, given as in Equation 13.12.

$$T_{P,Hot} = T_{Pinch} + \frac{1}{2}\Delta T_{min} \quad \forall hot \ streams \qquad (13.11)$$

$$T_{P,Cold} = T_{Pinch} - \frac{1}{2}\Delta T_{min} \quad \forall cold \ streams \qquad (13.12)$$

Example 13.2 Algebraic and automated targeting

Revisit the chemical manufacturing process in Example 13.1. Use MS Excel to solve the following:

1. Determine the minimum hot and cold utility targets algebraically with PTA, for a ΔT_{min} value of 10°C.
2. Use ATM to determine the minimum utility targets for ΔT_{min} values of 30°C and 40°C.

SOLUTION

1. Determine the minimum hot and cold utility targets with PTA, for a ΔT_{min} value of 10°C.
   The five-step procedure is followed to determine the minimum utility targets.
   i. *Determine the shifted temperatures (T^*) of all streams.*
      Shifted temperatures of hot and cold streams are determined by Equations 13.3 and 13.4, respectively. All shifted temperatures are located in column 3 of Table 13.3 for the supply temperatures, while those for the target temperatures are found in the last column.
   ii. *Setting up the shifted temperature scale.*
      The shifted temperature scale is set up, by arranging the temperatures in descending order. This is shown in column C of the heat cascade table in MS Excel (see Figure 13.11). Next,

TABLE 13.3

Shifted Temperatures of Hot and Cold Streams

Streams	T_S (°C)	T_S^* (°C)	T_T (°C)	T_T^* (°C)
H1	230	225	60	55
H2	168	163	60	55
C3	40	45	220	225
C4	140	145	175	180
C5	40	45	55	60

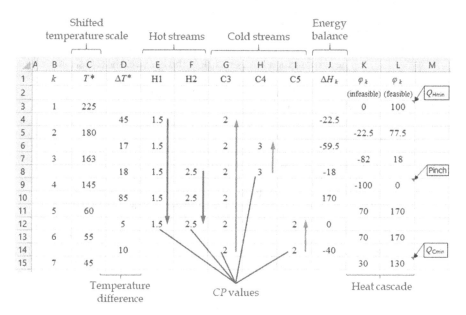

FIGURE 13.11
Solving PTA in MS Excel for Example 13.2.

the temperature difference for each interval is calculated using Equation 13.5, and located in column D. Between columns E and I, hot streams (pointing downwards) and cold streams (pointing upwards) are then located within their shifted supply and target temperatures. The CP values are inserted in the cells, within their respective temperature intervals.

iii. *Perform the energy balance within each temperature interval*

In column J, the net enthalpy values are calculated within interval k following Equations 13.6 and 13.7. As shown in Figure 13.11, negative enthalpy values are found in the first three intervals, as well as in the last one, indicating that energy deficits are experienced in these intervals. On the other hand, excess energy with positive enthalpy values is observed in the fourth and fifth intervals.

iv. *Perform heat cascade across the temperature intervals*

v. *Determine the actual pinch temperatures.*

Heat cascade is conducted in column K, with a negative value observed in between temperature levels of 180°C and 145°C. In other words, this is an infeasible heat cascade. The largest negative cascaded heat is identified at 145°C, i.e., −100 MW. Its absolute value of this negative cascaded heat is placed at the highest level (φ_0) so to restore the feasibility of the heat cascade, this feasible heat cascade is carried out in column L. The minimum hot and cold utilities are identified in the first and last entries, i.e., 100 (Q_{Hmin}) and 130 MW (Q_{Cmin}). The pinch temperature (T_{Pinch}) where the cascaded heat diminishes is identified as 145°C.

The actual pinch temperature of the hot streams ($T_{P,Hot}$) is calculated as 150°C following Equation 13.11, while the cold stream pinch temperature ($T_{P,Cold}$) is calculated as 140°C following Equation 13.12.

2. Solve the ATM to determine the minimum utility targets for ΔT_{min} values of 30°C and 40°C.

Steps 1–3 are repeated as described earlier. Note that the shifted temperatures in step 1 are different than those in the earlier case. In step 4, the heat cascade is conducted using ATM. The model is solved using MS Excel, with the objective in Equation 13.10 implemented using Solver (see setting in Figure 13.12). For ΔT_{min} values of 30°C, the minimum utility targets are identified as 176 (Q_{Hmin}) and 206 MW (Q_{Cmin}), respectively, as shown in Figure 13.13. The actual pinch temperatures are then calculated as 168°C ($T_{P,Hot}$) and 138°C ($T_{P,Cold}$) using Equations 13.11 and 13.12.

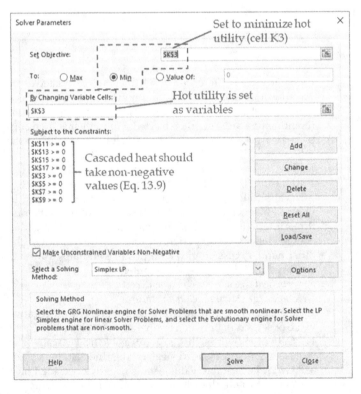

FIGURE 13.12

Setting of Excel Solver for solving ATM in Example 13.2.

A	B	C	D	E	F	G	H	I	J	K	L
1	k	T*	ΔT*	H1	H2	C3	C4	C5	ΔHₖ	φₖ	
2											Q_{Hmin}
3	1	235								176	
4			20			2			-40		
5	2	215								136	
6			25	1.5		2			-12.5		
7	3	190								123.5	
8			35	1.5		2	3		-122.5		
9	4	155								1	
10			2	1.5		2			-1		Pinch
11	5	153								0	
12			83	1.5	2.5	2			166		
13	6	70								166	
14			15	1.5	2.5	2		2	0		
15	7	55								166	
16			10	1.5	2.5				40		Q_{Cmin}
17	8	45								206	

FIGURE 13.13
Results of ATM in Example 13.2 ($\Delta T_{min} = 30°C$).

The same procedure is repeated for the case with ΔT_{min} of 40°C, the minimum utility targets are observed to increase to 196 (Q_{Hmin}) and 226 MW (Q_{Cmin}), respectively, as shown in Figure 13.14 The actual pinch temperatures are then calculated as 168°C ($T_{P,Hot}$) and 128°C ($T_{P,Cold}$). It can be observed that the HEN will have higher hot and cold utility targets with the larger ΔT_{min} value (as compared to thatin Example 13.1).

A	B	C	D	E	F	G	H	I	J	K	L
1	k	T*	ΔT*	H1	H2	C3	C4	C5	ΔHₖ	φₖ	
2											Q_{Hmin}
3	1	240								196	
4			30			2			-60		
5	2	210								136	
6			15	1.5		2			-7.5		
7	3	195								128.5	
8			35	1.5		2	3		-122.5		
9	4	160								6	
10			12	1.5		2			-6		Pinch
11	5	148								0	
12			73	1.5	2.5	2			146		
13	6	75								146	
14			15	1.5	2.5	2		2	0		
15	7	60								146	
16			20	1.5	2.5				80		Q_{Cmin}
17	8	40								226	

FIGURE 13.14
Results of ATM in Example 13.2 ($\Delta T_{min} = 40°C$).

13.4 Basics of the Pinch Design Method (PDM)

To synthesize an HEN, a convenient tool for use is the *grid diagram* (Figure 13.15). As shown, the grid diagram displays the hot and cold streams without the process units. All hot streams are represented by an arrow point from right to the left, while cold streams in the reverse direction[2]. Their supply and target temperatures are indicated by values above the stream. When a heat exchanger is to be located on the grid diagram, it is represented by a pair of cycles linked with a vertical line. The energy recovered from the hot stream to the cold stream is indicated by the value below the heat exchanger. Note that the pinch temperature divides the HEN into regions above and below the pinch. Hence, it is convenient to carry out the HEN design in these regions separately.

To ensure the HEN achieves the utility targets (identified by HRPD or PTA), the PDM has the following important features (Linnhoff and Hindmarsh, 1983; Smith, 2016):

- *Start design at the pinch*—The pinch is the most constrained area in the HEN. Hence, pairing of hot and cold streams should first be carried out at this region, as their choices are limited. Once heat exchangers are located to pair a hot and cold streams at the pinch, we can then move away to pair other streams.

- *CP inequality*—An observation of the HRPD in Figure 13.16a reveals that in the region below the pinch, segment of the cold composite curve are generally steeper than those of the hot composite curve. This is the main rational behind this important feature. As the slope of segment corresponds to the inverse of *CP* value (readers may want to prove this), this leads to the inequality rule in Equation 13.13.

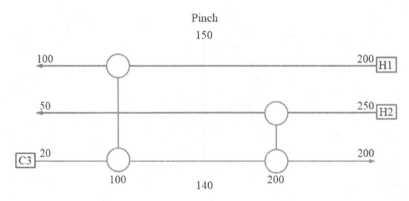

FIGURE 13.15
The grid diagram for HEN synthesis.

---

[2] Note that hot and cold streams are commonly drawn in reverse direction in other HEN books, but the calculation steps remain identical.

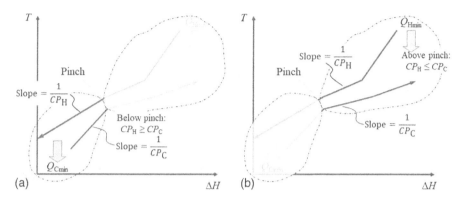

FIGURE 13.16
CP inequality for regions: (a) below, and (b) above the pinch.

Hence, the latter should be followed, where hot stream of higher CP value is to be paired with cold stream (of lower CP value).

$$CP_H \geq CP_C \qquad (13.13)$$

On the other hand, Figure 13.16b shows that in the region above the pinch, segment of the hot composite curve tend to be steeper than those of the cold composite curve. Hence, hot stream to be paired to the cold stream in this region should have a steeper slope (i.e., lower CP value) than the cold stream, following the inequality in Equation 13.14.

$$CP_H \leq CP_C \qquad (13.14)$$

Note that both inequality rules in Equations 13.13 and 13.14 are only applicable for heat exchanger matches at the pinch, i.e., where both its hot and cold streams are touching the pinch temperature. For heat exchangers away from the pinch (either or both of its hot and cold streams are not touching the pinch temperature), these inequality rules are not necessary, as there is more flexibility for placing the match, and with temperature difference bigger than ΔT_{min}.

- To facilitate the compliance of the CP inequality rules, it is useful to make use of the CP *table* during the pairing of hot and cold streams at the pinch. CP values of hot and cold streams to be paired (at the pinch) are arranged in descending order in the CP table, so that the CP inequality rules in Equations 13.13 and 13.14 are observed during the pairing. An example of the CP table is shown in Figure 13.17, where H1 with a larger CP value (than H2) is paired with C3, in order to observe Equation 13.14.

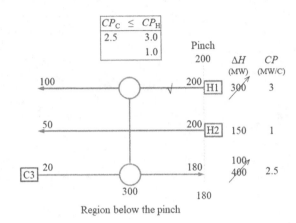

FIGURE 13.17
Use of *CP* table and the "tick-off" heuristic.

- *The tick-off heuristic*—In synthesizing the HEN, it is important to keep the cost to the minimum. Hence, it is desired to have a smaller number of heat exchanger units. This can be achieved by recovering the maximum heat when a heat exchanger is paired on the hot and cold streams. However, due to the enthalpy availability (hot stream) and requirement (cold stream), the heat to be recovered should be maximized, but cannot be bigger than that possessed by the two streams that are paired. For instance, in Figure 13.17, hot stream H1 has an enthalpy of 300 kW. When H1 is to be paired with cold stream C3 with enthalpy of 400 kW, the paired heat exchanger can only transfer a total of 300 kW. Doing so allows H1 to reach its target temperature. Hence, H1 is now ticked off, as its enthalpy is fulfilled completely.

The basic steps of PDM are given as follows:

1. Draw the grid diagram, with pinch temperature dividing the HEN into regions below and above pinch.
2. Pair hot and cold streams that are located at the pinch with heat exchangers, guided by *CP* inequality and *CP* table. Apply tick-off heuristic to maximize heat recovery for the two streams to be paired.
3. Calculate the intermediate temperature for stream where its enthalpy is partially satisfied, using the following equation:

$$\Delta H = CP(T_{HE} - T_{CE}) \tag{13.15}$$

where T_{HE} and T_{CE} are stream temperatures at the hot and cold ends, respectively. For instance, cold stream C3 in Figure 13.17 has a total enthalpy requirement of 400 MW. Upon the pairing with H1, its enthalpy is partially fulfilled, with the heat exchanger transferring a total heat of 300 MW. Hence, the intermediate temperature at the exchanger cold end is calculated using Equation 13.15 as 140°C.

Steps 2 and 3 are to be repeated for all streams at the pinch before step 4 is attempted.

4. Pair other hot and cold streams away from the pinch; apply tick-off heuristic. In region below the pinch, all cold streams are to be exhausted. In other words, their enthalpy requirements are fulfilled by hot streams through heat exchange, as no heating (with hot utility) is permitted in this heat source region that has excess energy (see discussion in Section 13.2). On the other hand, all hot streams in region above the pinch are to be exhausted, i.e., their enthalpies are fully transferred to cold stream in this region. Doing this will ensure no cooling (with cold utility) is carried out in this heat sink region that experiences energy deficit. Once a heat exchanger is paired to recover heat between hot and cold streams, Equation 13.15 is used to determine the intermediate temperatures for stream where its enthalpy is partially satisfied.

5. Use hot or cold utility to satisfy the remaining enthalpy requirement of the streams, so that they can reach their respective target temperatures. Note that cold utility is only permitted in region below the pinch (heat source), while hot utility in region above the pinch (heat sink). The total hot/cold utility used for the HEN should match the utility targets identified using the HRPD or PTA.

6. Form a complete HEN by joining both regions below and above the pinch. Perform a temperature profiles check to ensure all heat exchangers have feasible temperature profiles for their hot and cold streams.

The overall procedure of the PDM is shown in Figure 13.18, and next demonstrated with an example.

Example 13.3

The chemical manufacturing process in Example 13.1 is revisited. The minimum hot and cold utility targets have been identified as 100 and 130 MW, respectively, for a ΔT_{min} of 10°C (see Examples 13.1 and 13.2). Design the HEN that achieves the utility targets using the PDM.

SOLUTION

1. *Draw the grid diagram.*
   The grid diagram for this problem is shown in Figure 13.19. As shown, the pinch temperature divides the HEN into regions above and below the pinch. Both hot streams H1 and H2, as well as cold stream C3 are found in both regions; in other

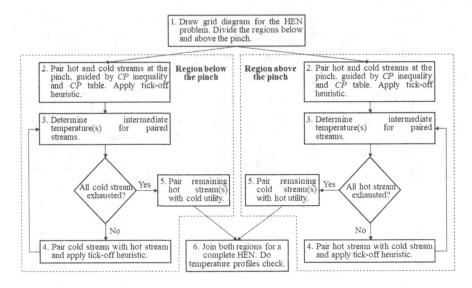

FIGURE 13.18
Overall procedure of the PDM.

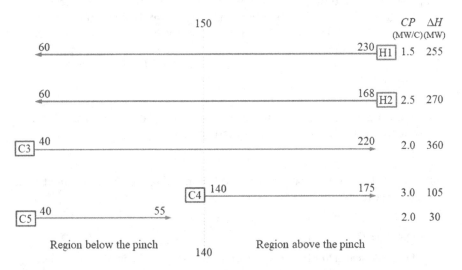

FIGURE 13.19
Grid diagram of the HEN problem of the chemical manufacturing process (all temperatures are given in °C).

words, they cut across the pinch temperatures. However, cold stream C4 starts from cold pinch temperature (140°C) and is only found in the region above the pinch. On the other hand, cold stream C5 is found in region below the pinch; note also that this stream does not reach the pinch temperature. CP and

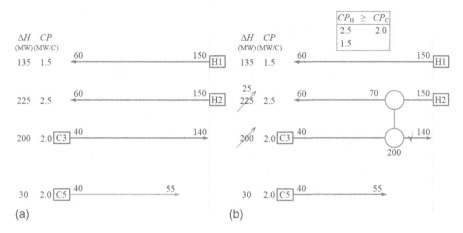

FIGURE 13.20
(a) Grid diagram for region below the pinch (step 1); (b) pairing of H2 and C3 (steps 2 and 3; temperatures are given in °C).

ΔH values of all streams are given at the right side of the grid diagram.

We first perform stream matching for region below the pinch. The grid diagram for this region is given in Figure 13.20a. Note that the ΔH values of all streams are recalculated using Equation 13.1. Since only part of the HEN is involved, the ΔH values of streams are generally smaller than those in the complete grid diagram (Figure 13.19). Exceptional case is observed for C5, as this stream is only found in this region below the pinch.

We may perform a simple enthalpy balance using the grid diagram in Figure 13.20a. As shown, two hot streams are found in this region, i.e., H1 and H2 with a total enthalpy of 360 MW (= 135 + 225 MW). This amount of enthalpy has to be removed so that the hot streams can reach their target temperatures of 60°C. Besides, two cold are observed, i.e., C3 and C5; they require a total enthalpy of 230 MW (= 200 + 30 MW) in order to reach their target temperatures. In other words, there is a net excess enthalpy of 130 MW (= 360 – 230 MW) in this region, if the hot streams can fully recover their heat to the cold streams. Cold utility is necessary for the removal of the excess enthalpy of 130 MW, so that all streams will reach their target temperatures.

2. *Pair hot and cold streams at the pinch with heat exchangers, guided by CP inequality and CP table. Apply tick-off heuristic.*

3. *Calculate intermediate temperature for stream where its enthalpy is partially satisfied.*

Following the PDM, hot and cold streams at the pinch are to be paired in priori, since it is the most constrained area in the HEN. As shown in Figure 13.20a, these stream candidates to be paired include H1, H2, and C3. Note that C5 is excluded from this step, as it is not located at the pinch (i.e., it does not reach the pinch temperature).

In Figure 13.20b, *CP* table is added above the grid diagram, with descending order of *CP* values of all stream candidates. The *CP* table indicates that cold stream C3 (*CP* = 2.0 MW/°C) can only be paired with hot stream H2 (*CP* = 2.5 MW/°C), and not with H1 that has a lower *CP* value (1.5 MW/°C). Hence, a heat exchanger unit is added to recover heat from H2 to C3. This heat exchanger will recover 200 MW of heat, i.e., smaller among the enthalpy values of the two candidates. Doing this allows us to "tick off" C3, which means that its enthalpy requirement is completely fulfilled (see Figure 13.20b). We then update the new enthalpy of H2 (225 − 200 = 25 MW), and make use of Equation 13.15 to calculate its intermediate temperature at the cold end of the heat exchanger, i.e., 70°C (= 150 − 200/2.5°C).

4. *Pair other hot and cold streams away from the pinch; apply tick-off heuristic.*

3. *Calculate intermediate temperature for stream where its enthalpy is partially satisfied.*

Since the stream candidates at the pinch are paired, we next move to analyze other streams away from the pinch. Figure 13.20b shows that both H1 and H2 consist of heat that are to be removed, while C5 requires heating. Note that in this region below the pinch, no hot utility should be used to provide heating of C5. Hence, its enthalpy requirement is to be fulfilled by H1 and/or H2. Among the two candidates, H1 is a better choice to be paired with C5, as it has sufficient enthalpy (135 MW) needed by the latter (30 MW). Hence, a heat exchanger is paired to recover 30 MW of heat from H1 to C5, which allows us to tick off C5 (Figure 13.21a). We then update the enthalpy of H1, i.e., 105 MW (= 135 − 30 MW). Equation 13.15 is next used to calculate the intermediate temperature of H1 at the cold end of the exchanger, i.e., 130°C (= 150 − 30/1.5; see Figure 13.21a).

5. *Use hot or cold utility to satisfy the remaining enthalpy requirement of the streams.*

Since all cold streams in the region are exhausted, the remaining enthalpy of H1 (105 MW) and H2 (25 MW) are to be removed using cold utility. Two coolers (i.e., cooling exchanger) are added at the left most of the grid diagram (Figure 13.21b), with their duties corresponding to the remaining enthalpy of these streams (Figure 13.21b). Doing this allows H1 and H2 to be ticked off, as their enthalpy is fully fulfilled. Note that Equation 13.15 may be used to verify the cooling duty of these coolers. For instance, the cooling duty of the H1 cooler can be calculated

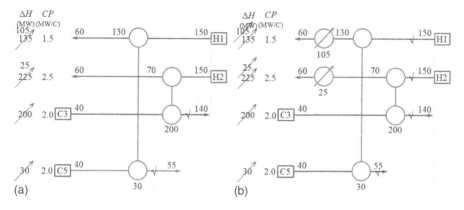

FIGURE 13.21
(a) Pairing of heat exchangers away from the pinch (step 4); (b) Use cold utility to satisfy the remaining enthalpy requirement of the streams (step 5; all temperatures given in °C).

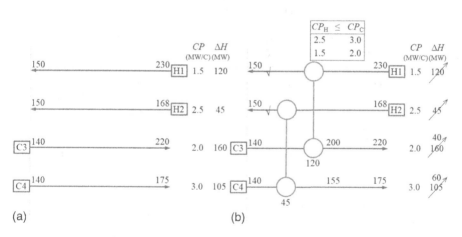

FIGURE 13.22
(a) Grid diagram for region above the pinch (step 1); (b) pairing of H1-C3 and H2-C4 (steps 2 and 3; temperatures are given in °C).

as 105 MW (= 1.5 × (130 – 60)). Similar calculations are applicable for H2 cooler.

We next move to analyze region above the pinch. Figure 13.22a shows that this region has two hot streams, i.e., H1 and H2 with a total enthalpy of 165 MW (= 120 + 45 MW). Besides, two cold streams, i.e., C3 and C4 are also observed, with a total enthalpy of 265 MW (= 160 + 105 MW). It can be observed that the total enthalpy of the hot streams is insufficient to fulfill the enthalpy requirement of the cold streams. Hence, hot utility of 100 MW (265 – 165 MW) is needed in this region.

Another important observation is that the enthalpy of each stream in this region may be added to those in the region below the pinch to match their total values in the complete grid diagram. For instance, H1 has an enthalpy of 120 MW above the pinch (Figure 13.22a) and 135 MW below the pinch (Figure 13.21a); these values are added to 255 MW (= 120 + 135 MW), matching the total enthalpy of H1 in Figure 13.20.

2. *Pair hot and cold streams at the pinch with heat exchangers, guided by CP inequality and CP table. Apply tick-off heuristic.*
3. *Calculate the intermediate temperature for stream where its enthalpy is partially satisfied.*

Similar to earlier case, hot and cold streams at the pinch are first to be paired with heat exchanger(s), since it is the most constrained area in the HEN. As shown in Figure 13.22a, all H1, H2, C3, and C4 are stream candidates that will be considered, as they are either reaching (H1 and H2) or departing from (C3 and C4) the pinch temperatures.

The *CP* table in Figure 13.22b indicates that H1 is to be paired with C3, while H2 with C4. In other words, two heat exchangers are paired at the pinch. Tick-off heuristic is applied for H1 and H2, as their available heat is recovered completely to the cold streams. The enthalpy values of C3 (160 – 120 = 40 MW) and C4 (105 – 45 = 60 MW) are updated, while Equation 13.15 is used to calculate the intermediate temperatures at the hot side of these exchangers, i.e., 200°C (C3) and 155°C (C4), as shown in Figure 13.22b.

4. *Pair other hot and cold streams away from the pinch; apply tick-off heuristic.*
5. *Use hot or cold utility to satisfy the remaining enthalpy requirement of the streams.*

Since the enthalpy of both H1 and H2 are removed completely, step 4 is not necessary. In step 5, hot utility is used to fulfill the enthalpy requirement of the cold streams. As shown in Figure 13.23, two heaters (i.e., heating exchangers) are added at the rightmost of the grid diagram, which allows both streams to be ticked off. For instance, C3 requires a heating duty of 40 MW, which is now supplied by the heater so that it can reach its target temperature of 220°C. Note that Equation 13.15 may be used to verify the heating duty of these heater (i.e., 2.0 × (220 – 200) = 40 MW). Similar calculations are applicable for C4.

6. *Form a complete HEN and perform a temperature profile check*

Finally, grid diagrams of both regions are joined to form a complete HEN, as shown in Figure 13.24. A temperature profile check is performed for all heat exchangers, to ensure that the

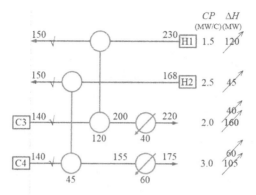

FIGURE 13.23
Use hot utility to satisfy the remaining enthalpy requirement of the streams for region above the pinch (step 5; all temperatures given in °C).

ΔT_{min} is observed, especially for heat exchangers at the pinch. For instance, exchanger H2-C4 has a ΔT_{min} of 10°C at its cold end, as it is located at the pinch. At its hot end, the ΔT_{min} is observed to be larger, i.e., 13°C (= 168 – 155°C). On the other hand, exchanger H1-C5 has a much larger ΔT_{min}, i.e., 95°C (= 150°C – 55°C; hot end) and 90°C (= 130°C – 40°C; cold end) due to its location that is away from the pinch.

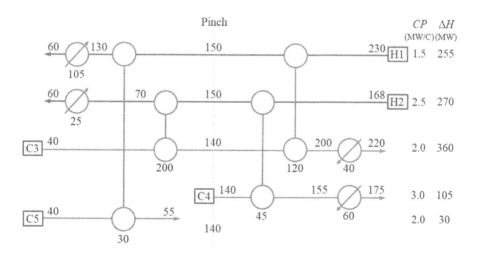

FIGURE 13.24
A complete HEN for Example 13.3 (all temperatures given in °C).

13.5 Stream Splitting in PDM

The PDM discussed earlier is useful in ensuring the hot and cold utility targets be met while the HEN is synthesized. However, in some cases, the above outlined procedure may be infeasible in generating an HEN that fulfills all criteria. For such cases, stream splitting may be considered. Illustrative cases are shown next.

In Figure 13.25, cold stream C3 has to be paired with hot streams H1 and/or H2 in order to raise its temperature. However, its *CP* value (3.5 kW/°C) is larger than those of H1 (3.2 kW/°C) and H2 (1 kW/°C). In other words, the *CP* inequality constraint in Equation 13.14 is not fulfilled. If one were to pair H1 and H2 with C3, temperature infeasibility will result for the paired heat exchangers. Note also that in region below the pinch, no hot utility is permitted for the heating of cold streams. Hence, to fulfill its enthalpy requirement, C3 is split in order to be paired with H1 and H2 in a parallel configuration. One possible solution is shown in Figure 13.25, where the enthalpy of C3 and H2 (160 kW) are simultaneously fulfilled, while H1 will have its remaining enthalpy (32 kW) to be removed using cooling utility. The *CP* values of the C3 branches may be calculated using Equation 13.15. The C3 branch that is paired with H2 will have *CP* value of 1.0 kW/°C (= 160/(180 − 20) kW/°C), while the branch that is paired with H1 will have *CP* value of 2.5 kW/°C (= 400/(180 − 20) kW/°C).

Apart from the constraint of *CP* inequality, stream splitting is also necessary when the number of streams is limited at the pinch region. Figure 13.26 shows such a situation. As shown, cold stream C3 has *CP* value larger than

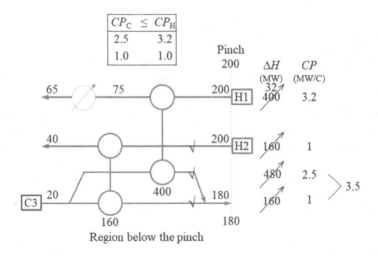

FIGURE 13.25
Splitting of cold stream C3 below the pinch (all temperatures given in °C, enthalpy in MW).

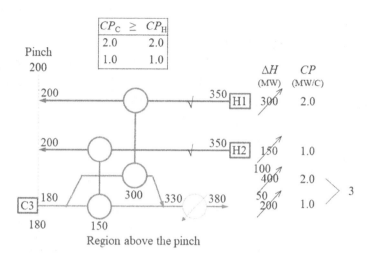

FIGURE 13.26
Splitting of cold stream C3 above the pinch (all temperatures given in °C, enthalpy in MW).

both hot streams H1 and H2. To maintain temperature feasibility, C3 is split to be paired with H1 and H2. One possible solution is shown in Figure 13.26, where the branches of C3 will have equal *CP* values with H1 and H2. Doing this fulfill the enthalpy requirement of the hot streams, while the remaining duty of C3 (150 kW) will be fulfilled using heater.

A generic guide for the splitting of streams at the pinch region is shown in Figure 13.27.

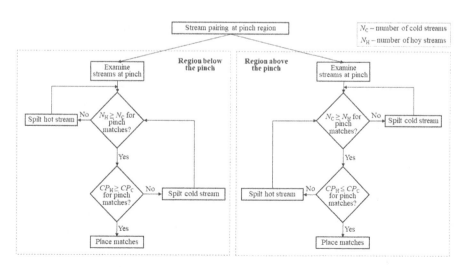

FIGURE 13.27
Flowchart for splitting of streams at the pinch region.

TABLE 13.4

Data for Additional Hot Streams for Example 13.4

Streams	Supply Temperature T_S (°C)	Target Temperature T_T (°C)	Mass Capacity Flowrate, CP (MW/°C)	Enthalpy Change, ΔH (MW)
H3	230	150	0.5	40

Example 13.4

Example 13.3 is revisited. However, in this case, the chemical manufacturing process is assumed to have an additional hot stream (e.g., boiler flue gas) where its heat may be recovered, with data given in Table 13.4. Targeting shows that the HEN will have a lower hot utility target of 60 kW, while its cold utility will remain at 130 kW. The hot and cold pinch temperatures are identical as before, i.e., 150°C and 140°C. Re-design the HEN in order to fulfill its hot and cold utility targets.

SOLUTION

As the new hot stream H3 has a higher temperature than the hot pinch temperature (150°C), this stream is located in the region above the pinch. As shown in Figure 13.28, there are three hot streams in this region, while only two cold streams are eligible for pairing. Hence, cold stream C3 is split so as to be paired with H1 and H3. Doing this fulfill the enthalpy requirement of these streams.

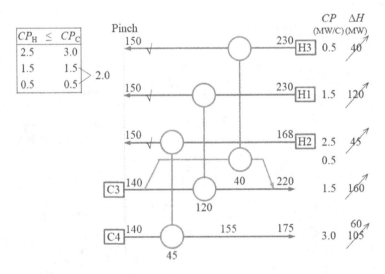

FIGURE 13.28

Splitting of cold stream C3 to be paired with H1 and H3 (Example 13.4; all temperatures given in °C, enthalpy in MW).

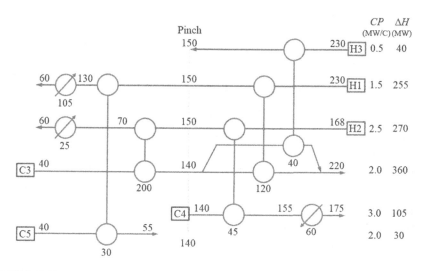

FIGURE 13.29
A complete HEN for chemical manufacturing process in Example 13.4 (all temperatures given in °C, enthalpy in MW).

The remaining heat exchangers are identical to those in Example 13.3. The complete HEN is shown in Figure 13.29. As shown, the HEN has hot and cold utility targets of 60 kW and 130 kW.

13.6 Utility Selection Using Grand Composite Curve (GCC)

In the process plants, hot utilities such as steam and hot oil are commonly used to provide heating needs. On the other hand, cold utility such as cooling water is commonly used for heat removal in the plant. In earlier sections, HRPD and PTA were introduced to determine the utility targets for a HEN problem. However, these tools are not appropriate in the selection of hot and cold utilities. A better tool for utility selection is the *grand composite curve* (GCC). The GCC is constructed by plotting shifted temperatures versus the cascaded heat in the PTA.

A generic GCC is shown in Figure 13.30. A few features of the GCC are worth mentioning. As shown, the hot utility target is given by the opening at the top section of the GCC, while cold utility is indicated by the opening at the bottom. Note that the hot and cold utility targets are identical to those determined by the HRPD and PTA. The pinch is identified at the point where the GCC touches the *y*-axis. When the GCC segments are moving to the right, it means that excess heat is generated from the process, while heat is consumed when the segments are moving to the left. Hence, for sections with

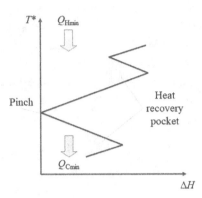

FIGURE 13.30
A generic GCC.

segments moving to the right and to the left, the "heat recovery pocket" is formed, where process-to-process heat recovery takes place (see Figure 13.30). Procedure for using GCC for utility selection is given as follows:

1. Construct the GCC by plotting shifted temperatures (T^*) versus the cascaded heat (φ_k) in the PTA (Figure 13.30).

2. Determine the shifted temperatures for hot and/or cold utilities. Since the GCC is plotted using shifted temperatures, hot and/or cold utilities have to be plotted using the shifted temperatures too. In this regard, hot utility is treated as hot stream. Hence, their shifted temperature is colder by half of the ΔT_{min} value, which is calculated using Equation 13.3. On the other hand, shifted temperatures of the cold utility will be hotter by half of the ΔT_{min} value, calculated using Equation 13.4.

3. To determine the usage of hot/cold utility:

   a. Steam as hot utility—saturated steam of lower pressure and temperature is preferred than those at higher temperature, due to cost and safety considerations. In other words, low pressure steam (LPS) is preferred than medium (MPS), and the latter is preferred over high-pressure steam (HPS). To determine the minimum amount of steam use, the preferred steam (e.g., LPS) is first placed at its shifted temperature as a horizontal segment, and is being maximized until it touches the GCC or its heat recovery pocket, as shown in Figure 13.31a. This steam level will provide heating to the opening of the GCC beneath it. The same procedure is repeated for the next preferred steam (e.g., MPS), and will cover the remaining parts of the GCC where heating is needed (see Figure 13.31a). The heating load requirement(s) is given by the horizontal distance of the steam segment(s).

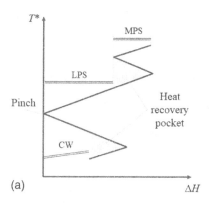
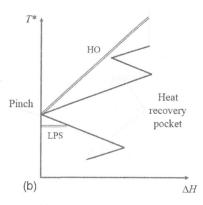

FIGURE 13.31
GCC for (a) steam and cooling water placement; (b) hot oil and steam generation.

    b. Heating medium such as hot oil may be used to heat process units that require higher operating temperature (than steam). To determine the minimum flowrate of the heating medium, the latter is plotted as a supply line that starts its supply temperature (Figure 13.31b). The latter is used as a pivot, where the heating medium line is drawn with the steepest slope until it touches and stay above the GCC entirely (or bound by its minimum return temperature). The minimum flowrate of the heating medium may be calculated from the inverse slope of the heating medium line.

    c. Cold utility—higher temperature utility such as cooling water (CW) is usually preferred than those at lower temperature (e.g., refrigeration) due to cost consideration. To determine the minimum usage, the preferred utility is plotted as a supply line from its supply temperature, which serves as the pivot point. The supply line is drawn with the steepest slope and stays below the GCC entirely (but bound by its maximum return temperature). The minimum usage of the cooling water is determined from the inverse slope of the supply line (Figure 13.31a).

    d. Steam generation—if the rejected heat below the pinch has a high temperature (typically higher than 110°C), steam generation may be considered, instead of rejecting the heat to cooling utility. If saturated steam is to be generated, the steam level is placed below the GCC as a horizontal segment (at its shifted temperature), and is being maximized until it touches the GCC segment, or its heat recovery pocket (Figure 13.31b). The amount of load that may be used for steam generation is given by the horizontal distance of the steam segment.

The use of GCC for utility selection is demonstrated using the next example.

Example 13.5

For the chemical manufacturing process (Example 13.3), the utility targets have been identified as 100 and 130 MW for its hot and cold utility targets. Use the GCC for utility selection, and modify the HEN configuration with insights from GCC.

1. For hot utility, available for service are HPS at 240°C (34 bar), and MPS at 185°C (11.3 bar). Determine the load of the steams, with priority given to steam of lower pressure. For cold utility, available for service is cooling water at 35°C. Determine the minimum flowrate of cooling water given that its return temperature of 55°C and a heat capacity of 4.2 kJ/kg.°C.
2. Instead of steam, hot oil is used as heating medium. Determine the minimum flowrate and return temperature of the hot oil, given that it has a supply temperature of 285°C and a heat capacity of 2 kJ/kg.°C. Apart from using cooling water, it is desired to generate LPS at 120°C (2 bar) from this process. Determine the amount of LPS that may be generated, given that its enthalpy is 2202 kJ/kg.

SOLUTION

1. For hot utility, available for service are HPS at 240°C (34 bar), and MPS at 185°C (11.3 bar). Determine the load of the steams, with priority given to steam of lower pressure. For cold utility, available for service is cooling water at 40°C. Determine the minimum flowrate of cooling water given that its return temperature of 60°C and a heat capacity of 4.2 kJ/kg.°C.

   i. *Construct the GCC*

   The PTA results are shown in Figure 13.11. It is used to construct the GCC by plotting its shifted temperatures (T^*; column C) versus the cascaded heat (φ_k; column L). The resulting GCC is shown in Figure 13.32. As shown, the hot utility target (Q_{Hmin}) is determined as 100 MW, while cold utility target (Q_{Cmin}) is identified as 130 MW. The shifted pinch temperature is identified at 145°C. The "heat recovery pocket" is also labeled on the GCC in Figure 13.32.

   ii. *Determine the shifted temperatures for hot and/or cold utilities.*

   iii. *To determine the usage of steam and cooling water.*

   The shifted temperatures of HPS and MPS are first determined using Equation 13.3, i.e., 235°C (=240°C − 5°C) and 180°C (=185°C − 5°C), respectively. On the other hand, shifted temperatures of the cooling water are calculated using Equation 13.4 as 40°C (= 35°C + 5°C) and 60 (= 55°C + 5°C), for its supply and target temperatures.

   Since MPS is preferred over HPS, the steam level of the former is placed above the GCC at its shifted temperature of 180°C, and is being maximized until it touches the GCC.

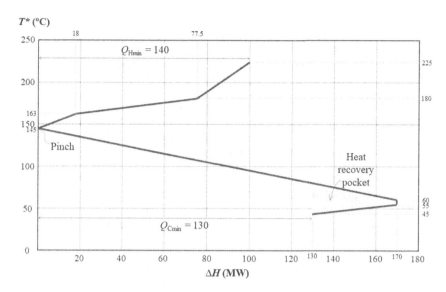

FIGURE 13.32
GCC for the chemical manufacturing process in Example 13.5.

As shown in Figure 13.33, the heating load served by MPS is identified as 77.5 MW. Next, the steam level of HPS is placed at its shifted temperature of 235°C, and is being maximized until covering the remaining opening of the GCC. Figure 13.33, shows that it serves a total heating load of 22.5 MW (=100 – 77.5 MW).

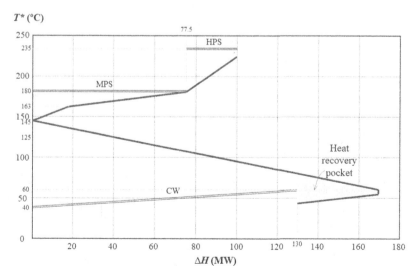

FIGURE 13.33
GCC for optimum use of steam and cooling water in Example 13.5.

On the other hand, the cooling water is placed at its shifted supply (40°C) and target temperatures (60°C) below the GCC, and covers the entire cooling load of 130 MW. It is then determined that the cooling water flowrate may be calculated as 1547.6 kg/s (= 130,000/4.2/20 kg/s).

2. Instead of steam, hot oil is used as heating medium. Determine the minimum flowrate and return temperature of the hot oil, given that it has a supply temperature of 285°C and a heat capacity of 2.0 kJ/kg.°C. Apart from using cooling water, it is desired to generate LPS at 120°C (2 bar) from this process. Determine the amount of LPS that may be generated, given that its enthalpy is 2202 kJ/kg.

    i. *Construct the GCC.*

    ii. *Determine the shifted temperatures for hot and/or cold utilities.*

    iii. *To determine the usage of hot oil, steam generation, and cooling water.*

The same GCC in Figure 13.32 may be used. Next, the shifted supply temperature of the hot oil (HO) is determined as 280°C (= 285°C – 5°C) using Equation 13.3. On the other hand, as LPS is to be generated using the access heat, it is treated as a cooling utility, and its shifted temperature is calculated using Equation 13.4 as 125°C (= 120°C + 5°C), while the shifted temperatures of cooling water remain identical.

We first examine the use of HO as hot utility. As shown in Figure 13.34, the HO supply covers the entire heating requirement of 100 MW from its shifted supply temperature (280°C), which is used as a pivot. In order to keep its flowrate to the minimum, it is drawn with the steepest slope but

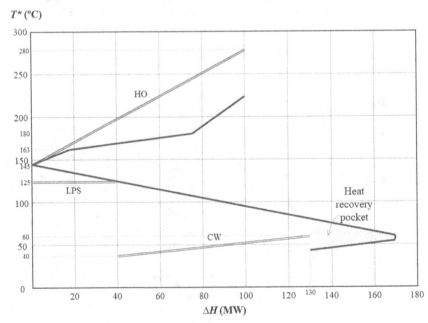

FIGURE 13.34
GCC for optimum use of hot oil, steam generation, and cooling water in Example 13.5.

yet stays above the GCC entirely. As shown in Figure 13.34, the HO supply line touches the pinch, and hence has a shifted return temperature of 145°C (or actual temperature of 150°C). The minimum flowrate of the heating medium may be calculated from the inverse slope of the supply line as 370.4 kg/s (= 100,000/2/(280 – 145) kg/s).

In the region below the pinch, LPS generation is preferred over the use of cooling water, as it may be sold (e.g., to neighbor plant) for profit, or used as preheating sources (e.g., for boiler feed water) for cost saving. The LPS is first located at its shifted supply temperature of 125°C, and is being maximized until it touches the GCC. As shown in Figure 13.34, the load that may be used for LPS generation is identified as 40 MW, which corresponds to LPS of 18.2 kg/s (= 40,000/2202 kg/s).

Next, cooling water (CW) is placed at its shifted temperatures of 40°C and 60°C, and will cover the remaining cooling requirement of the GCC. Figure 13.34 shows that this cooling load corresponds to 90 MW (= 130 – 40 MW). This corresponds to a flowrate of 1071.4 kg/s (= 90,000/4.2/(60–40) kg/s). In other words, lower flowrate of cooling water is needed for this case, as compared to earlier case without steam generation (Figure 13.33).

It is useful to re-design the HEN in Figure 13.24 by incorporating the heating and cooling exchangers. Attention should be paid to the temperature profiles of the heat exchangers. The HEN for the case with hot oil (HO), cooling water, and steam generation is shown in Figure 13.35. As shown, C4 is split in order to be paired with H2 and hot oil supply, in order to ensure feasible heat transfer. The final process flow diagram with a complete HEN structure is given in Figure 13.36.

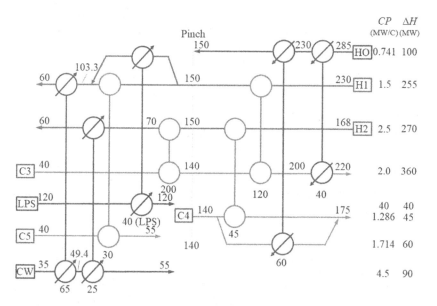

FIGURE 13.35
HEN design with the pairing of heating and cooling heat exchangers.

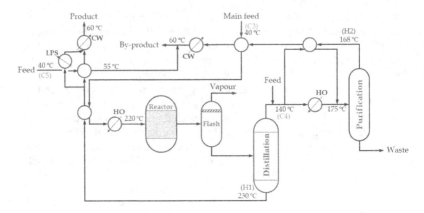

FIGURE 13.36
Process flow diagram with HEN structure for chemical manufacturing example.

13.7 Process Changes and Effects on Hot and Cold Utilities

In the process plant, it is common to have changes in process operation on a regular basis. If process changes affect the temperature and/or enthalpy of hot or cold streams, the utility consumption of the HEN may be affected. In some cases, higher/lower utility may result, depending on where the stream is located. The concept on process change evaluation on their utility effect is guided by the famous "plus-minus principles" (Linnhoff et al., 1982; Smith, 2016). In general, the following strategies should be adopted if one were to reduce utility consumptions for the HEN:

  i. Larger enthalpy for cold stream(s) below the pinch
 ii. Larger enthalpy for hot stream(s) above the pinch
iii. Smaller enthalpy for hot stream(s) below the pinch
 iv. Small enthalpy for cold stream(s) above the pinch

Strategies (i) and (ii) are termed as the "plus principles", while strategies (iii) and (iv) are "minus principle" and are illustrated in Figure 13.37. Operational changes other than the above listed will lead to higher utility consumption which is undesired.

Example 13.6

Revisit the HEN of the chemical manufacturing process in Figure 13.35. Identify the new utility targets and also its revised HEN structure if stream H1 will have a new supply temperature of 240°C, instead of its original temperature of 230°C.

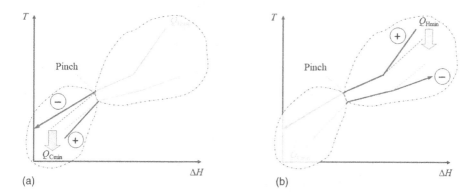

FIGURE 13.37
Plus-minus principles: (a) for region below the pinch; (b) for the region above the pinch.

SOLUTION

The HRPD of the HEN problem is replotted in Figure 13.38, with hot composite curve having higher supply temperature of 240°C, instead of 230°C (Figure 13.9). As shown, the hot utility target (Q_{Hmin}) is reduced to 85 MW, while the heat recovery target (Q_{REC}) is increased to 410 MW. Note that the minimum cold utility (Q_{Cmin}) remains identical at 130 MW. The pinch temperatures are also identical to before, i.e., 150°C and 140°C.

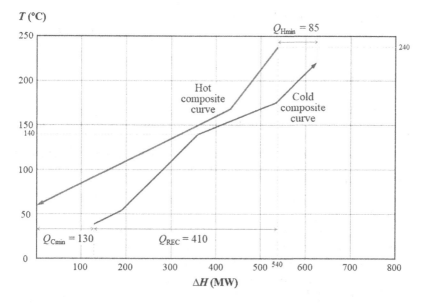

FIGURE 13.38
HRPD for revised temperature of H1 (Example 13.6).

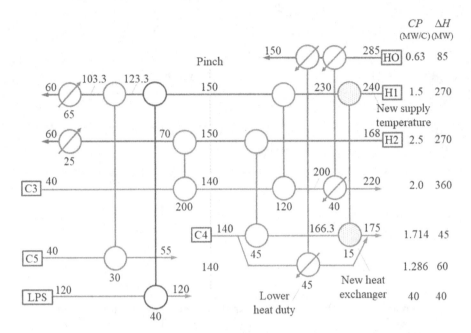

FIGURE 13.39
Grid diagram for revised temperature of H1 (Example 13.6).

The grid diagram in Figure 13.36 is also revised, and is given in Figure 13.39. As shown, a new heat exchanger is added between streams H1 and C4, with heat duty of 15 MW. Note also that the heat duty of exchanger pairing hot oil and C4 is reduced to 45 MW, which is smaller than its original value of 60 MW (Figure 13.35).

ADDITIONAL READINGS

In synthesizing an HEN, there are cases where *threshold problems* is encountered, i.e., only hot/cold utility is needed for the HEN. It is advised to check the shape of the HRPD before the HEN is constructed (Smith, 2016). In addition, various energy-intensive processes (e.g., distillation, evaporator, reactors) may be integrated with the background process to reduce its overall utility consumption. Readers may check the details in Smith, 2016.

EXERCISES

1. Table 13.5 shows the data of four process streams that will undergo heating and cooling operations in a chemical process (Smith, 2016). Solve the following tasks:

   a. Use HRPD to determine the minimum hot and cold utility targets for the process, for a ΔT_{min} value of 10°C, 20°C,

TABLE 13.5

Stream Data for Problem 13.1

Stream	Supply Temperature T_S (°C)	Target Temperature T_T (°C)	Enthalpy Change, ΔH (MW)
1	20	180	32.0
2	250	40	31.5
3	140	230	27.0
4	200	80	30.0

and 30°C. What conclusion can you make based on the observation?

  b. Use PTA to determine the minimum hot and cold utility targets, for a ΔT_{min} value of 10°C.
  c. Design the HEN with PDM, for a ΔT_{min} value of 10°C.
2. Table 13.6 shows the data of four process streams that will undergo heating and cooling operations (Linnhoff et al., 1982). Solve the following tasks:

  a. Use HRPD to determine the minimum hot and cold utility targets for the process, for a ΔT_{min} value of 10°C, 20°C, 30°C, and 40°C. What conclusion can you make based on the observation?
  b. Use PTA to determine the minimum hot and cold utility targets, for a ΔT_{min} value of 10°C.
3. Table 13.7 shows the stream data of four process streams that will undergo heating and cooling operations (Smith, 2016). Solve the following tasks:

  a. Use PTA to verify that for a ΔT_{min} value of 20°C, the minimum hot and cold utility targets for this process are 15 and 26 MW, respectively, and its pinch temperatures are 120°C and 100°C.
  b. Design the HEN with PDM.

TABLE 13.6

Stream Data for Problem 13.2

Stream	Supply Temperature T_S (°C)	Target Temperature T_T (°C)	Heat Capacity Flowrate, CP (kW/°C)
1	20	135	2.0
2	170	60	3.0
3	80	140	4.0
4	150	30	1.5

TABLE 13.7

Stream Data for Problem 13.3

Stream	Supply Temperature T_S (°C)	Target Temperature T_T (°C)	Heat Capacity Flowrate, CP (MW/°C)
1	400	60	0.3
2	210	40	0.5
3	20	160	0.4
4	100	300	0.6

TABLE 13.8

Cold Stream Data for Heat Exchangers on an Offshore Platform in Problem 13.4

Stream	Supply Temperature T_S (°C)	Target Temperature T_T (°C)	Enthalpy Change, ΔH (kW)
1	40	65	9,850
2	30	65	11,860
3	65	90	10,860

4. Revisit the HEN design of Example 13.4 (Figure 13.29), and determine if it is possible to have an alternative matching between the newly added hot stream with the cold streams.

5. Table 13.8 shows the data of three cold streams that require process heating on an offshore platform (Foo et al., 2022). Hot water is used as a heating medium for these cold streams. Solve the following tasks:

   a. Construct a cold composite curve for the cold streams to determine its total heating requirement. Verify the hot utility target with a PTA.

   b. Construct a GCC from PTA.

   c. Use GCC to determine the minimum flowrate of hot water that is used as hot utility. Assuming that hot water will be supplied at 140°C. The minimum temperature difference (ΔT_{min}) is taken as 10°C. Determine also the targeted return temperature of hot water.

6. Revisit the HEN design of Example 13.5 (Figure 13.35), determine if it is possible to have an alternative configuration for its heat exchanger pairing with utility streams:

   a. Pairing of hot streams H1 and H2 with cooling water (CW).

   b. Pairing of cold streams C3 and C4 with hot oil (HO).

7. For Example 13.6, the cold stream C5 will have a revised target temperature of 70°C. Solve the following tasks:
   a. Revise the HRPD in Figure 13.38 to examine this process change on its utility targets.
   b. Modify the HEN in Figure 13.39 to match the new utility targets in (a).

References

Foo, D. C. Y., Diban, P., and Ooi, R. E. H. (2022). Modelling and optimisation of separation and heating medium systems for offshore platform, in Foo, D. C. Y. (Eds.), *Chemical Engineering Process Simulation*, 2nd Ed. Amsterdam, Elsevier.

Hohmann, E. C. (1971). *Optimum Networks for Heat Exchange*, PhD Thesis. University of Southern California, Los Angeles.

Linnhoff, B. and Flower, J. 1978a. Synthesis of heat exchanger networks. I. Systematic generation of energy optimal networks, *AIChE Journal*, 24, 633–642.

Linnhoff, B. and Flower, J. 1978b. Synthesis of heat exchanger networks. II. Evolutionary generation of networks with various criteria of optimality, *AIChE Journal*, 24, 642–654.

Linnhoff, B. and Hindmarsh, E. 1983. The pinch design method of heat exchanger networks, *Chemical Engineering Science*, 38, 745–763.

Linnhoff, B., Townsend, D. W., Boland, D., Hewitt, G. F., Thomas, B. E. A., Guy, A. R., and Marshall, R. H. 1982. *A User Guide on Process Integration for the Efficient Use of Energy*. Rugby: IChemE.

Papoulias, S. A. and Grossmann, I. E. (1983). A structural optimization approach in process synthesis – II: Heat recovery networks. *Computers and Chemical Engineering*, 7(6): 707–721.

Smith, R. (2016). *Chemical Process: Design and Integration*, 2nd Ed. West Sussex, England: John Wiley.

14

Combined Heat and Power (CHP)

Combined heat and power (CHP), or *cogeneration*, is commonly used in the process plants to improve their thermal efficiency. Apart from electricity generation, various levels of steam may be generated via CHP for heating and non-heating purposes (e.g., stripping agent, blanking) in the plant. In this chapter, graphical, algebraic and mathematical programming methods for CHP units are introduced, aiming to assist process engineers to better optimize energy resources in the plant.

14.1 Shortcut Model for CHP

The conventional model for turbine requires detailed calculations for individual turbine, in order to evaluate its outlet enthalpy at constant entropy, along with its steam flowrate. A shortcut model has been proposed by El-Halwagi et al. (2009) for the identification of cogeneration potential of a CHP system, where detailed calculations can be bypassed. It is assumed that demand of steam of a process plant is known (with given enthalpy level and quantity), steam of high enthalpy level will be generated and let down through turbine (where power is generated), in order to fulfill the steam demand. Hence, power is generated as a side benefit apart from fulfilling the steam demand of the process plant.

For instance, in the steam system in Figure 14.1, high- (HPS), medium- (MPS) and low-pressure steam (LPS) are needed in the plant. Very high pressure (VHPS) and high-pressure steam (HPS) are generated in the CHP system through boilers or heat recovery steam generator (HRSG; when waste heat is available), which are then let down in order to fulfill the demand of HPS, MPS and LPS of the plant. Besides, power generation through boilers or HRSG also leads to cost savings; this makes the CHP an attractive energy-saving alternative.

In the shortcut model of El-Halwagi et al. (2009), the *extractable energy* (e_{Head}) is based on header levels where the turbine operates in between. As shown in Equation 14.1, the extractable energy is calculated from the specific enthalpy (H_{Head}) with known temperature and pressure, along with the efficiency of the header (η_{Head}):

$$e_{\text{Head}} = \eta_{\text{Head}} H_{\text{Head}} \tag{14.1}$$

DOI: 10.1201/9781003173977-16

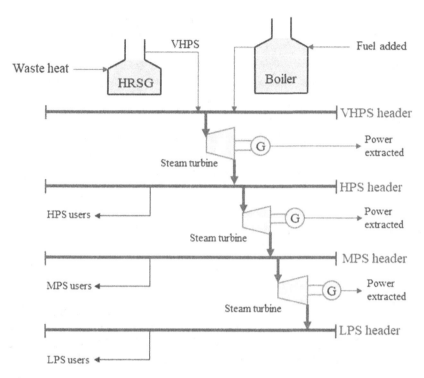

FIGURE 14.1
Steam sources and demand at different levels.

Next, the term of *extractable power* of header (E_{Head}) was also introduced (El-Halwagi et al. (2009), given as in Equation 14.2 that follows:

$$E_{Head} = F_{STM}e_{Head} = F_{STM}\eta_{Head}H_{Head} \qquad (14.2)$$

where F_{STM} is the steam flowrate. More details of this shortcut model are found in El-Halwagi et al. (2009) and El-Halwagi (2017).

14.2 Graphical Targeting Method

To determine the cogeneration potential of a CHP system, a graphical tool called the *cogeneration pinch diagram* (CPD; El-Halwagi et al., 2009) is useful. Procedure for plotting the CPD is given as follows:

1. Arrange all steam sources in ascending order of their pressure or specific enthalpy level. Do the same for all steam demand.

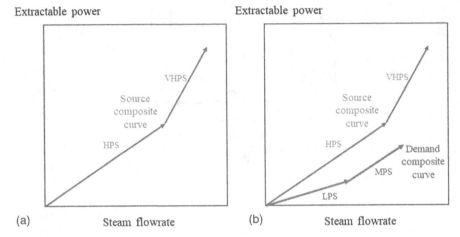

FIGURE 14.2
Construction of CPD: (a) source composite curve; (b) demand composite curve.

2. All steam sources are plotted from the origin on an extractable power-versus-flowrate diagram, to form a *source composite curve*. The individual sources are connected by linking the tail of a latter source to the head of the earlier one (Figure 14.2a).

3. Step 2 is repeated for all steam demand to form the *demand composite curve*, on the same extractable power-versus-flowrate diagram. This forms the CPD, as shown in Figure 14.2b. For a feasible CPD, all higher pressure steam sources must stay directly above lower pressure steam demand.

4. To maximize the cogeneration potential, the demand composite curve is shifted upwards along the source composite curve, until both composite curves are aligned vertically at their terminal points, as shown in Figure 14.3. The vertical gap between the terminal points of the composite curves is the total cogeneration potential (P_T) of the CHP system without condensing turbine. On the other hand, the horizontal gap on the left of the CPD is the excess steam (F_{XS}) from the system.

5. For cases where high-pressure steam demand is located below a lower pressure steam, such as that shown in Figure 14.4a, the CPD is considered infeasible. To restore feasibility, the segment(s) of the higher pressure steam demand(s) is to be shifted upward along the source composite curve until it stays entirely below the higher pressure steam segment(s), such as that shown in Figure 14.4b. Similarly, the horizontal gap on the left of the CPD indicates the excess steam (F_{XS}) from the system. The total cogeneration potential (P_T) for a non-condensing turbine system is given by the summation of the vertical gaps between

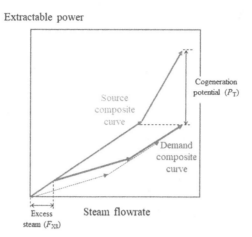

FIGURE 14.3
CPD with maximum cogeneration potential (without condensing turbine).

the composite curve segments. At the right end of the CPD, when the demand composite segment extends beyond the source composite curve, it means that the CHP system is experiencing steam deficit. To restore the feasibility, the steam demand has to be reduced or larger amount of higher pressure steam has to be generated in order to remove the deficit, such as that shown in Figure 14.5.

6. For CPD with excess steams (e.g., Figures 14.3 and 14.4), different alternatives are available for the use of the excess steam. To maximize cogeneration potential, condensing turbine may be installed so

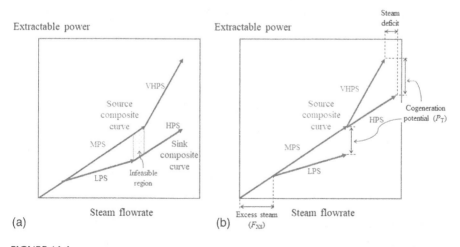

FIGURE 14.4
Infeasible CPD where (a) high-pressure steam demand located below lower pressure steam; (b) with steam deficit.

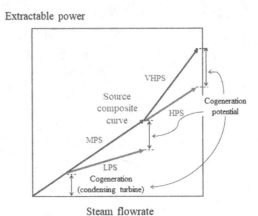

FIGURE 14.5
Feasible CPD with condensing turbine.

that more power can be generated from the CHP system. For such a
case, the vertical gap at the left end of the CPD also contributes to the
total cogeneration potential (P_T) of the CHP system (see Figure 14.5).
Alternatively, the excess steam may be sold to nearby plants that are
in need of lower pressure steam. If no excess steam is to be generated,
flowrate of the lower pressure steam source may be reduced.

The configuration of the CHP system can be derived from the CPD by
inspection. For instance, for the CPD in Figure 14.5, two turbines are used to
let down the VHPS and MPS, to fulfill the demands of HPS and LPS, respectively. Besides, a condensing turbine is used to generate more power from
the excess steam for the CHP. The CHP configuration is given in Figure 14.6.

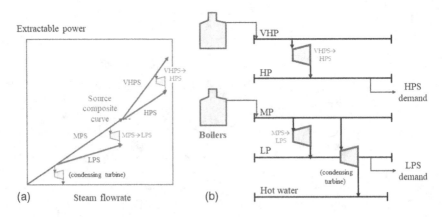

FIGURE 14.6
CHP configurations for CPD in Figure 14.5 (with condensing turbine): (a) turbine placement in
the CPD; (b) steam network.

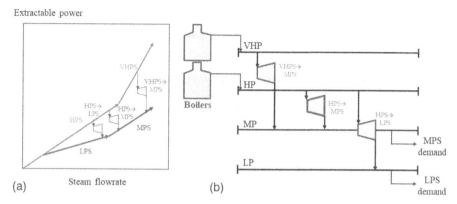

FIGURE 14.7
CHP configurations for CPD in Figure 14.3 (without condensing turbine): (a) turbine placement in the CPD; (b) steam network.

Similar inspection may be carried out for the CHP system without condensing turbine in Figure 14.3, with its configuration given in Figure 14.7. Note that two turbines are used to let down the VHPS and HPS in order to fulfill the demand for MPS. Note also that the amount of steam that passes through the turbine can be observed from the horizontal distance of the segments in the CPD (see the example that follows).

Example 14.1

In a CHP system, VHPS of 41 bar and HPS of 10 bar are to be generated. The plant requires MPS of 5.5 bar and an LPS of 3 bar, apart from some power demand. Data for the steam sources and demands are given in Table 14.1. For a header efficiency (η_{Head}) of 0.7, the extractable power for all headers has been calculated using Equation 14.2, and shown in Table 14.1. Determine the cogeneration potential of the CHP system using CPD, for cases with and without condensing turbine.

TABLE 14.1

Data for Example 14.1

Steam Source, i	Flowrate, F_{Sri} (kg/h)	Specific Enthalpy (kJ/kg)	Extractable Power, E_{SRi} (×1,000 kW)*
HPS	40,000	3,215	25
VHPS	60,000	3,430	40
Steam Demand, j	**Flowrate, F_{SKj} (kg/h)**	**Specific Enthalpy (kJ/kg)**	**Extractable Power, E_{SKj} (×1,000 kW)***
LPS	32,000	2,732	17
MPS	50,000	2,880	28

* *Note:* kJ/h = 1/3600 Kw.

SOLUTION

To identify the cogeneration potential, the procedure for plotting the CPD is followed:

1. *Arrange all steam sources in ascending order of their pressure or specific enthalpy level. Do the same for all steam demand.*
2. *All steam sources are plotted from the origin on an extractable power-versus-flowrate diagram, to form a source composite curve. The individual sources are connected by linking the tail of a latter to the head of the earlier source.*
3. *Step 2 is repeated for all steam demand to form the demand composite curve, on the same extractable power-versus-flowrate diagram. This forms the CPD.*

   The steam sources and demands in Table 14.1 have been arranged in ascending order of their specific enthalpy values. The source and demand composite curves are then plotted to form an initial CPD in Figure 14.8. As shown, the vertical gap between the composite shows a cogeneration potential (P_T) of 8,000 kW.

4. *To maximize the cogeneration potential, the demand composite curve is shifted upwards along the source composite curve, until both composite curves are aligned vertically at their terminal points*

   The demand composite curve is shifted along the HPS segment of the source composite curve, until their terminal points are aligned vertically. As shown in Figure 14.9, the vertical gap

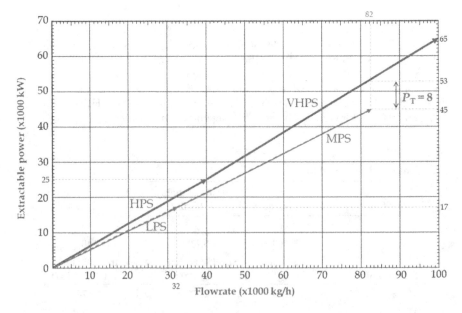

FIGURE 14.8
Initial CPD for Example 14.1.

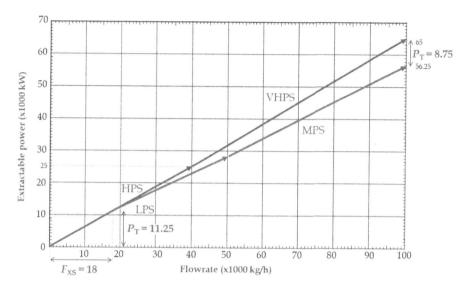

FIGURE 14.9
CPD for CHP with condensing turbine for Example 14.1.

between the composite curves shows that at the right a cogeneration potential (P_T) of 8,750 kW. At the left end of the CPD, an excess of 18,000 kg/h of HPS (F_{XS}) is observed.

6. *For CPD with excess steams, different alternatives are available for the use of the excess steam. To maximize cogeneration potential, condensing turbine may be installed so that more power can be generated from the CHP system. Alternatively, the excess steam may be sold to nearby plants that are in need of lower pressure steam. If no excess is to be generated, flowrate of the lower pressure steam source may be reduced.*
   If condensing turbine will be installed, an extra cogeneration potential of 11,250 kW will be resulted, given by the vertical gap at the left end of the CPD. In other words, the CHP system will have a total cogeneration potential of 20,000 kW (= 8,750 + 11,250 kW), as shown in Figure 14.9.
   If no condensing turbine will be installed, the access steam may be sold to neighbor plant for extra income. The cogeneration potential of the CHP system will remain identical to that determined in Step 4, i.e., 8,750 kW.
   For the case where no excess steam is to be generated (F_{XS} = 0), the HPS flowrate may be reduced to 22,000 kg/h (= 40,000 − 18,000 kg/h). Note that the cogeneration potential (P_T) will remain as 8,750 kW, as shown in the CPD in Figure 14.10.

As described, one may use the insight of the CPD to derive the CHP configuration. For the CPD in Figure 14.10, the overlapping of the steam segments allows us to identify the let-down of higher pressure to lower pressure steam. For instance, since HPS has a flowrate of 22,000 kg/h,

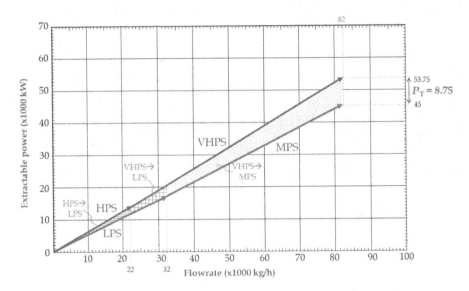

FIGURE 14.10
CPD without excess steam nor condensing turbine for Example 14.1.

it is let down completely to fulfill the demand of the LPS (region as labeled as "HPS → LPS" in Figure 14.10). The latter has a demand of 32,000 kg/h. Hence, the remaining demand of 10,000 kg/h (= 32,000 – 22,000 kg/h) is to be supplied by VHPS (region of "VHPS → LPS" in Figure 14.10). Finally, the remaining flowrate of the VHPS (82,000 – 32,000 = 50,000 kg/h) is to be let down to fulfill the demand of MPS (region of "VHPS → MPS" in Figure 14.10). The configuration of the CHP system is given in Figure 14.11a. One may use the same insight to derive the configuration for the CHP system with condensing turbine in Figure 14.9, with its configuration given in Figure 14.11b.

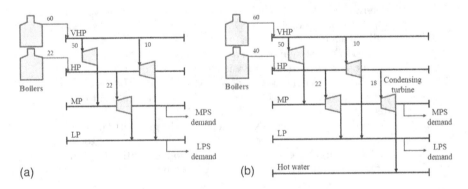

FIGURE 14.11
CHP configuration for Example 14.1: (a) with condensing turbine; (b) without condensing turbine[1] (steam flowrate in 1,000 kg/h).

---

[1] As an exercise, readers may make use of the CPD insights to derive this CHP configuration.

14.3 Algebraic Targeting Method

The graphical method of CPD is a useful tool in determining cogeneration potential for a CHP system, it suffers with inaccuracy issue and being cumbersome like other graphical methods. Hence, the algebraic targeting method known as *steam cascade analysis* (SCA) developed by Ng et al. (2017) serves as a good alternative tool for calculating the exact cogeneration target.

A generic framework for SCA is given in Table 14.2, while its procedure for construction is given as follows.

1. Arrange the specific enthalpy (H_k) of all steam sources and demand in descending order in columns 1 and 2.

2. In column 3, locate the flowrates of steam source ($F_{SR,k}$) at their respective enthalpy levels. Do the same for all steam demand flowrates ($F_{SK,k}$) in column 4.

3. In column 5, calculate the net steam flowrate at every enthalpy level k ($F_{Net,k}$) with Equation 14.3. A positive value indicates net steam surplus, while a negative value means net steam deficit.

$$F_{Net,k} = F_{SR,k} - F_{SK,k} \quad \forall k \qquad (14.3)$$

4. Perform *steam cascade* in column 6, where steam from higher level is accumulated to the next level (δ_k), given as in Equation 14.4. Last entry in this column indicates the excess steam (F_{XS}) of the CHP system.

$$\delta_k = \delta_{k-1} + F_{Net,k} \quad \forall k \qquad (14.4)$$

TABLE 14.2

A Generic Framework for SCA

k	H_k	$F_{SR,k}$	$F_{SK,k}$	$F_{Net,k}$	δ_k	E_k	P_k
1	H_1	$F_{SR,1}$	$F_{SK,1}$	$F_{Net,1}$			$P_1 = 0$
					δ_1	E_1	
2	H_2	$F_{SR,2}$	$F_{SK,2}$	$F_{Net,2}$			P_2
					δ_2	E_2	
$\vdots$	$\vdots$	$\vdots$	$\vdots$	$\vdots$	$\vdots$	$\vdots$	$\vdots$
k	H_k	$F_{SR,k}$	$F_{SK,k}$	$F_{Net,k}$			P_k
					δ_k	E_k	
$\vdots$	$\vdots$	$\vdots$	$\vdots$	$\vdots$	$\vdots$	$\vdots$	$\vdots$
$n-1$	H_{n-1}	$F_{SR,n-1}$	$F_{SK,n-1}$	$F_{Net,n-1}$			P_{n-1}
					δ_{n-1}	E_{n-1}	
n	H_n						P_n

5. Determine the cogeneration potential (E_k) generated across every enthalpy interval in column 7, following Equation 14.5.

$$E_k = \eta \delta_k (H_k - H_{k+1}) \ \forall k \qquad (14.5)$$

6. Perform the *power cascade* in column 8, where the cogeneration potential (E_k; column 7) is accumulated in column 8, following Equation 14.6.

$$P_k = \begin{cases} P_k = 0 \ k = 1 \\ P_{k-1} + E_{k-1} \ k = 2,...n \end{cases} \qquad (14.6)$$

For a CHP with condensing turbine, the total cogeneration potential of the CHP system is given by the last entry in column 8, i.e., $P_n = P_T$. For cases where no condensing turbine will be installed, the cogeneration potential is given by the second last entry in column 8, i.e., $P_{n-1} = P_T$.

7. There may be cases where negative δ_k value(s) is observed in column 6, which means that steam deficit is experienced, such as that shown in Figure 14.4b. For such cases, flowrate(s) of the steam source (same or at higher level) is to be increased, so that steam deficit is removed. In other words, all δ_k values in column 6 have to be non-negative, as given by Equation 14.7. Note that non-negative δ_k values also ensure all E_k and P_k in columns 7 and 8 to be non-negative values.

$$\delta_k \geq 0 \ \forall k \qquad (14.7)$$

Example 14.2

Determine the cogeneration potential for the problem in Example 14.1 using SCA. For both cases with and without condensing turbine, excess steam is allowed. Derive the CHP configuration based on insights of the SCA.

SOLUTION

The procedure of SCA is following to determine the cogeneration potential:

1. *Arrange the specific enthalpy (H_k) of all steam sources and demand in descending order in columns 1 and 2.*
2. *In column 3, locate the flowrates of steam source ($F_{SR,k}$) at their respective enthalpy levels. Do the same for all steam demand flowrates ($F_{SK,k}$) in column 4.*
3. *In column 5, calculate the net steam flowrate at every enthalpy level k ($F_{Net,k}$).*

TABLE 14.3

Cascade Table for Example 14.2 (with Condensing Turbine)

k	H_k (kJ/kg)	$F_{SR,k}$ (kg/h)	$F_{SK,k}$ (kg/h)	$F_{Net,k}$ (kg/h)	$F_{C,k}$ (kg/h)	E_k (kW)*	P_k (kW)*
1	3,430	60,000		60,000			
					60,000	2,508	
2	3,215	40,000		40,000		(E_1)	2,508
					100,000	6,514	
3	2,880		50,000	−50,000		(E_2)	9,022
					50,000	1,439	
4	2,732		32,000	−32,000		(E_3)	10,461
					18,000	9,562	
5	0				(F_{XS})	(E_4)	20,023
							(P_T)

* *Note:* kJ/h = 1/3600 Kw.

The specific enthalpy values are arranged in descending order in columns 1 and 2 of the cascade table (Table 14.3). Flowrates of all steam sources and demands are located at their respective enthalpy levels, in columns 3 and 4. In column 5, the net steam flowrate at every enthalpy level is calculated. As shown, surplus steam is observed at levels 1 and 2, while steam deficit is observed in levels 3 and 4.

4. *Perform steam cascade in column 6.*
5. *Determine the cogeneration potential (E_k) generated across every enthalpy interval in column 7.*
6. *Perform the power cascade in column 8, where the cogeneration potential (P_k) is accumulated in column 8.*

We next perform steam cascade, with results shown in column 6. The last entry in column 6 indicates that there is an excess steam (F_{XS}) of 18,000 kg/h. Next, cogeneration potential across every interval is determined in column 7. Finally, the cogeneration potential is calculated for every level in column 8. The last entry in this column indicates that the CHP system has a total cogeneration potential (P_T) of 20,023 kW, if condensing turbine is to be installed. This is the same solution as in Figure 14.9, but with higher numerical accuracy. Note however that the SCA cannot locate cogeneration potential for CHP without condensing turbine.

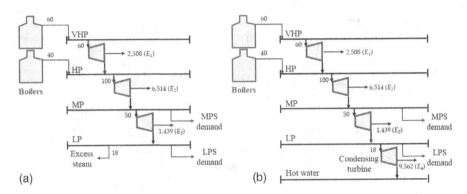

FIGURE 14.12
CHP configuration based on SCA results of Table 14.3 (steam flowrate in 1000 kg/h; power in kW).

14.4 Automated Targeting Model (ATM)

Apart from graphical and algebraic approaches, optimization-based approach is also useful in evaluating the CHP system. One such approach is the *automated targeting model* (ATM) that is conceptually similar to the algebraic approach.

The basic framework of the ATM is shown in Figure 14.13, while its constraints are given as those in Equations 14.3–14.7.

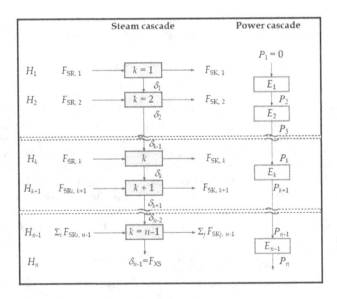

FIGURE 14.13
Basic framework of the ATM for CHP.

With mathematical-based approach, different optimization objectives may be set, depending on case-specific requirements. The simplest form of objective is to maximize the cogeneration potential (P_T) of the CHP system, with the objective in Equation 14.8.

$$\text{Maximize } P_T = P_n \tag{14.8}$$

The objective may also be set to maximize the *total saving and profit (TSP)* of the CHP system, which is contributed mainly by cost saving from power (CS_P) and steam (CS_{STM}), as well as the profit of selling excess steam (PF_{XS}), given as in Equation 14.9. In cases where waste material is used for energy generation (i.e., waste-to-energy project), saving from waste disposal (CS_{Waste}) may also be included.

$$\max TSP = CS_P + CS_{STM} + PF_{XS} + CS_{Waste} \tag{14.9}$$

Alternatively, one may also set the objective to minimize the total annualized cost (*TAC*) of the CHP system, which considers the annualized capital cost (*ACC*) of all equipment (i.e., boilers and turbines), as well as the operating cost (*OC*), given as in Equation 14.10.

$$\min TAC = ACC + AOT(OC) \tag{14.10}$$

where *AOT* is the annual operating time of the CHP system. Note that the cost terms in Equation 14.9 and Equation 14.10 are case-specific (see Example 14.3 and case study).

For the calculation of *ACC* of the CHP units, the "big M formulation" in Equation 14.11 is often used.

$$B_{eqp} \geq \left. PR_{eqp} \middle/ M \right. \tag{14.11}$$

where PR_{eqp} and B_{eqp} are parameter and binary variable for the unit in concern, while M is an arbitrary large value[2]. The binary variable B_{eqp} can be incorporated into cost correlation for calculation of *ACC* (see Example 14.3 that follows).

Note also that all basic constraints (Equations 14.3–14.7) of ATM linear, solving the objective to maximize the cogeneration potential (Equation 14.8) or *TSP* (Equation 14.9) will make the ATM a linear program (LP) problem; the latter can be solved to obtain a global solution, if it exists. For an LP problem, MS Excel (with Solver add-in) is a convenient tool. For more complex formulations with non-linearity (i.e., non-linear programming—NLP), or binary variables (with Equation 14.11), the ATM will become a mixed integer linear program—MILP, or mixed integer non-linear program—MINLP), other commercial optimization software such as LINGO and GAMS should be used. Note that the ACC and TAC correlations presented here are meant for performance targeting purposes, with accuracy of approximately 70-80%. Detailed cost calculation is necessary before final decision is to be made.

---

[2] See Chapter 9 for the use of big-M formulation in calculating the piping cost.

Example 14.3

Resolve the CHP problem in Example 14.1 with ATM for the following tasks:

1. Use MS Excel to solve the ATM by maximizing its cogeneration potential (P_T), for a CHP system with and without condensing turbine.
2. Use MS Excel to maximize the *TSP* of the CHP system, when no excess steam is allowed. The cost-saving terms in Equation 14.9 are described by Equation 14.12 and Equation 14.13, respectively (PF_{XS} and CS_{Waste} are omitted). The unit costs of power, MPS, and LPS are given as 0.08 \$/kWh, 0.01 \$/kg, and 0.007 \$/kg, respectively. Derive also its CHP configuration based on the insights of ATM.

$$CS_P\ (\$/h) = CT_P P_T \qquad (14.12)$$

$$CS_{STM}\ (\$/h) = \sum_s CT_{STM,s} F_{STM,s} \qquad (14.12)$$

where CT_P and $CT_{STM,s}$ are unit costs of power and steam of level s, while $F_{STM,s}$ (kg/h) is the flowrates of steam of level s.

3. Use LINGO to minimum the *TAC* of the CHP system, when no excess steam is allowed. Determine the results for a different scenario when the maximum steam flowrate constraint is removed. The *ACC* term of the CHP system in Equation 14.10 is contributed by the individual units, i.e., boilers for the generation of VHPS (ACC_{VHPB}, Equation 14.14) and HPS (ACC_{HPB}, Equation 14.15), as well as the turbines across different steam levels ($ACC_{Turb,s}$; Equation 14.16). Note that the cost correlation in Equations 14.14–14.16 was linearized based on the non-linear model of Bruno et al. (1998). The operating cost term (*OC*) for the CHP is mainly contributed by that of boilers, where different fuel type f is used, as described by Equation 14.17 and Equation 14.18. Note that the *OC* correlation in Equation 14.17 considers 30% of other factors such as pumping power, sewer charges of blowdown, etc. (Bruno et al., 1998). The *AOT* is assumed as 8000 h.

$$ACC_{BL-VHPS}\ (\$/y) = 26,659 + 3.512\ F_{VHPS} \qquad (14.14)$$

$$ACC_{BL-HPS}\ (\$/y) = 12,340 + 1.626\ F_{HPS} \qquad (14.15)$$

$$ACC_{Turb,s}\ (\$/y) = 81,594 + 18\ E_{Turb,s} \qquad \forall s \qquad (14.16)$$

$$OC\ (\$/h) = 1.3 \sum_f F_{Fuel,f} CT_f \qquad (14.17)$$

$$F_{Fuel,f}\ (kg/h) = \frac{F_{STM,s} H_s}{LHV_f \eta_{boil}} \quad \forall f\ \forall s \qquad (14.18)$$

where CT_f and LHV_f are the unit cost ($/kg) and lower heating value (kJ/kg) of the fuel, F_{VHPS} and F_{HPS} are the respective flowrates of VHPS and HPS (both in kg/h), H_s is the specific enthalpy of steam (kJ/kg), and η_{boil} is the boiler efficiency, which is assumed at 85%. For this case, it is assumed that natural gas is used as fuel, which has a unit cost (CT_f) of 0.3 $/kg and LHV of 55,100 kJ/kg.

SOLUTION

1. *Use MS Excel to solve the ATM by maximizing its cogeneration potential (P_T), for a CHP system with and without condensing turbine.*

    The cascade table is a convenient way to construct the ATM in MS Excel. Hence, the ATM has the same structure as the cascade table in Table 14.3, along with Solver configuration, as shown in Figure 14.14. As shown, the objective is set to maximize the cogeneration potential (cell I13), subject to the non-negativity constraints in Equation 14.7. Note that constraints described by Equations 14.3–14.6 have been incorporated into the Excel setting. For instance, Equation 14.3 is described by column F. The steam cascade in Equation 14.4 is set up in column G. The calculation of cogeneration potential in Equation 14.5 and Equation 14.6 is found in columns H and I.

    The flowrates of VHPS and HPS are used as optimization variables for this case, which have their available flowrate given as in Table 14.1. Hence, additional constraints are added in the

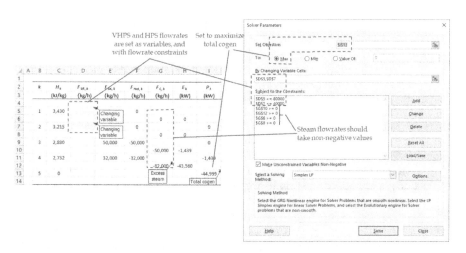

FIGURE 14.14
Setting of MS Excel and Solver for Example 14.3.

ATM for their flowrate availabilities, following Equation 14.19 (see Figure 14.14).

$$F_{STM,s} \leq F_{STM,s}^{max} \quad \forall s \qquad (14.19)$$

The ATM result in Figure 14.15 shows that the maximum cogeneration potential (cell I13) of the CHP system is identified as 20,023 kW, for the case where condensing turbine is installed. The results also show that both VHPS and HPS are utilized to their maximum availability. Where no condensing turbine is installed for the CHP system, cell I11 shows that the CHP will have a cogeneration potential of 10,461 kW, while 18,000 kg/h of excess steam is reported. All these targets are identical to the solutions as reported by the algebraic technique in Table 14.3. The CHP configurations are hence similar to those reported for the algebraic technique, i.e., without condensing turbine (Figure 14.12a), and with condensing turbine (Figure 14.12b).

2. Use MS Excel to *maximize the TSP of the CHP system, when no excess steam is allowed.*

   The setting in Excel Solver is similar to that in task 1, except that the objective is set to maximize the *TSP* (Equation 14.9). The latter has its detailed calculation given below the cascade table (see Figure 14.16). Note that an additional constraint in Equation 14.20 is added in the Solver so that no excess stream is to be produced from the CHP system in cell G12 (level $k = 4$). The configuration of Solver is shown in Figure 14.16.

$$\delta_4 = 0 \qquad (14.20)$$

   The ATM results in Figure 14.16 indicate that a *TSP* of 1,426 $/h is resulted, with total cogeneration potential of 8,771 kW. Note that the ATM result indicates that the HP steam has a reduced flowrate of 22,000 kg/h, as compared to the solution with excess flowrate (Figure 14.15). The configuration of the CHP system can be derived similarly to that using the algebraic method and is given in Figure 14.17.

3. Use LINGO to *minimum the TAC of the CHP system. Determine the results if the steam flowrate constraints are removed.*

To enable the calculation of *ACC* of boilers and turbines, the correlations in Equations 14.14–14.16 are used together with the big *M* formulation (Equation 14.11), given in Equations 14.21–14.30 that follow.

$$ACC_{BL-VHPS}(\$/y) = 26,659B_{BL-VHPS} + 3.512\, F_{VHPS} \qquad (14.21)$$

$$ACC_{BL-HPS}(\$/y) = 12,340B_{BL-HPS} + 1.626\, F_{HPS} \qquad (14.22)$$

$$ACC_{TB-VHPS}(\$/y) = 81,594B_{TB-VHPS} + 18\, E_{TB-VHPS} \qquad (14.23)$$

	A	B	C	D	E	F	G	H	I	J	K
1											
2		k	H_k	$F_{SR,k}$	$F_{SK,k}$	$F_{Net,k}$	$F_{C,k}$	E_k	P_k		
3			(kJ/kg)	(kg/h)	(kg/h)	(kg/h)	(kg/h)	(kW)	(kW)		
4											
5		1	3,430	60,000	Changing variable	60,000					
6							60,000	2,508			
7		2	3,215	40,000	Changing variable	40,000			2,508		
8							100,000	6,514			
9		3	2,880		50,000	-50,000			9,022		
10							50,000	1,439			
11		4	2,732		32,000	-32,000			10,461	Total cogen (without condensing turbine)	
12							18,000	9,562			
13		5	0				Excess steam		20,023	Total cogen (with condensing turbine)	
14											
15											

FIGURE 14.15
Results of ATM for maximizing cogeneration potential for Example 14.3.

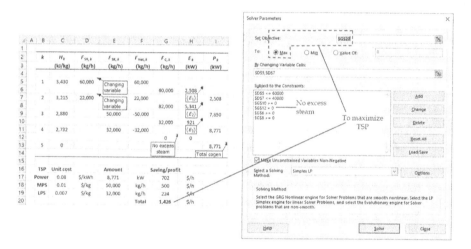

FIGURE 14.16
Results of ATM for maximizing the *TSP* for Example 14.3.

$$ACC_{TB-HPS}(\$/y) = 81,594 B_{TB-HPS} + 18\, E_{TB-HPS} \qquad (14.24)$$

$$ACC_{TB-MPS}(\$/y) = 81,594 B_{TB-MPS} + 18\, E_{TB-MPS} \qquad (14.25)$$

$$B_{BL-VHPS} \geq F_{VHPS}/M \qquad (14.26)$$

$$B_{BL-HPS} \geq F_{HPS}/M \qquad (14.27)$$

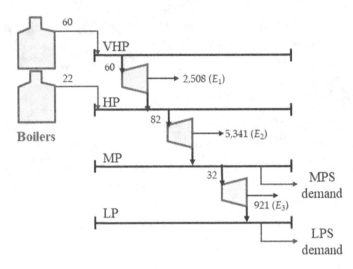

FIGURE 14.17
CHP configuration for maximum *TSP* for Example 14.3.

$$B_{\text{TB-VHPS}} \geq E_{\text{TB-VHPS}}\big/M \qquad\qquad (14.28)$$

$$B_{\text{TB-HPS}} \geq E_{\text{TB-HPS}}\big/M \qquad\qquad (14.29)$$

$$B_{\text{TB-MPS}} \geq E_{\text{TB-MPS}}\big/M \qquad\qquad (14.30)$$

The use of binary variables makes the formulation an MILP problem. The ATM model is solved using LINGO, with objective in Equation 14.10 (minimize *TAC*) and constraints in Equations 14.3–14.7 and Equations 14.17–14.30. The LINGO formulation for the above model is written as follows:

Model

```
!OBJECTIVE FUNCTION;
Min = TAC;
TAC = ACC + AOT*OC;

!Capital cost (all in $/y);
ACC = ACC _ BL _ VHPS + ACC _ BL _ HPS + ACC _ TB _ VHPS + ACC _
TB _ HPS + ACC _ TB _ MPS;
ACC _ BL _ VHPS = 26659*B _ BL _ VHPS + 3.512*VHPS;
ACC _ BL _ HPS  = 12340*B _ BL _ HPS  + 1.626*HPS;
ACC _ TB _ VHPS = 81594*B _ TB _ VHPS + 18*E _ VHPS;
ACC _ TB _ HPS  = 81594*B _ TB _ HPS  + 18*E _ HPS;
ACC _ TB _ MPS  = 81594*B _ TB _ MPS  + 18*E _ MPS;
```

```
!Binary variables & big M formulation;
B _ BL _ VHPS >= VHPS/1000000;
B _ BL _ HPS >= HPS/1000000;
B _ TB _ VHPS >= E _ VHPS/1000000;
B _ TB _ HPS >= E _ HPS/1000000;
B _ TB _ MPS >= E _ MPS/1000000;

@BIN(B _ BL _ VHPS);
@BIN(B _ BL _ HPS);
@BIN(B _ TB _ VHPS);
@BIN(B _ TB _ HPS);
@BIN(B _ TB _ MPS);

!Operating cost;
OC = 1.3*CT _ NG*FNG;
FNG = FNG _ VHPS + FNG _ HPS;
FNG _ VHPS = VHPS*H _ VHPS/LHV _ NG/EFF _ BOIL;
FNG _ HPS = HPS*H _ HPS/LHV _ NG/EFF _ BOIL;
H _ VHPS = 3430; !kJ/kg;
H _ HPS = 3215;  !kJ/kg;
CT _ NG = 0.3;   !$/kg;
LHV _ NG = 55100;!kj/kg;
EFF _ BOIL = 0.85;
AOT = 8000;      !h/y;

!Steam cascade (kg/h);
D0 = 0;
D1 = D0 + VHPS;  !k = 1;
D2 = D1 + HPS;   !k = 2;
D3 = D2 - 50000; !k = 3;
D4 = D3 - 32000; !k = 4;
D4 = 0;          !No excess steam;
D1 >= 0; D2 >= 0; D3 >= 0; D4 >= 0; !Steam cascade must be
positive values;
VHPS <= 60000;   !Flowrate constraint;
HPS <= 40000;    !Flowrate constraint;

!Power cascade (kW);
P1 = 0;                   !k = 1;
P2  =  P1  +  E _ VHPS;  E _ VHPS  =  D1*(3430  -  3215)*EFF _
HEAD/3600;                !k = 2;
P3  =  P2  +  E _ HPS;   E _ HPS   =  D2*(3215  -  2880)*EFF _
HEAD/3600;                !k = 3;
P4  =  P3  +  E _ MPS;   E _ MPS   =  D3*(2880  -  2732)*EFF _
HEAD/3600;                !k = 4;
P5 = P4;                  !k = 5;
P5 = COGEN;
EFF _ HEAD = 0.7;
END
```

Solving the LINGO code leads to the minimum *TAC* of 18.8 million \$/y[3], with the ATM solution given in Figure 14.18. As shown, the

---

[3] LINGO solution is omitted here due to space constraint. Readers may download the LINGO model from book support website in order to generate the solution.

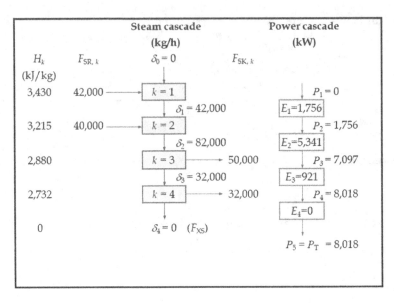

FIGURE 14.18
ATM solution for Example 14.3 (with steam constraint).

total cogeneration potential (P_T) is determined as 8,018 kW (last entry in the power cascade). For this scenario with steam flowrate constraint, the ATM determines that the HPS flowrate is to be maximized at 40,000 kg/h, while the VHPS flowrate is operated at 42,000 kg/h which is lower than its maximum flowrate (60,000 kg/h).

It is worth noting that the CHP configuration can also be derived from the ATM through inspection. In the first enthalpy interval of the ATM in Figure 14.18, the power cascade shows that extractable power of 1,756 kW is generated (E_1); this corresponds to the letdown of 42,000 kg/h (δ_1) of VHPS to the HPS level. Similarly, 40,000 kg/h (δ_2) of HPS is letdown to the MPS level and generated 5,341 kW (E_2) of extractable power. Finally, 921 kW of power (E_3) is generated when 32,000 kg/h (δ_3) of HPS is letdown to the LPS level The cascaded values of the extractable power hence make up a total of 8,018 kW of total cogeneration potential (P_T) in the last level. With these observations, the CHP configuration for this scenario can be derived and shown in Figure 14.19a.

On the other hand, when the steam flowrate constraint (Equation 14.19) is omitted, the following LINGO codes are to be used instead:

```
!VHPS <= 60000;
!HPS  <= 40000;
```

Solving the LINGO code leads to a lower *TAC*, i.e., 18.0 million $/y, with the ATM solution given in Figure 14.20. Without the steam constraint, the ATM determines that no generation of VHPS is necessary, while a total of 82,000 kg/h HPS is to be generated for the CHP system.

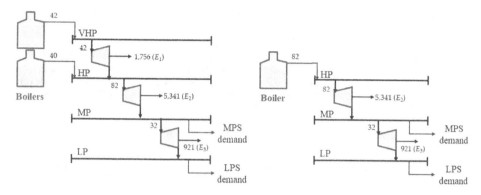

FIGURE 14.19

CHP configuration for scenarios: (a) steam flowrate constraints; (b) without steam flowrate constraint (steam flowrate in 1,000 kg/h; power in kW) for Example 14.3.

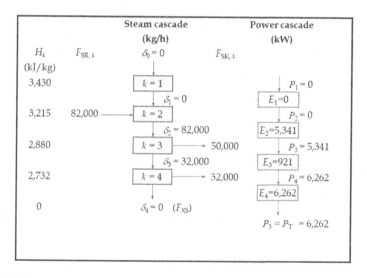

FIGURE 14.20

ATM solution for Example 14.3 (without steam constraint).

In other words, the *ACC* of VHPS boiler and turbine is avoided, which leads to reduced *TAC*. Note also that Figure 14.20 shows that the total cogeneration potential (P_T) is reduced to 6,262 kW. The CHP configuration for this scenario may be derived through inspection, as shown in Figure 14.19b.

A summary of both scenarios in this example, along with that in Example 14.1 (Figure 14.11a)[4] is given in Table 14.4. The latter shows that

---

[4] Readers may try to redo the calculation with Equation 10 (*TAC*), Equation 13 - Equation 15 (*ACC* for boilers and turbines) and Equation 16 (*OC*).

TABLE 14.4

Summary of Economic Performance for Several Scenarios

	Maximum Cogeneration (Figure 14.11a)	Minimum TAC; with Steam Flowrate Constraints (Figure 14.19a)	Minimum TAC; Without Steam Flowrate Constraints (Figure 14.19b)
OC ($/y)	2,303	2,270	2,195
ACC ($/y)	688,175	640,651	421,581
TAC ($/y)	19,109,730	18,804,400	17,983,780
Steam flowrates (kg/h)	60,000 (VHPS); 22,000 (HPS)	42,000 (VHPS); 40,000 (HPS)	82,000 (HPS)
Power generation (kW)	8,750	8,018	6,262

when the CHP is set to maximize its cogeneration potential at 8,750 kW (Figure 14.11a), it has the highest *TAC* value of 19.11 million $/y. This is due to its higher *OC* and *ACC* for having two boilers and three turbines. When the ATM is solved to minimize for the *TAC*, the latter is reduced to 18.80 million $/y (Figure 14.19a). The lowest *TAC* of 17.98 million $/y is achieved when the steam flowrate constraints are removed from the ATM, with the CHP system having only one boiler and two turbines (Figure 14.19b).

14.5 Heat Recovery and CHP

CHP may be integrated with *heat exchanger network* (HEN) for the overall reduction of utility in the process plant. Both CHP and HEN require energy source for their operation for their operation. For non-integrated system (Figure 14.21a), hot utility (Q_{Hmin}) is used for HEN, while energy source (Q_{CHP}) is used for CHP. These energy sources may be reduced if the CHP system is integrated with the HEN. As shown in Figure 14.21b, when the CHP is integrated in the region below the pinch of the HEN, the excess heat that was originally rejected to the cold utility (Q_{Cmin}) can now be sent to the CHP. Doing this leads to overall saving of hot and cold utilities of the plant. Similar saving is also achievable when CHP is placed in the region above the pinch, where steam generated from the CHP can reduce the hot utility of the HEN.

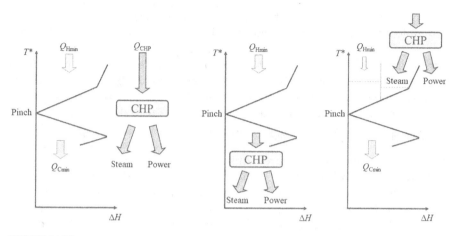

FIGURE 14.21
(a) No integration between HEN and CHP; (b) overall reduction of hot and cold utilities when CHP in region below the pinch; (c) overall reduction of hot and cold utilities when CHP is integrated in region above the pinch..

14.5 Industrial Case Study

A CHP project is evaluated in a sulfuric acid recovery (SAR) plant, which is located within an industrial park in the UK (Yeo and Foo, 2019). The plant is built to recover waste acid effluents from nearby plants, and to supply fresh sulfuric acid (H_2SO_4; of 98% purity) to the adjacent clients. The process flow diagram of the SAR plant is shown in Figure 14.22. As shown, the waste acid

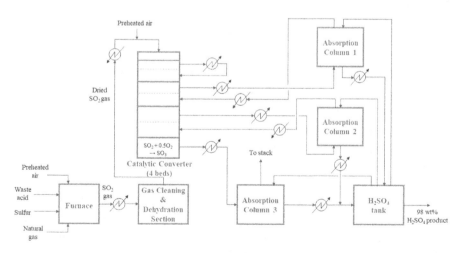

FIGURE 14.22
Process flow diagram of the SAR plant. (From Yeo and Foo, 2019.)

stream is first mixed with preheated air and natural gas prior to being sent to a high-temperature furnace operated at 1000°C. Elemental sulfurs are added to ensure that the final product in meeting its specification. Waste acid is first converted into sulfur dioxide (SO_2) gases in the furnace. Effluent of the latter is a gas stream of high temperature, which is cooled before being sent to the gas cleaning and dehydration section. The dehydrated gas is then fed with the pre-heated air to the catalytic converter, where sulfur trioxide (SO_3) gas is produced through oxidation process. The catalytic converter consists of four beds and inter-bed cooling/heating is required to ensure that the inlet gases is fed to each bed at their optimal temperatures. Absorption columns are used to remove the SO_3 product, where it is reacted to form liquid H_2SO_4 product (98% purity).

Upon the analysis of the HEN, the *grand composite curve* (GCC)[5] indicates that the SAR plant has a high pinch temperature, i.e., shirted temperature (T^*) of 990°C, with a large amount of energy to be rejected to the cooling utility in the region below the pinch (see Figure 14.23). It is desired to make use of the available waste heat for a new CHP system. A heat recovery steam generator (HRSG) may be installed so that the waste heat can be used for the co-generation of power and steam. The generated power may be used internally by the SAR plant, while the steam produced may be sold to other nearby plants within the industrial park. Three steam users were identified, with data given in Table 14.5.

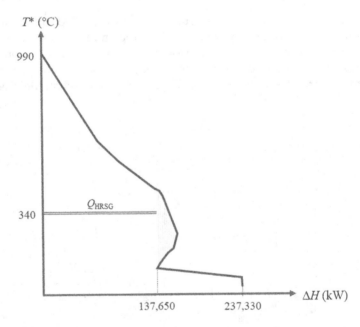

FIGURE 14.23
GCC of the SAR plant (for inlet H_2SO_4 concentration of 69%).

---

[5] Readers may refer to Chapter X for details of the GCC.

TABLE 14.5

Steam Requirements of Neighboring Plants

User	Steam Type	Demand (kg/h)	Pressure (bar)	Enthalpy (kJ/kg)	Unit Cost (£/kg)
1	HPS	35,000	36	2973	0.011
2	MPS	30,000	15	2899	0.009
3	LPS	27,000	4	2753	0.006

Source: Yeo and Foo (2019).

For this case, the objective (Equation 14.31) is set to maximize the profit of the CHP system (PF_{CHP}), which account for revenue from steam sales (REV; Equation 14.32), cost of the power consumed by the SAR plant (Equation 14.33; power from the CHP unit is insufficient), as well as the ACC of the CHP units, i.e., HRSG (Equation 14.34) and turbines (Equation 14.35). It may be assumed that all turbines will be operated at the header efficiencies (η_{Head}) of 70%, while the annual operating time (AOT) is given as 7920 h. The unit price for power is taken as 0.04 £/kWh (Yeo and Foo, 2019), while those for the steam are given in the last column of Table 14.5.

$$\max PF_{CHP}\left(£/y\right) = REV - COST_P - ACC_{HRSG} - ACC_{Turb} \tag{14.31}$$

$$REV\left(£/y\right) = AOT \sum_s CT_{STM,s} F_{STM,s} \tag{14.32}$$

$$COST_P\left(£/y\right) = CT_P(P_{SAR} - P_T)AOT \tag{14.33}$$

$$ACC_{HRSG}\left(£/y\right) = 5,317 + 0.152\, F_{VHPS} \tag{14.34}$$

$$ACC_{Turb}\left(£/y\right) = \sum_k 62,011 + 13.7\, E_k \tag{14.35}$$

As mentioned earlier, waste heat is to be recovered using the HRSG from the region below the pinch (see GCC in Figure 14.23). Due to the availability of the waste heat, an additional constraint is added to limit the maximum generation of the VHPS, given as in Equation 14.36.

$$F_{VHPS} \le \frac{Q_{HRSG}\eta_{HRSG}}{H_{VHPS}} \tag{14.36}$$

where η_{HRSG} is the efficiency of the HRSG, which may be assumed at 85%.

For the inlet concentration of 69% H_2SO_4, the power consumption of the SAR plant (E_{SAR}) was reported as 23,100 kW, while the available heat (Q_{HRSG}) is reported as 137,650 kW (see GCC in Figure 14.23). This heat may be used to produce VHPS at 80 bar (with a specific enthalpy of 2989 kJ/kg).

TABLE 14.6

ATM Solution for SAR Plant for inlet concentration of 69% H2SO4

k	H_k (kJ/kg)	$F_{SR,k}$ (kg/h)	$F_{SK,k}$ (kg/h)	$F_{Net,k}$ (kg/h)	$F_{C,k}$ (kg/h)	E_k (kW)	P_k (kW)
1	2,989	140,903		140,903			
		(VHPS)			140,903	448	
2	2,973		35,000	−35,000			448
					105,903	1,524	
3	2,899		30,000	−30,000			1,972
					75,903	2,155	
4	2,753		27,000	−27,000			4,127
					48,903	26,178	(P_T)
5	0				(F_{XS})		30,305

The ATM is solved with the objective in Equation 14.31, subject to the constraints in Equations 14.3–14.7 and Equations 14.32–14.36, with an optimum solution given in Table 14.6. The ATM model determines that a profit of 190,563 £/y the CHP system is reported as, where 140,903 kg/h of VHPS is generated to fulfil the steam demand of the three users, while 4,127 kW of generated power consumed internally within the SAR plant.

As the SAR plant recovers waste acid effluents from different plants in the nearby region, the inlet concentration of the sulfuric acid tends to fluctuate in the range of 63–69 wt%. Consequently, its power consumption also varies due to its operation. Both of these values are summarized in Table 14.7.

Re-solving the ATM with the excess heat and power consumption data in Table 14.7 yields different sets of solutions, given in Table 14.8. The cogeneration potential (P_T) and profits for different inlet concentrations are shown in Figure 14.24. As shown, the CHP unit will have higher profit with higher inlet concentrations. It is also obvious that the CHP should avoid its operation when the inlet concentration is lower than 65%, due to its diminishing profit.

TABLE 14.7

Excess Heat and Power Consumption for Different Inlet H_2SO_4 Concentrations of the SAR Plant

Inlet Concentration (wt%)	Excess Heat (kW)	Power Consumption (kW), P_{SAR}
63	140,480	23,890
65	138,940	23,370
67	138,470	23,250
69	137,650	23,100

TABLE 14.8

Solution Summary for Different Inlet H_2SO_4 Concentrations of the SAR Plant

Inlet Concentration (wt%)	REV (£/h)	$COST_P$ (£/h)	ACC (£/y)
63	817	785	271,600
65	817	767	270,334
67	817	763	269,958
69	817	759	269,302

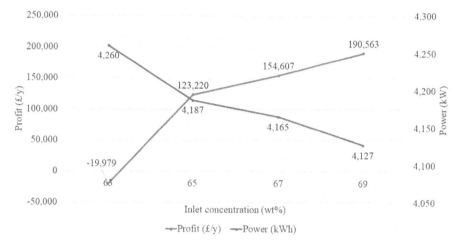

FIGURE 14.24

Cogeneration potential and profits for different inlet H_2SO_4 concentrations of the SAR plant. (Adapted from Yeo and Foo, 2019.)

14.6 Further Readings

The various methods for CHP evaluation in this chapter may be extended to cater to more complex cases. For instance, one may consider the installation of CHP for a *total site*; the latter refers to an industrial area where different processes/plants are located at a close distance. For such a case, CHP may be operated using excess heat from energy-intensive processes, with a similar concept as the integration of CHP with heat recovery presented in this chapter. Doing so helps to reduce both fuel requirement and CO_2 emission (due to fuel reduction). Several techniques for such integration have been reported and evaluated (see the work of Soh and Foo (2020)). One may also extend the ATM for CHP integration within a total site by including binary variables; this allows the automated selection of steam levels for an optimal design (Kho and Foo, 2021).

In evaluating waste-to-energy project, the ATM may be extended to consider storage constraints for combustible waste (to reduce fire risk), such as that reported by Chua and Foo (2021). The latter also extended the ATM into multi-period planning of CHP system, where different amount of combustible waste is available for a given period.

In a later work by Ang and Foo (2022), gas turbine and combined cycle were evaluated using the ATM. The combined cycle improves the efficiency of the CHP system, with higher output per unit of input energy.

PROBLEMS

1 Redo the SCA in Example 14.2. For cases where excess steam is not allowed, determine the revised cogeneration potential by reducing the flowrate of VHPS or HPS. Compare the solution with the results of Example 14.3 (part 2).

2 Table 14.9 shows data for two steam sources, i.e., VHPS and HPS that are considered in a CHP system of a process plant. The plant requires MPS and LPS, apart from some power demand. The header efficiency (η_{Head}) may be assumed as 0.7. If condensing turbine will not be installed (i.e., no access steam), determine the cogeneration potential by reducing the HPS flowrate. Determine also the cogeneration potential if the VHPS flowrate is to be reduced.

3 For the SAR plant case study (Section 14.5), determine the *TSP* (Equation 14.9) for the following cases. For both cases, the excess heat available for steam generation has been provided in Table 14.7. The unit cost of steam and power has been given in the case study.

a. For the case with an inlet concentration of 65 wt% H_2SO_4, the excess LPS from the CHP will be sold to a new customer.

b. For the case with inlet concentration of 67 wt% H_2SO_4, no excess steam should be generated. Observe the reduction of its cogeneration potential (P_T).

TABLE 14.9

Data for Problem 14.2

Steam Source, i	Flowrate, F_{Sri} (kg/h)	Pressure, P_{Sri} (bar(a))	Specific Enthalpy, H_{SRi} (kJ/kg)
VHPS	54,213	41.4	3,289
HPS	46,119	11.0	3,087

Steam Demand, j	Flowrate, F_{SKj} (kg/h)	Pressure, P_{SKj} (bar(a))	Specific Enthalpy H_{SKj} (kJ/kg)
MPS	49,947	5.5	2,852
LPS	27,224	4.1	2,747

Source: El-Halwagi (2017).

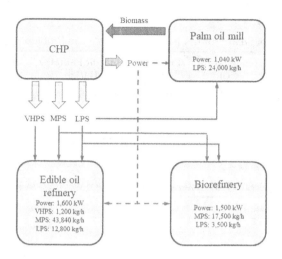

FIGURE 14.25
A palm oil processing complex that consists of a CHP plant, palm oil mill, biorefinery, and edible oil refinery.

4 A CHP project is evaluated for a palm oil processing complex (Figure 14.25); the latter consists of a palm oil mill, biorefinery, and edible oil refinery (Ng et al., 2017). The CHP system is to be fed with palm biomass from the palm oil mill, where VHPS, HPS, and power are to be generated for use in the complex (see steam details in Table 14.10). For cost calculation, *TAC* of the boilers and turbines are given in Equations 14.14–14.16. Since the palm biomass is considered waste material from the mill, no *OC* is necessary for this case. The efficiency of turbines is given as 70%. No condensing turbine is to be used in this case. Solve the following tasks:

a. Determine the minimum *TAC* for this CHP project.
b. Determine the amount of biomass needed, given that the biomass has an average LHV of 18,000 kJ/kg and moisture content of 40%, and boiler efficiency is assumed as 75%.

TABLE 14.10

Details of Steam for the Palm Oil Processing Complex in Problem 14.4

Steam	Temperature (°C)	Pressure (bar)	Specific Enthalpy (kJ/kg)	Demand (kg/h)
VHPS	350	65	3,031	1,200
HPS	300	40	2,962	-
MPS	250	12	2,936	61,340
LPS	150	4	2,753	40,300

TABLE 14.11

Details of Steam for the Edible Oil Refinery in Problem 14.5

Steam	Temperature (°C)	Pressure (bar)	Specific Enthalpy (kJ/kg)	Unit Price ($/kg)
VHPS	300	60	2,886	0.08
HPS	250	30	2,857	
MPS	200	15	2,796	0.03

TABLE 14.12

Important Parameters for CHP System in Problem 14.5

Parameter	Value
AOT	8,640 hours
Boiler efficiency	85 %
Isentropic efficiency	70 %
LHV of natural gas	55,100 kJ/kg
Unit price of natural gas	0.145 $/kg
Unit price of power	0.10 $/kW

5 A cogeneration project is evaluated for an edible oil refinery (Chua and Foo, 2021), with a power demand of 3,914 kW. Besides, the refinery has two steam demands, i.e., 5,421 kg/h of VHPS and 35,350 kg/h of MPS (see steam details in Table 14.11). A by-product known as glycerin pitch is generated from the refining process at a flowrate of 450 kg/h, which is disposed to landfill periodically with the current practice. Note that the glycerin pitch poses high LHV (14,640 kJ/kg), and hence is evaluated for its possibility of generating HPS and power through a CHP system (important parameters are given in Table 14.12). Doing this helps to reduce the consumption of natural gas of the plant, and hence lower its operating cost and CO_2 emission. To allow the combustion of the glycerin pitch, a new boiler is to be installed for the generation of HPS. As the glycerin pitch has limited supply, additional VHPS has to be generated through the existing boiler that fueled by natural gas (and let down through turbine for plant use; see). Note that no LPS will be generated from the CHP, as the excess MPS generated from the CHP can be sold to nearby plants (with unit price in Table 14.11).

Solve the following tasks:

1 Determine the minimum *TAC* (Equation 14.10) of the CHP system. *ACC* of the new HPS boiler may be calculated by the correlation in Equation 14.15, while that for the turbine in Equation 14.16. Besides, *OC* of the CHP system is mainly contributed by the natural gas boiler

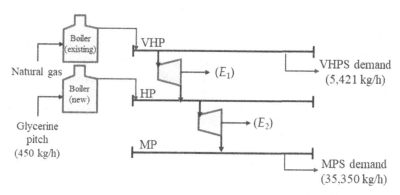

FIGURE 14.26
CHP system for an edible oil refinery. (From Chua and Foo, 2021.)

that is used to generate additional VHPS, given in Equation 14.17. To determine the *OC*, flowrate of the natural gas can be calculated using Equation 14.18 based on the VHPS flowrate (HPS generation from glycerin pitch is free).

2 To determine the actual financial benefit (*FINBEN*) for the CHP project, objective in Equation 14.37 is used, which is the difference between the *TSP* (calculating using Equation 14.9) and *TAC* values (Equation 14.10) of the project. Besides, the avoided cost of the waste landfill also contributed to the TSP, with a unit cost of 0.12 $/kg. Construct a chart to show the relationship between *FINBEN* and cogeneration potential that covers 20, 30, 40, and 50% of the total power consumption of the plant (3,914 kW).

$$\min FINBEN\ (\$/y) = TSP - TAC \qquad (14.37)$$

References

Ang, J. C. and Foo, D. C. Y. 2022. Optimisation of cogeneration system design with extended automated targeting model (ATM), *Applied Thermal Engineering*, 217, 119148.

Bruno, J. C., Fernandez, F., Castells, F., and Grossmann, I. E. 1998. A rigorous MINLP model for the optimal synthesis and operation of utility plants, *Chemical Engineering Research and Design*, 76(3), 246–258.

Chua, X. Y. and Foo, D. C. Y. 2021. Optimisation of cogeneration system and fuel inventory with automated targeting model, *Clean Technologies and Environmental Policy*, 23(8), 2369–2383.

El-Halwagi, M. M. (2017). Sustainable Design Through Process Integration, 2nd ed. Waltham, U.S.: Elsevier.

El-Halwagi, M. M., Harell, D., and Spriggs, H. D. 2009. Targeting cogeneration and waste utilization through process integration, *Applied Energy*, 86, 880–887.

Kho, W. Y. and Foo, D. C. Y. 2021. Extended automated targeting model (ATM) for total site analysis (TSI), *Asia Pacific Journal of Chemical Engineering*, 16(6), 2717.

Ng, R. T. L., Loo, J. S. W., Ng, D. K. S., Foo, D. C. Y., Kim, J.-K., and Tan, R. R. 2017. Targeting for cogeneration potential and steam allocation for steam distribution network, *Applied Thermal Engineering*, 113, 1610–1621.

Soh, W. L. and Foo, D. C. Y. 2020. Evaluation of cogeneration potential with different total site integration (TSI) techniques, *Process Integration and Optimization for Sustainability*, 4, 91–105.

Yeo, Z. M. and Foo, D. C. Y. 2019. Targeting waste heat utilisation with cogeneration, *Process Integration and Optimization for Sustainability*, 3(3), 413–421.

15

Extended Application: Synthesis of Heat-Integrated Water Network*

Water and energy are two important resources for the process plants. The recovery of these resources has been dealt with separately in the past chapters. In this chapter, the recovery of these resources is carried out simultaneously, through the synthesis of *heat-integrated water network* (HIWN).

As shown in Chapters 2–12, one will have to consider both flowrate and impurity constraints in synthesizing a *water network*, while temperature consideration has been ignored. However, temperature constraints are important in many process operations. In particular, some processes are operated at a fixed temperature range, and hence water streams that are fed to these units (for both recycled water and freshwater) have to undergo heating/cooling operations in order to reach their desired temperature. To reduce utility consumption, these water streams may undergo heat recovery through a *heat exchanger network* (HEN). In other words, two separate networks are to be synthesized in a HIWN, i.e., water network and HEN. One may also view the HIWN problem as a special case of these network synthesis problems.

In this chapter, a hybrid model based on mathematical programming is introduced for the synthesis of HIWN, which is built on the fundamentals of water network and HEN syntheses[1]. The method will assist process engineers to better optimize energy and water resources in the plant. Following the philosophy of process integration, the performance targets of the HIWN, i.e., minimum freshwater flow rate, minimum hot utility target, or minimum operating cost (*OC*) are first determined using the hybrid model. Once the performance targets are identified, the HEN of HIWN can then be constructed using the *pinch design method**.

---

* Readers are encouraged to read Chapter 13 for the basics of energy recovery and Chapter 12 for the basics of water network synthesis with superstructural model prior to reading this chapter.

[1] See Example 9.1 for a similar problem.

15.1 Simultaneous Superstructure-Targeting Model—A Hybrid Approach

In this section, a hybrid model for simultaneous superstructure and targeting of HIWN is presented. The hybrid model was originally proposed by George et al. (2011) for HIWN synthesis, which has been simplified here for direct water reuse/recycle with single hot and cold utilities. Although water network and HEN may be viewed as two separate subnetworks, they interact with each other to form the HIWN (see Figure 15.1).

As the name implies, the hybrid model performs the following tasks:

- Matching of water sinks and sources of the water network through superstructure model.
- Targeting of utility of HEN through transshipment model.

Steps to construct the hybrid model are given as follows:

1. Construct a superstructure for direct water reuse/recycle network (see Figure 15.1a). The model has been described in Chapter 12 and is reproduced here. The constraint in Equation 15.1 describes that the flowrate requirement of sink j is to be fulfilled by process source i ($F_{SRi,SKj}$), and/or the fresh water r ($F_{FWr,SKj}$). Equation 15.2 describes that the impurity load sent from process sources and freshwater feed(s) must stay lower than the maximum limit acceptable by the process sinks. Equation 15.3 indicates that each process source i may allocate its flowrate to sink j, while the unutilized flowrate will be discarded

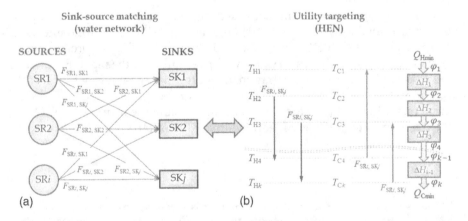

FIGURE 15.1
Two subnetworks of HIWN are solved using: (a) superstructure model for water sinks-sources matching; (b) transshipment model for utility targeting.

as wastewater ($F_{SRi,WW}$). Equation 15.4 indicates that all flowrate variables of the superstructural model should take non-negative values:

$$\sum_i F_{SRi,SKj} + \sum_r F_{FWr,SKj} = F_{SKj} \ \forall j \tag{15.1}$$

$$\sum_i F_{SRi,SKj} C_{SRi} + \sum_i F_{FWr,SRj} C_{FWr} \le F_{SKj} C_{SKj} \ \forall j \tag{15.2}$$

$$\sum_j F_{SRi,SKj} + F_{SRi,WW} = F_{SRi} \ \forall i \tag{15.3}$$

$$F_{SRi,SKj}, F_{FWr,SKj}, F_{SRi,WW} \ge 0 \ \forall i, j \tag{15.4}$$

2. From the water network superstructure, determine the water streams that will undergo heating and cooling operations, due to the difference in their supply (T_S) and target (T_T) temperatures. Note that all streams have to be considered, i.e., freshwater feed(s), reuse/recycle streams between sinks and sources, as well as wastewater discharge from all sources. These streams will be included in the HEN section of the HIWN, as they may undergo heat recovery, prior to the usage of external hot and cold utilities.

3. Construct the transshipment model to identify the utility targets for the HEN section. A simplified transshipment model that is conceptually identical to the *Problem Table Algorithm* (see Section 13.3) is shown in Figure 15.1b. As shown, all *hot streams* (that will undergo cooling operation) are located within their respective intervals in the hot temperature scale, while all *cold streams* (that will undergo heating operation) are in the cold temperature scale. Note that the hot and cold temperature scales are constructed using the supply (T_S) and target (T_T) temperatures of the individual hot/cold streams, along with their corresponding temperatures at the cold (T_{CS}) or hot side (T_{HS}); all temperatures are arranged in ascending order. To ensure feasible heat transfer, *minimum temperature difference* (ΔT_{min}) is incorporated to generate the corresponding temperature at the hot/cold side levels, following Equations 15.5 and 15.6, respectively. For instance, Equation 15.5 is used to determine the corresponding hot side temperature of T_{C1} (target temperature of the first cold stream), i.e., T_{H1}. Similarly, Equation 15.6 is used to determine the corresponding cold side temperature of T_{H2} (supply temperature of first hot stream), i.e., T_{C2}:

$$T_{HS} = T_{Cold} + \Delta T_{min} \ \forall \ \text{cold streams} \tag{15.5}$$

$$T_{CS} = T_{Hot} - \Delta T_{min} \ \forall \ \text{hot streams} \tag{15.6}$$

Within every temperature interval k, the net enthalpy values of hot ($\Delta H_{k,hot}$) and cold streams ($\Delta H_{k,cold}$) are calculated with Equations 15.7 and 15.8, respectively. The latter require the use of mass flowrate of the hot (m_{Hot}) and cold streams (m_{Cold}), as well as the temperature difference across interval k of the hot ($\Delta T_{hot,k}$) and cold ($\Delta T_{cold,k}$) temperature scales. For both Equations 15.7 and 15.8, it is assumed that the water heat capacity (C_p) stays at a constant value:

$$\Delta H_{k,hot} = \sum_{Hot} m_{Hot,k} C_p \Delta T_{Hot,k} \ \forall k \tag{15.7}$$

$$\Delta H_{k,cold} = \sum_{Cold} m_{Cold,k} C_p \Delta T_{Cold,k} \ \forall k \tag{15.8}$$

where

$$\Delta T_{Hot,k} = T_{Hot,k} - T_{Hot,k+1} \ \forall k \tag{15.9}$$

$$\Delta T_{Cold,k} = T_{Cold,k} - T_{Cold,k+1} \ \forall k \tag{15.10}$$

Equations 15.11 and 15.12 are next used to perform *heat cascade* across the temperature intervals by taking into consideration of the *net enthalpy* of the hot streams ($\Delta H_{k\,hot}$) and cold streams ($\Delta H_{k,\,cold}$), respectively. Excess heat at higher temperature intervals may be cascaded (φ_k) to lower temperature interval following Equation 15.11. For a feasible heat transfer, the cascaded heat can only take a positive value, as described by Equation 15.12:

$$\varphi_k = \varphi_{k-1} + \Delta H_{k,hot} - \Delta H_{k,cold} \ \forall k \tag{15.11}$$

$$\varphi_k \geq 0 \ \forall k \tag{15.12}$$

The cascaded heat entering the first level (φ_1) indicates the minimum hot utility (Q_{Hmin}) of the HEN, while the cascaded heat leaving the last level indicates its minimum cold utility (Q_{Cmin}). Note that the temperature set where the cascaded heat diminishes ($\varphi_k = 0$) is identified as the hot ($T_{P,Hot}$) and cold pinch temperatures ($T_{P,Cold}$) of the HEN problem.

Note also that the simplified transshipment model used here can only cater to single levels of hot and cold utilities. A more generic transshipment model for targeting multiple utilities may be found in Papoulias and Grossmann (1983).

4. Solve the hybrid model with either of the following objectives:

i. Minimum operating cost (*OC*), given as in Equation 15.14.

$$\text{Minimize } OC = \sum_r F_{FW_r} CT_{R_r} + F_{WW} CT_{WW} + Q_{Hmin} CT_{HU} + Q_{Cmin} CT_{CU} \tag{15.14}$$

where CT_{R_r}, CT_{WW}, CT_{HU} and CT_{CU} are unit costs of fresh resource r, wastewater, hot utility and cold utility, respectively.

ii. Minimum freshwater flow rate and minimum hot utility with the two-stage optimization approach. In stage 1, the freshwater flow-rate (F_{FW}) is minimized (with Equation 15.15), subject to the constraints in Equations 15.1–15.4. Once the minimum fresh resource flowrate is identified, it is then included as a new constraint in stage 2 (Equation 15.16), where hot utility (Q_{Hmin}) is minimized with the objective in Equation 15.17, subject to the additional constraints in Equations 15.5–15.12. As the model is a linear program (LP), any solution found is globally optimal, if it exists.

$$\text{Minimize } F_{FW} = \sum_r F_{FWr,SKj} \tag{15.15}$$

$$F_{FW,min} = F_{FW} \tag{15.16}$$

$$\text{Minimize } \varphi_1 = Q_{Hmin} \tag{15.17}$$

Once the performance targets of the HIWN are identified, the *pinch design method* may be used to design the HIWN; the latter has been described in detail in Chapter 13.

Example 15.1 Single contaminant case

A heat-integrated water network is to be synthesized, with its limiting data given in Table 15.1 (Bagajewicz et al., 2002). Its freshwater supply is assumed to have zero impurity and a temperature of 20°C, while its wastewater discharge is restricted to a maximum temperature of 30°C. The minimum temperature difference (ΔT_{min}) is assumed as 30°C, while water heat capacity is taken as 4.2 kJ/kg.°C.

Use the four-step procedure to solve for the minimum fresh and wastewater flowrates, followed by the minimum hot utility needed for the HIWN. Design the HEN for the HIWN that fulfills the targets.

TABLE 15.1

Limiting Water Data for Example 15.1

SK_j	Flowrate, F_{SKj} (kg/s)	Concentration, C_{SKj} (ppm)	Temperature, T_{SKj} (°C)
SK1	100	50	100
SK2	40	50	75
SK3	166.67	800	100
SR_i	Flowrate, F_{SRi} (kg/s)	Concentration, C_{SRi} (ppm)	Temperature, T_{SRi} (°C)
SR1	100	100	100
SR2	40	800	75
SR3	166.67	1100	100

Source: Bagajewicz et al. (2002).

SOLUTION

The four-step procedure is followed to determine the minimum fresh and wastewater water flowrates, followed by the minimum hot and cold utilities.

1. *Construct a superstructure for a direct water reuse/recycle network.*

   A superstructure for the water network is set up in MS Excel, as shown in Figure 15.2. Note that it is similar to the RCN examples solved in Chapter 9[1], except that all water sinks and sources will have to consider their temperature profiles, apart from flowrates and concentrations. Flowrate variables for the HIWN are also highlighted in Figure 15.2. These are the variables that will be optimized in order to achieve the minimum flowrate and energy targets of the HIWN.

2. *From the water network superstructure, determine the water streams that will undergo heating and cooling operations.*

   From Figure 15.2, some water sinks and sources are observed to have the same temperatures, e.g., sources SR1 and SR3, as well as sinks SK1 and SK3 (all at 100°C). Hence, recovery streams among these sinks and sources do not need to undergo any heating/cooling operation. In other words, these streams are excluded from HEN consideration, as shown in the simplified superstructure in Figure 15.3a.

   On the other hand, if water recovery will take place among sources and sinks of different temperatures, these recovery streams will undergo heating (e.g., SR2–SK1 or SR2–SK3) or cooling operations (e.g., SR1–SK2 or SR3–SK2) in the HEN. Note that the HEN will also include all freshwater streams (20°C) that are fed to the water sinks (FW–SK1, FW–SK2, FW–SK3) of higher temperature, as well as the water sources that are

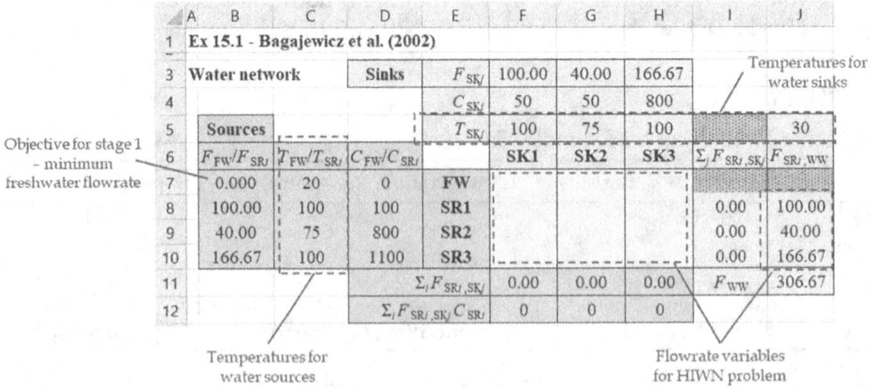

FIGURE 15.2
Superstructure for water network (all flowrates in kg/s; concentration in ppm; temperature in °C).

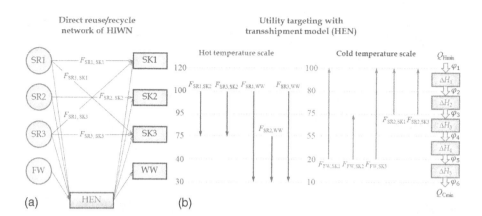

FIGURE 15.3
(a) Simplified superstructure showing steams connecting sinks and sources of identical temperatures are excluded from HEN; (b) water streams that will undergo heating and cooling operations in the HEN are located within their temperature intervals of the transshipment model for utility targeting.

discarded as wastewater (SR1–WW, SR2–WW and SR3–WW) at a lower temperature (30°C).

3. *Construct the transshipment model.*

The supply and target temperatures of the hot and cold streams are used to set up the hot and cold temperature scales of the transshipment model. Note that hot/cold corresponding temperatures have to be determined (using Equations 15.5 and 15.6) prior to their inclusion in the temperature scales. For instance, the highest value of the hot temperature scale is determined from the target temperature of several cold streams with Equation 15.5, i.e., 100 + 20 = 120°C. Similarly, the second value of the cold temperature scale is determined using Equation 15.6, i.e., 100 − 20 = 80°C. Once the hot and cold temperature scales are set up (in descending order of temperature), all streams are located within their respective temperature intervals, with hot streams on the left, and cold streams on the right side (Figure 15.3b).

Within every temperature interval k, the net enthalpy of hot steams ($\Delta H_{k,\,hot}$) and cold streams ($\Delta H_{k,\,cold}$) are calculated with Equations 15.7–15.12. A setup of this transshipment model in MS Excel is shown in Figure 15.4.

4. *Solve the hybrid model.*

The two-stage optimization model is used to determine the minimum freshwater and minimum hot utility. In stage 1, the model is set to minimize the freshwater flowrate (F_{FW}) with Equation 15.15, subject to the constraints in Equations 15.1–15.4. Setting in Excel Solver is shown in Figure 15.5a. The model determines that the freshwater flowrate is 77.274 kg/s.

Once the minimum freshwater flowrate is identified, it is included as a new constraint in stage 2 optimization (Equation

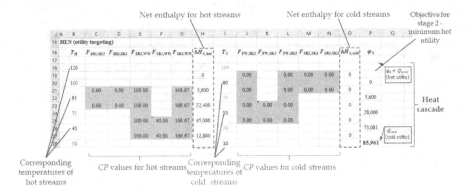

FIGURE 15.4
Setup of transshipment model in MS Excel (all temperatures in °C; flowrates in kg/s; heat flow in kW).

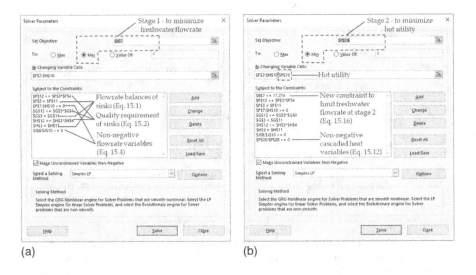

(a)   (b)

FIGURE 15.5
Setup of Excel Solver: (a) first stage—to minimize freshwater; (b) second stage—to minimize hot utility.

15.16), where hot utility (Q_{Hmin}) is minimized with the objective in Equation 15.17, subject to the constraints in Equations 15.5–15.12. As shown in Figure 15.5b, two new constraints are added in Excel Solver to limit the freshwater flowrate to 77.274 kg/s (Equation 15.16), along with the non-negative constraint for the cascade heat flow (Equation 15.12). The model is solved again to determine the minimum hot utility as 7,473 kW, with the results shown in Figure 15.6. The latter also determines that the cold utility of the HIWN (Q_{Cmin}) is identified as 4,227 kW. Note also

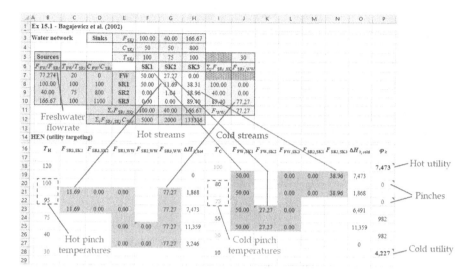

FIGURE 15.6

HIWN with minimum freshwater and hot utility target (all flowrates in kg/s; concentration in ppm; temperature in °C; heat flow in kW).

that the HEN has two sets of pinch temperatures, i.e., 100/80°C and 95/75°C. This information is important for HEN design.

Now the performance targets are identified, i.e., 77.274 kg/s fresh water and 7473 kW of hot utility, the HIWN can then be synthesized using the *pinch design method* (PDM). The latter have been discussed in detail in Chapter 13[2], and hence only important key steps are described in this section.

Important hot and cold stream data are first compiled and summarized in Table 15.2. As shown, two hot streams ($F_{SR1,SK2}$, $F_{SR1,WW}$) and three cold streams ($F_{FW,SK1}$, $F_{FW,SK2}$, $F_{SR2,SK3}$) are identified from the solved hybrid model in Figure 15.6. Their mass capacity flowrates (CP; column 5 of Table 15.2) are next

TABLE 15.2

Data for Hot and Cold Streams for HEN in Example 15.1

Streams	Supply Temperature T_S (°C)	Target Temperature T_T (°C)	Mass Flowrate, m_{Hot}, m_{Cold} (kg/s)	Mass Capacity Flowrate, CP (kW/°C)	Stream Type
$F_{SR1,SK2}$	100	75	11.69	49.1	Hot
$F_{SR1,WW}$	100	30	77.27	324.5	Hot
$F_{FW,SK1}$	20	100	50.00	210	Cold
$F_{FW,SK2}$	20	75	27.27	114.5	Cold
$F_{SR2,SK3}$	75	100	38.96	163.6	Cold

[2] See Sections 13.4 – 13.5 for details of PDM for HEN synthesis.

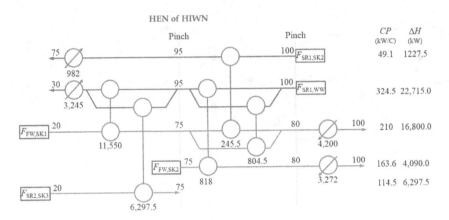

FIGURE 15.7
HEN of HIWN in Example 15.1 (all temperature in °C; heat flow in kW).

determined from the product of mass flowrate and water heat capacity (C_p), i.e., 4.2 kJ/kg °C. As the HEN has two sets of pinch temperatures, HEN designs are carried out independently in the regions above, below, and in between pinches. For the latter, no utility is used, as it is a region that is fully self-sufficient in energy, as shown in the grid diagram in Figure 15.7. The latter also shows two heaters are used in the region above the higher pinch temperatures (100/80°C), with a total hot utility of 7472 kW (= 4200 + 3272 kW). On the other hand, two coolers are used in regions below the lower pinch temperatures (95/75°C), with a total cold utility of 4227 kW (= 982 + 3245 kW). Both hot and cold utility consumptions are identical to the targets obtained in the transshipment model (Figure 15.6).

Example 15.2 Multiple contaminants case

A heat-integrated water network is to be synthesized, with its limiting data given in Table 15.3 (Dong et al., 2008). The freshwater supply is assumed

TABLE 15.3

Limiting Water Data for Example 13.2

SK_j	F_{SK_j} (kg/s)	$C_{SK_j,A}$ (ppm)	$C_{SK_j,B}$ (ppm)	$C_{SK_j,C}$ (ppm)	T_{SK_j} (°C)
SK1	30	0	0	0	100
SK2	40	50	40	15	75
SK3	20	50	50	30	35
SR_i	F_{SK_i} (kg/s)	$C_{SR_i,A}$ (ppm)	$C_{SR_i,B}$ (ppm)	$C_{SR_i,C}$ (ppm)	T_{SR_i} (°C)
SR1	30	100	80	60	100
SR2	40	150	115	105	75
SR3	20	125	80	130	35

Source: Dong et al. (2008).

to have zero impurity and a temperature of 20°C, while its wastewater is allowed to be discharged at 30°C. Synthesize a HIWN with minimum OC, assuming that the annual operating time of 8000 h. The unit costs of fresh water, wastewater, hot utility, and cold utility are given as 1 $/t, 1 $/t, 120 $/kW.y, and 10 $/kW.y, respectively (Kamat et al., 2019). To reduce energy consumption costs, the minimum temperature difference (ΔT_{min}) is assumed as 30°C, while water heat capacity is taken as 4.2 kJ/kg.°C.

SOLUTION

The four-step procedure is followed to determine the minimum OC of the HIWN

1. *Construct a superstructure for a direct water reuse/recycle network.*
2. *From the water network superstructure, determine the water streams that will undergo heating and cooling operations.*

   Since there are three impurities (A, B, and C) to be considered in this HIWN, the setup in MS Excel is similar to that in Example 13.1, with additional rows (cells H14-J16) to keep track of the load of each impurity.[3] As shown in Figure 15.8, flowrates in the matching-matrix (recovered water and freshwater feed) are set as changing variables (along with hot utility from transshipment model) in Solver.

   A simplified superstructure for the water network is given in Figure 15.9a, which shows that all direct water recycle streams (i.e., SR1–SK1, SR2–SK2, and SR3–SK3) are excluded from the HEN as their sinks and sources have the same temperatures.

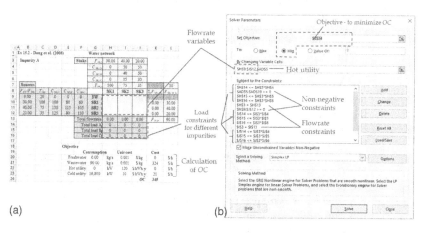

(a)   (b)

FIGURE 15.8
Setting for Example 15.2: (a) matching-matrix for water network; (b) Solver to minimize OC (heat cascade is omitted due to space limitation).

---

[3] See Example 9.2 for detailed explanation.

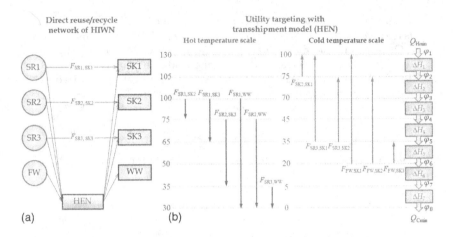

FIGURE 15.9
(a) Simplified water network superstructure and (b) HEN transshipment model for Example 13.2 (all flowrates in kg/s; concentration in ppm; temperature in °C; heat flow in kW).

All other streams (including freshwater feed and wastewater discharge) will undergo heating or cooling operations in the HEN, and hence be included in the transshipment model (see Figure 15.9b).

3. *Construct the transshipment model.*
4. *Solve the hybrid model.*

The transshipment model in Figure 15.9b shows all water streams that require heating and cooling are located within their respective temperature intervals in the hot and cold temperature scales. The latter are set-up using the supply and target temperatures of these streams, along with their corresponding temperatures (determined using Equations 15.5 and 15.6). The net enthalpy (ΔH_k) for all intervals is calculated with Equations 15.7–15.12. In the transshipment model, the hot utility (cascaded heat entering the first level (φ_1)) is also included as a changing variable in Solver, along with the flowrate variables (see Figure 15.8b). This is because, the objective is set to minimize the OC. The model is next solved, resulting in the minimum OC of 633 $/h. As shown in Figure 15.10, the freshwater of the HIWN is determined as 70 kg/s, while the hot and cold utilities are determined as 8,190 kW and 5,250 kW, respectively. Note that these performance targets obtained with the objective of minimum OC are identical to those obtained using the two-stage optimization approach (George et al., 2011). Once the minimum OC is determined, the HEN is designed using the PDM, as given in Figure 15.11.

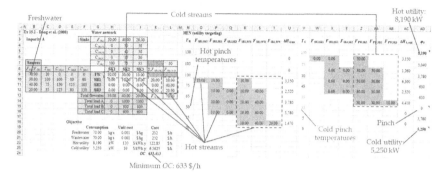

FIGURE 15.10
Results of a hybrid model for Example 15.2 (all flowrates in kg/s; concentration in ppm; temperature in °C; heat flow in kW).

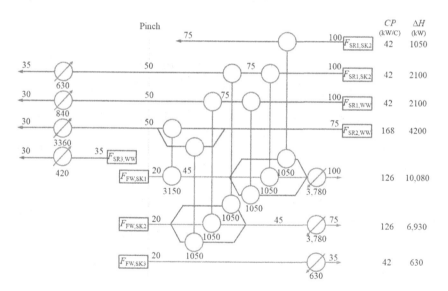

FIGURE 15.11
HEN of HIWN in Example 15.2 (all temperature in °C; heat flow in kW).

15.2 Further Readings

The method presented in this chapter can be extended to consider a water regeneration unit, i.e., *heat-integrated water regeneration network*. Such extensions were reported by Kamat et al. (2019) for single and multiple impurities problems. Besides, non-isothermal mixing of water sources has been reported to achieve lower utility requirements for some cases, as reported by George et al. (2011) and Kamat et al. (2019).

TABLE 15.4

Limiting Water Data for Problem 13.2

SK$_j$	F_{SKj} (kg/s)	C_{SKj} (ppm)	T_{SKj} (°C)
SK1	20	0	40
SK2	100	50	100
SK3	40	50	75
SK4	10	400	50
SR$_i$	F_{SRi} (kg/s)	C_{SRi} (ppm)	T_{SRi} (°C)
SR1	20	100	40
SR2	100	100	100
SR3	40	800	75
SR4	10	800	50

Source: Savulescu et al. (2005).

PROBLEMS

1 Redo Example 1 for minimum temperature difference (ΔT_{min}) of 10°C (hot and cold utilities are found to be 3736.28 kW and 490.44 kW; George et al., 2011).

2 A HIWN is to be synthesized, with its limiting water data given in Table 15.4. The freshwater supply is free from impurity and has a temperature of 20°C. Wastewater is allowed for discharge at 30°C. Synthesize a HIWN with a minimum temperature difference (ΔT_{min}) of 10°C with the following objective:

a. Minimum OC, with unit costs of fresh water, wastewater, hot utility, and cold utility given as 1 $/t, 1 $/t, 120 $/kW.y, and 10 $/kW.y, respectively (Kamat et al., 2019). The annual operating time may be assumed as 8000 h.

b. Minimum fresh water and wastewater flowrates, along with minimum hot and cold utilities (answers: 90 kg/s, 3780 kW, and 0 kW).

References

Bagajewicz, M., Rodera, H., and Savelski, M. 2002. Energy efficient water utilization systems in process plants, *Computers and Chemical Engineering*, 26(1), 59–79.

Dong, H. G., Lin, C. Y., and Chang, C. T. 2008. Simultaneous optimization approach for integrated water-allocation and heat-exchange networks, *Chemical Engineering Science*, 63, 3664–3678.

George, J., Sahu, G. C., and Bandyopadhyay, S. 2011. Heat integration in process water networks, *Industrial & Engineering Chemistry Research*, 50, 3695–3704.

Kamat, S., Sahu, G. C., Foo, D. C. Y., and Bandyopadhyay, S. 2019. Optimum synthesis of heat integrated water regeneration network, *Industrial & Engineering Chemistry Research*, 58, 1310–1321.

Papoulias, S. A., and Grossmann, I. E. 1983. A structural optimization approach in process synthesis—II. Heat recovery networks, *Computers and Chemical Engineering*, 7, 707–721.

Savulescu, L. E., Kim, J.-K., and Smith, R. 2005. Studies on simultaneous energy and water minimisation – Part II. Systems with maximum re-use of water, *Chemical Engineering Science*, 60, 3291–3308.

Appendix

In Chapter 3, Equations 3.1 and 3.2 are used for the calculation of *maximum load acceptable by sink j* (m_{SKj}) and *load possessed by source i* (m_{SRi}), respectively:

$$m_{SK_j} = F_{SK_j} q_{SK_j} \tag{3.1}$$

$$m_{SR_i} = F_{SR_i} q_{SR_i} \tag{3.2}$$

where F_{SKj} and F_{SRi} are flowrates of material sink j and source i, while q_{SKj} or q_{SRi} are their corresponding quality levels. Note that quality of water/hydrogen sinks and sources are usually addressed in impurity concentration (i.e., C_{SKj} or C_{SRi}). As both flowrate and quality parameters may exist in various units, some commonly used units of these parameters are documented in Table A.1, to assist readers for quick reference.

TABLE A.1

Units of Parameters in Equations 3.1 and 3.2

F_{SKj}/F_{SRi}	C_{SKj}/C_{SRi}	m_{SKj}/m_{SRi}
t/h	ppm	g/h
t/d	ppm	g/d
t/d	mg/L	g/d
t/h	wt%	t/h
kg/h	ppm	mg/h
kg/min	ppm	mg/min
kg/s	ppm	mg/s
kg/s	wt%	kg/s
mol/h	mol%	mol/h
mol/s	mol%	mol/s

Index

Note: Locators in *italics* represent figures and **bold** indicate tables in the text.

Printed in the United States
by Baker & Taylor Publisher Services